# Progress in
# OCEANOGRAPHY

## Volume 7

# PROGRESS IN
# OCEANOGRAPHY

## VOLUME 7

*Editor*

# MARY SWALLOW

*Institute of Oceanographic Sciences,*
*Wormley, Godalming, Surrey*

# PERGAMON PRESS

OXFORD · NEW YORK · TORONTO
SYDNEY · PARIS · FRANKFURT

| U.K. | Pergamon Press Ltd., Headington Hill Hall, Oxford OX3 0BW, U.K. |
| U.S.A. | Pergamon Press Inc., Maxwell House, Fairview Park, Elmsford, New York 10523, U.S.A. |
| CANADA | Pergamon of Canada, Suite 104, 150 Consumer's Road, Willowdale, Ontario M2J 1P9, Canada |
| AUSTRALIA | Pergamon Press (Aust.) Pty. Ltd., P.O. Box 544, Potts Point, N.S.W. 2011, Australia |
| FRANCE | Pergamon Press SARL, 24 rue des Ecoles, 75240 Paris, Cedex 05, France |
| FEDERAL REPUBLIC OF GERMANY | Pergamon Press GmbH, 6242 Kronberg-Taunus. Pferdstrasse 1. Federal Republic of Germany |

First Published in *Progress in Oceanography* Volume 7 Numbers 1–6, and supplied to subscribers as part of their subscription

Library of Congress Catalog Card No. 63–15353

ISBN 0 08 020329 9

# CONTENTS

# CONTENTS

# EDITORIAL NOTICE

I retire in July 1977; at my initial invitation and with the full support of Pergamon Press, Dr. Martin Angel and Professor James O'Brien have kindly agreed to become co-editors of *Progress in Oceanography*.

It is hoped that the availability of two editors may avoid delays necessarily occasioned by other commitments and by absence on leave, at sea or at meetings.

Manuscripts should be sent to:

DR. MARTIN ANGEL, Institute of Oceanographic Sciences, Wormley, Godalming, Surrey GU8 5UB, England. Telephone: Wormley 4141.

or to:

PROFESSOR JAMES J. O'BRIEN, Professor of Meteorology and Oceanography, Department of Meteorology, The Florida State Univeristy, Tallahassee, Florida 32306, U.S.A. Telephone (904) 644–581.

MARY SWALLOW

Prog. Oceanog., Vol. 7, p. 1, 1976. Pergamon Press. Printed in Great Britain.

# EDITOR'S NOTE 1976

*Progress in Oceanography*, in this new format, continues to present papers on diverse topics and in diverse styles. Review papers are welcome as the emphasis in the title should be on "progress" and review papers might well assess the present state of a subject. Nevertheless many other types of paper demonstrate how various marine topics have developed by focusing on some particular aspect; is it a truism to say that intense specialism needs an ever-broadening basis of knowledge? The submission of such papers is equally welcome. It may happen that various features of a paper, such as a detailed historical introduction, an unusually distinct description of instrumentation or a more exhaustive account of the observations, are too lengthy for publication in more usual journals, and yet by their very detail these features may be pertinent and their presentation may be necessary. If such oceanographic papers are significant then they are appropriate for publication in *Progress in Oceanography*.

My predecessors as editor, Dr Mary Sears and Dr Bruce Warren, indicated their willingness to consider original papers and this policy is being continued.

There is no bias towards any particular marine subject, the contents of each future volume will depend on the papers submitted and may thus be as eclectic as the word "oceanography". Volume 7 Number 1 is a paper on midwater fishes in the eastern North Atlantic, Number 2 furthers our knowledge of physical oceanography in the MEDOC area, other papers anticipated in Volume 7 are on growth conditions of manganese nodules and on some aspects of the physical chemistry of seawater.

Papers in *Progress in Oceanography* are now to appear as single parts of a volume. Such a manner of presentation should reduce delay in publication of an individual paper. I look forward to receiving further contributions.

*Institute of Oceanographic Sciences*  
*Wormley*  
*Godalming*  
*Surrey*

MARY SWALLOW

*Prog. Oceanog.* Vol. 7, pp. 3–58, 1976. Pergamon Press. Printed in Great Britain

# Midwater fishes in the eastern North Atlantic—I. Vertical distribution and associated biology in 30°N, 23°W, with developmental notes on certain myctophids

JULIAN BADCOCK* and N. R. MERRETT*

(*Received* 28 *March* 1975; *accepted* 20 *August* 1975)

Abstract—This paper is the first of a series describing the vertical distribution of midwater fishes in the eastern North Atlantic in areas approximately 10° of latitude apart near the 20° meridian. As such, it is concerned solely with collections made in 30°N, 23°W. The results are based primarily upon RMT 8 net samples collected using an opening–closing RMT 1+8 combination net. In all, 17,443 specimens were caught by RMT 8, divisible into 37 families, 66 genera with 98 species identified. Numerically the most abundant groups recognized were the Gonostomatidae (60% total catch), Macrorhamphosidae (23%) and Myctophidae (7%). Some definition of vertical limits is provided for the majority of species represented, but only 31 are considered in detail. Where possible, observed biological phenomena (e.g. development, sexual maturity, sexual dimorphism, etc.) and distribution limits have been correlated. In general, the characteristics of species' distributions, and the observed relations of distribution and migratory behaviour were as one would anticipate from past work. Among many species, a size–depth stratification was observed, and with migrants, migratory behaviour appeared dependent upon developmental state. The mode of species development, however, had no bearing upon ultimate migratory behaviour. Certain non-migratory elements of the population of an habitually migrant species, e.g. *Chauliodus danae*, could not be satisfactorily interpreted upon the basis of biological factors examined. Migrant species principally occupied 400–900 m depth by day, and 25–300 m by night, although they occurred at greater depths. Reverse migrations were only observed in *Macrorhamphosus*. The overall catches were dominated by non-migrants, and at depths greater than 200 m these comprised the most abundant species per depth at all times. *Cyclothone* constituted the most abundant genus sampled and provided greatest insight into distributional and biological detail. *C. braueri*, the most numerous species, is probably a single spawner, and the results demonstrate a size–depth stratification that may be correlated with sexual maturity stages. The olfactory structure in males is more complicated than previously described. The larger males develop a snout prolongation which would improve water flow through the nasal rosette and hence olfaction. Unlike *C. microdon, C. braueri* probably does not undergo sex-reversal. Developmental notes are included for the myctophid species *Notoscopelus resplendens, Benthosema suborbitale,* and *Hygophum hygomi* in the Appendix.

## CONTENTS

*Institute of Oceanographic Sciences, Wormley, Godalming, Surrey, U.K.

## INTRODUCTION

PRELIMINARY studies of the organization and structure of the vertical section of the ecosystem down to 950 m depth in a deep oceanic environment close to Fuerteventura (Canary Islands) made on the SOND* Cruise, 1965, revealed a marked ecological complexity at species and community level (ANGEL, 1969; FOXTON, 1969; CLARKE, 1969; BAKER, 1970; FOXTON, 1970a, b; BADCOCK, 1970; ROE, 1972a–d; ANGEL and FASHAM, 1973; PUGH, 1974). Among most taxa examined, well-defined patterns of depth distribution were apparent, and it was shown that many zooplanktonic and micronektonic species were restricted to relatively narrow depth ranges. Regular diel migrations of varying amplitude towards (dusk) and away from (dawn) the near-surface waters were made by a high proportion of the species present, although 'reverse' and non-migratory species also occurred. Thus, within the section, a broad spectrum of species' vertical distributional patterns and migratory behaviours was represented. These results, together with measurements of environmental temperature, salinity and underwater light intensity (CURRIE, BODEN and KAMPA, 1969; KAMPA, 1970), allowed tentative conclusions to be tendered regarding the interacting environmental and biological factors that underlay the observed ecological organization. Notably, morphological and structural variations in fish (BADCOCK, 1970) and decapod crustaceans (FOXTON, 1970a, b) were deduced to represent adaptations to the change in light regime with depth, although later work (FOXTON, 1971–1972) suggested that light was just one of many variables that differentially interact to limit distribution. For euphausiid and decapod crustaceans (BAKER, 1970; FOXTON, 1970b) the diel changes in the distribution of biomass were also demonstrated, emphasizing the role that migrant routes may play, as energy pathways, in the trophodynamics of the water column.

Although the overall SOND Cruise results represented a significant advance in our knowledge of the vertical section of ecosystems, it was clear that further investigations, extended to include a wider zoogeographic area and greater depth range,

could greatly improve our understanding of these systems. It was reasoned that investigations of the vertical structure of differing faunas from different areas using standard sampling techniques would emphasize regional variations and similarities in vertical patterns; which hopefully could be correlated with comparably scaled physico-chemical gradients in the environment, and thus make it possible to identify those factors which, either singly or in combination, determine the vertical limits of species and communities. Moreover, by making observations over a wide oceanic area, considerable information would be gained on horizontal distributions, faunal boundaries, and their possible relation to vertical zonation.

With the above objectives in mind, a detailed faunal survey has been made at each of six stations in the eastern North Atlantic. The sampling positions were sited at approximately 10° intervals of latitude between 11 and 60°N, close to the 20°W meridian. The collections made thus include diverse faunas, from tropical to boreal. At each locality the water column was sampled in detail down to depths between 2000 and 3000 m. Sampling was conducted as a day and night series of contiguous strata by an opening–closing net of novel design (BAKER, CLARKE and HARRIS, 1973), fished in a carefully controlled, routine manner. The data collected on zooplankton and micronekton, together with complementary temperature and salinity data, are immense in so far as studies of vertical distribution are concerned and have already produced results of interest (FOXTON, 1971–1972; CLARKE and LU, 1974; ANGEL and FASHAM, 1975). Several taxa were well represented in the collections, and among these, the meso- and bathypelagic fishes.

Our knowledge to date of the factors influencing distribution in both the vertical and horizontal planes is very conjectural. It has long been known that species' diurnal depth of occurrence varies from place to place, dependent upon local conditions (MURRAY and HJORT, 1912). HJORT (MURRAY and HJORT, 1912) noted that species'

---

*SOND, derived from the initial letter of the months Sept.–Dec., over which it was run.

depths increased southerly in the North Atlantic and correlated this with the underwater light regime. BODEN and KAMPA (1967) correlated daytime changes in depth of a scattering layer to concomitant changes in the light regime, and also demonstrated that particular scattering layers identify with particular isolumes during migration periods (see also CLARKE and BACKUS, 1964). CLARKE (1966) similarly showed that particular species follow particular isolumes during migration periods. On the basis of structural and morphological variation, BADCOCK (1970) speculated that light was a dominant factor influencing depth of occurrence of fish off Fuerteventura. FOXTON (1971–1972), however, showed that among certain decapod crustaceans diurnal and nocturnal distribution depths could be correlated with particular isotherms over several localities. He concluded that light may be an important local factor defining species' vertical limits, but that these could be modified by temperature.

It has been common practice to identify horizontal distributions of fishes, on the grand scale, in terms of water mass boundaries (e.g. HAFFNER, 1952; EBELING, 1962). As BACKUS, CRADDOCK, HAEDRICH and SHORES (1970) pointed out, however, zoogeographic distributions in the North Atlantic are far more complex than can be explained solely by the water mass hypothesis. In the eastern North Atlantic, the North Atlantic Central Water Mass has southerly limits at about 20°N (SVERDRUP, JOHNSON and FLEMING, 1942). BACKUS, MEAD, HAEDRICH and EBELING (1965) found an ichthyofaunal boundary within this region and FOXTON (1971–1972) conjectured that the transitional water lying between the North and South Atlantic Central waters may act as a southern faunal boundary for the typically northern decapod species. The northern limits of distribution, FOXTON pointed out, may, on the other hand, occur in areas where the species' optimum temperature level lay shallower than its optimum light level. PAXTON (1967) had developed a similar model for the lantern fishes of the Pacific. He surmised that northern distributions may be curtailed when the isotherm of minimum temperature tolerated by the species lay shallower

than its optimal zone of light intensity; and conversely, the southern limits were defined when the isotherms of maximum tolerance of the species lay deeper than the depth to which the species may migrate. ROBISON (1972), in his study of fishes from the Gulf of California, on the other hand, concluded that specific temperature and light levels may not be general determinants of diurnal depths. He suggested that levels of diurnal distribution may be defined by a physiological tolerance to a particular oxygen concentration; or, in the case of vertical migrators, by temperature differences existing across their diel vertical depth ranges. He also stated that the composition of the midwater fauna may be influenced by the nature of the surface waters. PICKFORD (1946) considered water density important in controlling the distribution of the octopod *Vampyroteuthis infernalis* Chun, but EBELING (1962) regarded it as unlikely to have much effect upon active nekton, citing the melamphaid fishes. HAEDRICH and CRADDOCK (1968), however, suggested that water density may have an influence in determining the distribution of the argentinoid fish *Winteria telescopa* Brauer, 1901.

The mass of present opinion suggests that ultimately the horizontal distribution of some midwater fishes is generally limited by water mass boundaries (see COHEN, 1973); but also that these distributions may be modified by other local environmental factors, both physico-chemical and biological (e.g. the degree of primary production). Although the determinants of vertical distribution are clearly not resolved, from the literature available, there does arise the suggestion that those factors modifying the zoogeography of a species within its characteristic water mass are closely related to those determinants of vertical distribution.

MATERIALS AND METHODS

Excluding the deepest tow (#80, 3000–2000 m, fished on 8/4/72), the series of collections was completed in 7 days (31/3–6/4/72). The limits of the area sampled were bounded by latitudes 29°37·7′N and 30°13.8′N and longitudes 22°55·4′W and 23°17·3′W; an area of almost 800 square miles.

Sampling was carried out with the National Institute of Oceanography (now Institute of Oceanographic Sciences) combination net (RMT 1 + 8) employing the standard technique for vertical distribution studies described by BAKER, CLARKE and HARRIS (1973). Thus, the upper 2000 m of the water column was divided into 16 strata, each of which was sampled separately by day and night. The RMT 1 alone was used to sample the upper 10 m. Departures from the normal procedure were the 1 rather than 2 h tows in 25–10 m by day, 50–25 m and 25–10 m by night; also the $\frac{1}{2}$ rather than the 1 h tow in 10–0 m by night (Table 1). In addition, the stratum from 3000 (estimated)–2000 m was fished by the RMT 1 + 8 by day only. Daytime was taken as extending from 1 h after sunrise to 1 h before sunset; night-time from 1 h after sunset to 1 h before sunrise. Samples were not taken between these periods. No allowance was made by night for the Moon, which during the sampling period was in the third quarter (full 29/3, last quarter 6/4). Moonrise varied between about 2000 and 0200 and moonset between about 0600 and 1200 GMT (subtract 1 h for local time). No account for cloud cover was made.

The fishing data are summarized in Table 1. The duration and time fished (GMT) is shown for each of the strata, together with the ship's course and the distance fished by the net as indicated by the 'flowmeter' on the net monitor. Maximum and minimum temperatures recorded by the net monitor are also given, which, although not calibrated against the salinity–temperature–depth probe (STD) used in the hydrographic observations, provided a comparison between day and night tows from similar strata. Agreement was to half a degree between the temperature recorded by the monitor and the STD in depths greater than 100 m (see Table 1 and Fig. 2). A different monitor was used in the upper 100 m, however, which was less accurate and accounts for the discrepancy apparent in Table 1. For consistency in sampling, a constant towing course was, in most cases, maintained but modified where necessary by the prevailing weather conditions. The net fished at a mean speed of 2·5 knots* through the

water by day and 2·4 knots by night. Since the angle of the net from the vertical was not measurable whilst sampling, it was not possible to use the 'flowmeter' to indicate the volume of water filtered by the net. The figure obtained for distance fished (Table 1) is of comparative value only, and has not been used to compensate catch numbers of fishes to 'unit flow'.

An approximate reconstruction of the trajectories taken by the RMT 1 + 8 whilst fishing is given in Fig. 1 for all strata from 10 to 2000 m depth. In the shallower strata the initial stages of a tow were more easily controlled. With increase in depth, and therefore length of warp required, uniformity of sampling across a stratum was more difficult to attain. Figure 1 also illustrates the variability of fishing effort with depth. The sampling programme was designed to provide maximum depth coverage in the minimum period of time. The selection of the various widths of stratum, therefore, resulted from both practical and scientific considerations. The upper 100 m were divided to provide a high resolution of distributional limits in a zone where complex stratification of nocturnal migrants occurs. The results of the SOND cruise suggested that 100 m strata were necessary to provide the requisite resolution from 100 to 1000 m, despite the increasing duration of tow with depth. Below 1000 m successively widening strata were fished for an increased time period, on the supposition that the bathypelagic fauna is relatively sparse and less finely stratified than the mesopelagic. For convenience, reference to the strata sampled is made in terms of standard strata (*sensu* BAKER, CLARKE and HARRIS, 1973) and quoted from the upper to lower limit, rather than in the reverse manner in which normally it was fished (e.g. see Table 1; #18, 203–102 m is referred to as 100–200 m).

On completion of the tow, the net was brought inboard and the cod-end liners were removed and transferred, in tubs of cooled seawater, to the ship's laboratory. There, the separate catches were fixed in a copious volume of 10% seawater

---

*1 knot = 0·51 m s⁻¹.

*Table 1.* *Resumé of fishing data, and the number of specimens and species per depth.*

| Depth Range (m) | Tow No. (•) | Course (°True) | Distance fished (km) | Temperature range (°C) min. | max | Minimum number species caught | Total number specimens | Duration of tows at depth (GMT) Times |
|---|---|---|---|---|---|---|---|---|
| **DAY** | | | | | | | | |
| 10-0[1] | 27 | 160 | - | - | - | - | - | 1458-1558 |
| 25-10 | 26 | 160 | 5.04 | 17.5 | 17.6 | 2 | 346 | 1330-1430 |
| 50-25 | 25 | 160 | 9.69 | 17.0 | 17.5 | 2 | 3 | 1100-1300 |
| 100-55 | 24 | 160 | 9.29 | 16.7 | 17.1 | 12 | 52 | 0842-1042 |
| 203-102 | 18 | 180 | 10.40 | 17.2 | 17.4 | 13 | 47 | 1528-1729 |
| 300-205 | 17 | 180 | 9.61 | 15.4 | 16.8 | 12 | 76 | 1251-1500 |
| 400-305 | 11 | 185 | 9.06 | 13.7 | 14.8 | 12 | 378 | 1542-1743 |
| 505-405 | 2 | 182 | 9.45 | 12.5 | 13.2 | 23 | 3186 | 0910-1110 |
| 600-500 | 16 | 180 | 10.08 | 11.2 | 12.7 | 24 | 1646 | 0959-1204 |
| 700-600 | 10 | 185 | 8.90 | 10.2 | 11.1 | 23 | 654 | 1244-1448 |
| 800-700 | 9 | 185 | 9.13 | 9.4 | 10.1 | 26 | 575 | 0924-1133 |
| 900-795 | 4 | 182 | 8.19 | 8.5 | 9.5 | 26 | 306 | 1534-1734 |
| 1000-910 | 3 | 182 | 8.43 | 8.0 | 8.8 | 19 | 127 | 1219-1419 |
| 1250-1000 | 50 | 008 | 22.21 | 7.5 | 8.5 | 17 | 107 | 0917-1317 |
| 1500-1250 | 51 | 180 | 17.32 | 6.2 | 7.4 | 13 | 97 | 1452-1853 |
| 2000-1500 | 54 | 000 | 17.80 | 3.9 | 5.3 | 12 | 40 | 0831-1234 |
| 3000-2010 | 80 | 135 | 27.88 | 2.8 | 3.7 | -[2] | -[2] | 0950-1550 |
| **NIGHT** | | | | | | | | |
| 10-0[1] | 40 | 160 | - | - | - | - | - | 0217-0247 |
| 25-10 | 39 | 140 | 4.72 | 17.5 | 17.6 | 9 | 245 | 0057-0157 |
| 50-25 | 38 | 140 | 4.88 | 16.9 | 17.1 | 23 | 447 | 2342-0042 |
| 100-50 | 37 | 140 | 10.24 | 16.6 | 17.0 | 32 | 532 | 2115-2315 |
| 200-100 | 22 | 145 | 10.08 | 17.9 | 18.0 | 35 | 3669 | 0324-0530 |
| 300-205 | 21 | 145 | 10.40 | 16.0 | 17.9 | 31 | 356 | 0042-0245 |
| 400-300 | 15 | 171 | 9.92 | 13.6 | 14.9 | 14 | 353 | 0357-0612 |
| 500-405 | 8 | 182 | 8.35 | 12.6 | 13.4 | 17 | 2068 | 0342-0542 |
| 600-500 | 20 | 145 | 10.71[3] | 11.2 | 12.0 | 20 | 958 | 2138-2348 |
| 700-600 | 14 | 171 | 9.21 | 10.3 | 10.8 | 20 | 305 | 0038-0244 |
| 800-700 | 13 | 171 | 8.82 | 9.6 | 10.0 | 15 | 330 | 2117-2323 |
| 895-800 | 7 | 182 | 8.66 | 8.6 | 9.0 | 18 | 177 | 0038-0238 |
| 1000-900 | 6 | 182 | 6.61 | 8.0 | 8.5 | 16 | 84 | 2055-2255 |
| 1250-1005 | 48 | 325 | 18.74 | 6.8 | 8.2 | 19 | 137 | 2114-0114 |
| 1500-1250 | 52 | 180 | 20.16 | 5.6 | 6.7 | 15 | 112 | 2206-0206 |
| 2010-1500 | 57 | 310 | - | 4.1 | 5.7 | 10 | 30 | 2129-0136 |

[1]RMT 1 only, catch not considered.
[2]The 14 specimens of eight species from this haul were excluded since several were obvious contaminants; also no comparable tow was made at night.
[3]Estimated.

formalin. The samples were sorted later, ashore, and preserved in an aqueous storage fluid of 10% propylene glycol, 1% formalin, 0·5% phenoxetol (based upon STEEDMAN, 1974).

The fishes from each sample were identified and the standard length (SL) of each was measured,

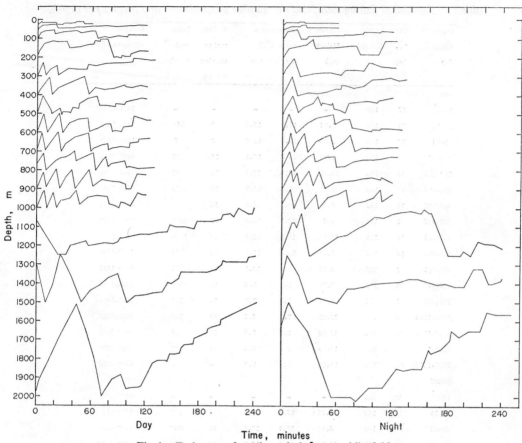

Fig. 1.    Trajectory of net through the water whilst fishing.

by projection on a ruler, to the nearest milli-metre, unless otherwise stated. Sex determination was effected through a slit, along either the flank or the ventral midline. A portion of the gonad was removed and examined under a binocular micro-scope. Selected gonads were sectioned and stained in haematoxylin and orange G for histological examination. The results indicated that the ovaries could be staged according to recognizable histo-logical changes associated with maturation in all species examined. The system of indices chosen (Table 2) was based upon that described by MERRETT (1971). Definitive stages in maturity of the testis were less easily recognized as some milt was present in all the 'immature' specimens. It was not possible even to determine the sex of many small and immature specimens, and only a negligible proportion of males was found to be

spent. Thus staging of the majority of males of the species examined could provide little additional evidence on the reproductive cycle to that obtained from the more precise staging of females. Consequently, testes were only staged in certain instances. Fecundity measurements were made on late Stage III/V females. The ovary pair was carefully removed from the body cavity and the total number of the most advanced generation of eggs was counted.

The RMT 8 (mesh size 4·5 mm) sampled more fish of a greater size range than did the RMT 1 (mesh size 0·32 mm). Consequently this analysis is based predominantly upon catches of the RMT 8, supplemented by some additional evidence from the RMT 1. The different pattern of sampling within the upper 100 m and also below 1000 m depth (see Fig. 1) in way no affects

Table 2. *Stages in ovarian maturity, modified from* MERRETT (1971).

| | | |
|---|---|---|
| Immature virgin: | Ovaries small, compact; yolk absent from oocytes. | (Stage I) |
| Recovering spent: | Ovaries somewhat flaccid, larger than Stage I; yolk absent from oocytes. | (Stage VII/II) |
| Yolk & chorion formation: | Ovaries firm, slightly enlarged; yolk granules present & chorion layer forming around oocytes | (Stage II) |
| Maturation of eggs: | Ovaries firm and distended. Eggs visible to the naked eye, opaque; yolk vesicles readily apparent, chorion fully formed. | (Stage III/V) |
| Ovulation: | Ovaries distended to such an extent that the transparent eggs, enlarged by the uptake of liquid and freed from the chorion, are shed with or without slight pressure on the flanks. | (Stage VI) |
| Spent: | Ovaries extremely flaccid, either an enlarged but empty sac or shrunken with an enlarged central lumen. | (Stage VII) |

the comparability of indicated species' vertical limits. It does, however, preclude direct quantitative comparisons of relative abundance and population structure outside the samples obtained at 100–1000 m depth. Numerical adjustment is possible to permit such analysis throughout the sampled water column, but its reliability is partly dependent upon sample size. The fish samples are not considered sufficiently large to justify this approach.

The stages in the life history of the species considered here are defined, unless otherwise stated, as follows: post-larval, specimens with the yolk sac completely absorbed but lacking the total complement of photophores; juvenile, specimens with a full complement of photophores but without identifiable gonads; sub-adult, specimens with identifiable gonads in Stage I of gonad maturity; adults, specimens in all the more advanced stages of gonad maturity including recovering spent (VII/II).

## RESULTS
*Limitations of sampling*

Any sample taken by a net is the product of an interaction between various biases. It is not the aim of this paper to consider these aspects in detail, but a brief discussion of some of the important factors influencing the results will be given, in an attempt to place the latter into their proper perspective.

A sampling programme designed with ambitions such as the present one should ideally be fished in each area within a single 24 h period, since catch anomalies may arise due to short term (non-diel) environmental changes over a longer period. The impracticality of this is obvious. Each locality, however, was fished within as short a time period as possible. In addition, STD dips were normally made at the beginning and end of each series, and usually midway through it, to monitor any environmental changes occurring over the sampling period.

A major drawback of the programme lay in the selected depth ranges each being fished once only per day and night. The catch numbers of many species were consequently low. This in itself, however, does not undermine the confidence expressed in the indicated vertical limits of distribution of the commoner species. Differential species' behaviour (e.g. patchiness of local distribution, relative activity with time, feeding and avoidance), on the other hand, will also influence the composition of a sample. Over a series of repeat hauls these effects may be normalized, but in a once-off-only programme the total effect may not be determined. It is tempting to consider the population structure of the more abundant species in quantitative terms. With the inherent impositions cited above, however, it is clear that only a tentative semi-quantitative approach is warranted. Furthermore, the results are based primarily upon the catches of one net, the RMT 8. It is accepted that this in itself is a severe restriction. HARRISSON (1967) showed that an increase in net size generally resulted in an increase in size of fish captured. Animals approaching the extreme limits of the catch size spectrum of the net are unlikely, therefore, to be present in the correct proportions of the population available. This further emphasizes the dangers of interpreting population fine structure in quantitative terms.

It has been pointed out that day and night depth pairs were fished to ensure as great a comparability as was practical. As Fig. 1 shows, however, a reasonable degree of repeatability is difficult to achieve over relatively broad depth bands. Handling difficulties, especially in the early stages of a haul at depth, in several cases, resulted in particular depth bands not being fished as well as would be desired (Fig. 1). It should also be pointed out that the day and night depth pairs fished in the upper 50 m are not comparable without a compensation calculation because of differences in time fished (p. 6).

The serious contamination problems apparently inherent in some open nets utilizing cod-end closing devices (BADCOCK, 1970; KRUEGER, 1972) have been eliminated by mouth opening–closing net systems. That some seepage of animals into the net occurs as it passes closed through respective species' distribution layers, however, is inevitable. The amount of contamination resulting from seepage is of a very low order. For example, at *Discovery* Sta. 8281 the RMT 1 + 8 was fished closed at about 2 knots in a continuous oblique manner from 0 to 1000 m depth and back to the surface for a duration of 1 h 31 min. The total catch in the RMT 8 comprised nine specimens of *Cyclothone braueri* and one snipe eel. The former species was the most numerous, and amongst the smallest, represented in other catches from this station, and, furthermore, was distributed vertically almost entirely within the upper 1000 m. The snipe eel was in very abraded condition and may have been hung up from a previous tow. In cases of the more abundant species, e.g. *Cyclothone*, such seepage probably accounts for the apparent distributional tails, and where these tails have been considered a product of contamination, indication is given in the results.

The depth bands fished, apart from those of the upper 100 m of the water column, were relatively coarse. Past work (e.g. PEARCY and LAURS, 1966; PAXTON, 1967; BADCOCK, 1970) has indicated that midwater fishes tend to have relatively discrete vertical distributions. Vertical distribution limits indicated for some species by the present survey, then, are likely to be exaggerated. Moreover, some species may undergo short diel vertical migrations which, because of the vertical coarseness of the fishing bands, may not necessarily be detectable from the results.

In conclusion, although several reservations have had to be made, the overall programme provides for the limits of vertical distribution to be determined per locality for a large number and variety of species with a reasonable degree of confidence. Although the indicated distributions are not absolute in themselves, they should create a sound basis for an inter-area comparison; from which, in conjunction with physical data supplied, a viable assessment of the factors influencing vertical and horizontal distributions may be made. A quantitative analysis of species' population structure is beyond the scope of the results, although a semi-quantitative assessment, approached tentatively, for the more abundant species is undoubtedly of value.

## Hydrography

Temperature, salinity and $\sigma_t$ relations for the three STD casts made during the sampling period are shown in Fig. 2. The variations expressed between the separate probes are within the limits of the normal expected dynamic variability. The profiles are as would be anticipated for the region and season. The isohaline layer is shown to extend from the surface to about 200 m depth, and early effects of summer warming are reflected in tem-

perature and density changes above the isothermal and isopycnal layer of approximately 60–200 m depth. North Atlantic Central Water (SVERDRUP, JOHNSON and FLEMING, 1942) dominates the water column between about 250 and 750 m depth, but below this the influence of Mediterranean Water can be detected. The influence of the latter water is most pronounced between 750 and 1300 m depth. At greater depths the salinity and temperature gradients are relatively smooth and in balance such that little change in density is reflected.

## Systematic account

In all, 17,443 fishes were caught by RMT 8,

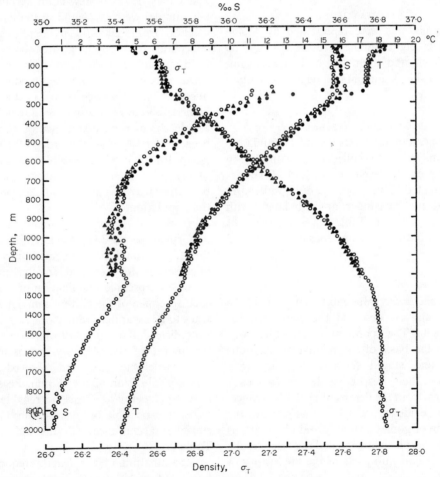

Fig. 2.   Relations of temperature, salinity and $\sigma_t$ with depth (solid triangles, 31/3/72; solid circles, 2/4/72; open circles, 6/4/72).

with a further 289 taken by RMT 1 at 10–0 m depth. The fishes were represented in 37 families, 66 genera, with 98 species identified. Fewer RMT 8 specimens were taken by day (7640 versus 9803), but the night catch was considerably swollen by the exceptionally large catch of *Macrorhamphosus* (see p. 46). The collections were dominated by members of the family Gonostomatidae, numerically represented by approximately 60% of the total catch. The Macrorhamphosidae (23%) followed by the Myctophidae (7%) were the next most abundantly sampled families. The systematic account is presented in three sections in which the more important species are discussed.

STOMIATOIDEI

The stomiatoids were represented by 11,456 specimens in eight families, 18 genera and 33 species. Catch and animal size data are summarized in Table 3, and data on depths of capture provided.

*Family Gonostomatidae (sensu* GREY, 1964)

The Gonostomatidae was the most abundantly caught family in the collections, with 10,386 specimens present. Seven genera and 17 species were represented. Seven species are considered in detail, and the remainder are tabulated, with depths of capture, in Table 3. A summary of vertical distribution for some species is presented in Fig. 3.

*Cyclothone braueri*

*Cyclothone braueri* was most numerous in the collections and so deserved the most detailed consideration. The population structure was studied on the basis of the relative proportions of males and staged females, together with analyses of their respective length–frequency distributions, from the depths of peak abundance. Additional evidence which arose from the examination of gonad histology and observations on sexual dimorphism augmented the basic results to give an insight into parts of the life history of the species.

The catches totalled 6479 specimens, of which 59% were caught by day and 41% by night. Post-larvae were not numerous in the collections; nine are among the day catches from the upper 100 m (Fig. 4). The depth distributions reflected by the day and night catches of adults and juveniles are shown in Fig. 5. Ninety-five per cent of the specimens were caught in 400–600 m depth throughout the 24-h period. Within this layer 70% of the total population occurred in 400–500 m and 25% in 500–600 m depth. The lack of any conspicuous change in this distribution by day or night indicates that *C. braueri* did not undergo an overall diurnal vertical migration, a conclusion common to the findings of other authors (BADCOCK, 1970; GOODYEAR, ZAHURANEC, PUGH and GIBBS, 1972). A few adult specimens (< 30 per sample) were incidental in the day and night catches from 200–400 m depth and in all collections from depths greater than 600 m (Fig. 4). The validity of the latter to indicate a downward extension of the depth distribution is questionable. Leakage into a closed RMT 8 hauled obliquely at normal speed for $1\frac{1}{2}$ h through the depths of peak abundance of *C. braueri* has been found to be of the order of 10 specimens (see p. 10). Thus it is impossible to assess the depth of origin of such comparatively small samples, but either way it is an insignificant proportion of the population.

It is known that *C. braueri* is a sexually dimorphic species. MARSHALL (1967) showed that the adults could be segregated into the larger microsmatic females and smaller macrosmatic males. Microscopic examination of the gonads of each specimen of the current samples allowed the sexes to be separated from about 12 mm. Further, it confirmed the reliability of recognizing males by the size of the olfactory organs in a sample of considerable size and showed that this was possible from about 16 mm. Because of the nature of spermatogenesis no reliable staging of testes was possible but ovarian development was staged in accordance with the indices given in Table 2.

Examination of the length–frequency distribution of males, females and specimens of indeterminate sex (Fig. 6) indicated several features

*Table* 3.   *Resumé of stomiatoid data.*

| Species | Catch Numbers | | | | Depths of Capture (m) | |
|---|---|---|---|---|---|---|
| | Day | Night | total | SL Range (mm) | Day | Night |
| Vinciguerria attenuata (Cocco, 1838) | 6 | 12 | 18 | 10-33 | 50-100[1]; 300-500.[4] | 25-100[1]; 50-200[2]; 200-300[6]; 500-600.[3] |
| V. nimbaria (Jordan & Williams, 1885) | 0 | 4 | 4 | 11-24 | - | 25-50[1]; 50-100.[4,5] |
| V. poweriae (Cocco, 1838) | 3 | 20 | 23 | 15-28 | 400-500.[3,5] | 50-100[1,2]; 200-300[3]; 50-200.[4,5] |
| Ichthyococcus ovatus (Cocco, 1838) | 0 | 1 | 1 | 29 | - | 200-300. |
| Cyclothone acclinidens Garman, 1899 | 5 | 1 | 6 | 31-39 | 800-1000. | 800-900. |
| C. alba Brauer, 1906 | 3 | 0 | 3 | 12-19 | 400-500. | - |
| C. braueri Jespersen & Tåning, 1926 | 3820 | 2659 | 6479 | 9-38 | 10-100; 200-2000. | 200-2000. |
| C. livida Brauer, 1906 | 3 | 3 | 6 | 16-31 | 500-800. | 500-800. |
| C. microdon (Günther, 1878) | 1786 | 924 | 2710 | 11-59 | 50-100; 500-2000. | 400-2000. |
| C. obscura Brauer, 1906 | 0 | 2 | 2 | 50-58 | - | 1500-2000. |
| C. pallida Brauer, 1906 | 47 | 65 | 112 | 16-63 | 600-1250. | 500-1500. |
| C. pseudopallida Mukacheva, 1964 | 112 | 110 | 222 | 14-45 | 400-1250. | 500-1250. |
| Cyclothone sp. post-larvae | 9 | 23 | 32 | 5-11 | 50-100; 400-500. | 25-100; 500-600. |
| Gonostoma bathyphilum (Vaillant, 1888) | 1 | 1 | 2 | 85-133 | 1250-1500. | 1250-1500. |
| G. elongatum Günther, 1878 | 13 | 73 | 86 | 14-150 | 100-200; 500-700. | 25-600; 1250-1500. |
| Bonapartia pedaliota Goode & Bean, 1895 | 23 | 17 | 40 | 12-47 | 100-300. | 100-500. |
| Margrethia obtusirostra Jespersen & Tåning, 1919 | 1 | 3 | 4 | 11-25 | 100-200. | 100-200. |
| Valenciennellus tripunctulatus (Esmark, 1871) | 177 | 459 | 636 | 9.0-30.5 | 100-500. | 100-500. |
| Argyropelecus aculeatus Cuvier & Valenciennes, 1849 | 80 | 75 | 155 | 7-62 | 300-500. | 100-500. |
| A. affinis group | 2 | 2 | 4 | 9-11 | 400-500. | 400-500. |
| A. hemigymnus Cocco, 1829 | 406 | 203 | 609 | 5-31 | 300-600. | 200-600. |
| Argyropelecus sp. post-larvae | 6 | 2 | 8 | 5-9 | | |
| Sternoptyx diaphana Herman, 1781 | 16 | 13 | 29 | 8-29 | 500-900. | 600-900. |
| S. pseudobscura Baird, 1971 | 1 | 1 | 2 | 17-19 | 900-1000. | 900-1000. |
| Sternoptyx sp. post-larvae | 21 | 38 | 59 | 6-10 | 500-700; 900-1000. | 500-800. |
| Chauliodus danae Regan & Trewavas, 1929 | 28 | 41 | 69 | 27-114 | 400-700. | 100-700. |
| C. sloani Bloch & Schneider, 1801 | 11 | 4 | 15 | 26-83 | 400-600. | 400-500; 600-1000. |
| Chauliodus sp. post-larvae | 20 | 60 | 80 | 15-33 | 50-300. | 25-200; 600-800. |
| Stomias sp. post-larva | 1 | 0 | 1 | 26 | 700-800. | - |
| Neonesthes capensis (Gilchrist & Von Bonde, 1924) | 0 | 2 | 2 | 26-124 | - | 50-100; 700-800. |
| Bathophilus vaillanti (Zugmayer, 1911) | 1 | 0 | 1 | 49 | 500-600. | - |
| Echiostoma barbatum Lowe, 1843 | 0 | 1 | 1 | 46 | - | 25-50. |
| Aristostomias grimaldii Zugmayer, 1913 | 1 | 1 | 2 | 47-121 | 900-1000. | 50-100. |
| A. xenostoma Regan & Trewavas, 1930 | 0 | 1 | 1 | 45 | - | 50-100. |
| Malacosteus niger Ayres, 1848 | 0 | 2 | 2 | 98-105 | - | 800-900. |
| Photostomias guernei Collett, 1889 | 4 | 5 | 9 | 62-115 | 600-900. | 200-500. |
| Idiacanthus fasciola Peters, 1877 | 7 | 14 | 21 | ♂37-48 ♀58-280 | 900-1250. 600-2000. | 900-1250. 25-200. |
| Totals | 6614 | 4842 | 11456 | | | |

Key: [1]post larvae; [2]prometamorphic stage; [3]midmetamorphic stage; [4]late postmetamorphic stage; [5]juvenile; [6]adult (Based on AHLSTROM and COUNTS, 1958).

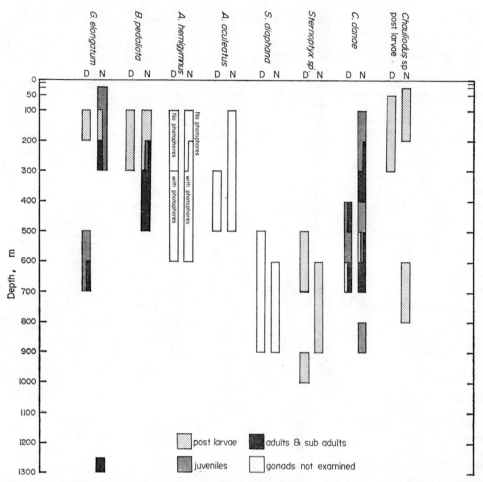

Fig. 3.  Vertical distribution of certain stomiatoid species. (Column width has no quantitative implication.)

of interest. The male population was unimodal, with a peak at 16–18 mm, which contrasted with the bimodal size-structure of females (peaks at 16–18 mm and 23–26 mm, Fig. 6). Consistent reduction occurred in the quantity of both sexes caught by night. This was most apparent among males of 17 and 18 mm and in females of 24–26 mm (Fig. 6), while the rest of the loss was spread throughout the size-ranges. The largest male caught was 26 mm, while females reached 38 mm. The latter measurement may be compared with a fish of maximum size (sex indeterminate) of 36 mm caught from a similar area (Stas. 34–39) on the *Michael Sars* Expedition (Koefoed, 1960). As expected, the highest frequency of specimens of indeterminate sex occurred at the lower end of the overall size-range. Considered as a whole, the length–frequency distribution of the total population is similar in form to that indicated by Jespersen and Tåning (1926, Fig. 10) for *C. braueri* in the Mediterranean, although the peak sizes of the present data are somewhat larger. Murray and Hjort (1912) indicated that two size-groups, but of larger fish, were present in at least the more northerly collections of *C. braueri* from the *Michael Sars* Expedition. As Koefoed (1960) pointed out, however, they (Murray and Hjort, 1912) showed that the average lengths were greater in the northern than in the southern section of the collections, a feature confirmed by hitherto unreported *Discovery* samples from higher latitudes.

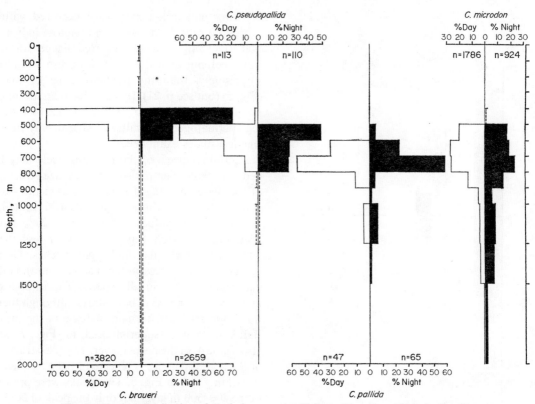

Fig. 4.   Vertical distribution of *Cyclothone* species. Values of less than 1 % in broken line.

There are marked differences in composition of the population of this species within the layers of peak abundance, which are consistent throughout the 24-h period (Fig. 5). The sex ratios of the total day- and night-time samples (1 ♂ : 3·1 ♀ and 1 ♂ : 2·3 ♀, respectively) bear little relation to the situations found within the 400–500 m and 500–600 m layers separately (Fig. 5). Both by day and night the sex ratio was closer to parity in the shallower layer (400–500 m; day, 1 ♂ : 2·5 ♀ and night 1 ♂ : 1·6 ♀, as against 500–600 m; day, 1 ♂ : 9·4 ♀ and night 1 ♂ : 5·3 ♀). Despite the reduced catches at night the relative change between the sex ratios within the two layers was similar. Disparity in composition of the ovarian stages represented also occurred between the two layers (Fig. 5) and again the situation by day and night was found to be similar. The vast majority of immature females (I) were present in the 400–500 m layer. Less than 1 % each of the

total day and night catches of females caught in 500–600 m depth were Stage I fish, while 34 % and 42 %, respectively, were taken from the 400–500 m layer. Similarly, Stage II females were more abundant in the shallower stratum examined (Fig. 5). Females with ovaries in the growth phase of oogenesis (Stage III/V) were the group most evenly distributed between the two layers. This Stage is the most protracted during development and consequently it is difficult to estimate accurately the proximity of the eggs to ovulation (Stage VI). However, in 400–500 m depth the numbers of specimens in Stage VI were negligible (Fig. 5), as were also spent fish (Stage VII). This implies that, if stratification is uniform, those within Stage III/V were possibly closer to Stage II in this depth range. By the same token the reverse situation may be expected among Stage III/V fish from 500–600 m depth, where also the bulk of the representatives of Stages VI and VII

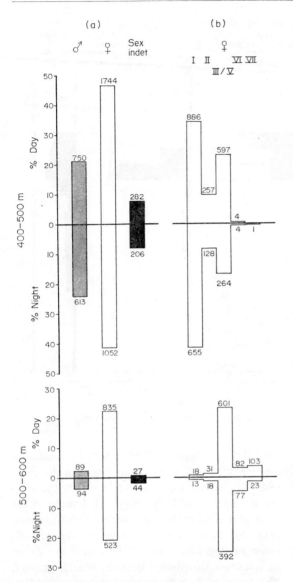

Fig. 5. *Cyclothone braueri.* Frequency, within the depth layers of peak abundance; (a) males, females and sexually indeterminate specimens as percentages of the total catches by day and night, respectively; (b) ovarian maturity stages as percentages of the total day and night catches, respectively.

were caught (Fig. 5). Although the considerably reduced numbers of *C. braueri* from the next deeper layer (600–700 m: day, 23; night, 28) have been omitted from this analysis, the high proportion of Stages III/V, VI and VII fish in these samples suggested that the lower horizon of the

breeding population may have occurred within this depth range. The evidence therefore indicates that spawning females occurred deeper in the water column than the less mature fish and also, surprisingly, than the preponderance of males (Fig. 5) (but see p. 21). It is notable that, although no quantitative method could be devised for staging spermatogenesis, milt was absent only in the smallest males examined.

Certain aspects of the breeding cycle may be deduced from comparison of the size-structure of the male and staged female components from 400–500 m with those from 500–600 m (Fig. 7). For clarity, values of less than 0·20% have been omitted from Fig. 7, a loss of less than 4·5% each of the day and night catches. Good agreement exits between the results from the day and night catches in all aspects. The sample of males was unimodal in both strata, although there was approximately a 1 mm increase in the mean standard length (superimposed in Fig. 7—see also Table 4) in the deeper layer. The size-structure of the staged females is reflected by the frequency situation given in Fig. 5. Two peaks were present in the 400–500 m layer but only the peak of larger-sized fish appeared in the deeper stratum. Similar size-stratification was reported by JESPERSEN and TÅNING (1926) and GOODYEAR, ZAHURANEC, PUGH and GIBBS (1972) for this species in the Mediterranean, and by BADCOCK (1970) from samples off the Canary Islands (in 28°N, 14°W). The shallower stratum peaks correspond with Stages I and III/V, respectively, and are linked by the intermediate Stage II females (Fig. 7 and cf. Fig. 5). The separation of the means of Stage I and III/V (see Fig. 7 and Table 4) is highly significant from both day and night data. Such findings accord with the observations of GOODYEAR, ZAHURANEC, PUGH and GIBBS (1972) that, among western Mediterranean specimens, smaller females generally had small eggs, while most large females had large, clearly visible ova. The single peak of females in 500–600 m depth was comprised predominantly of Stage III/V fish, but others in Stages VI and VII were present (Fig. 7 and cf. Fig. 5). In contrast to the well-separated means of female stages in the samples

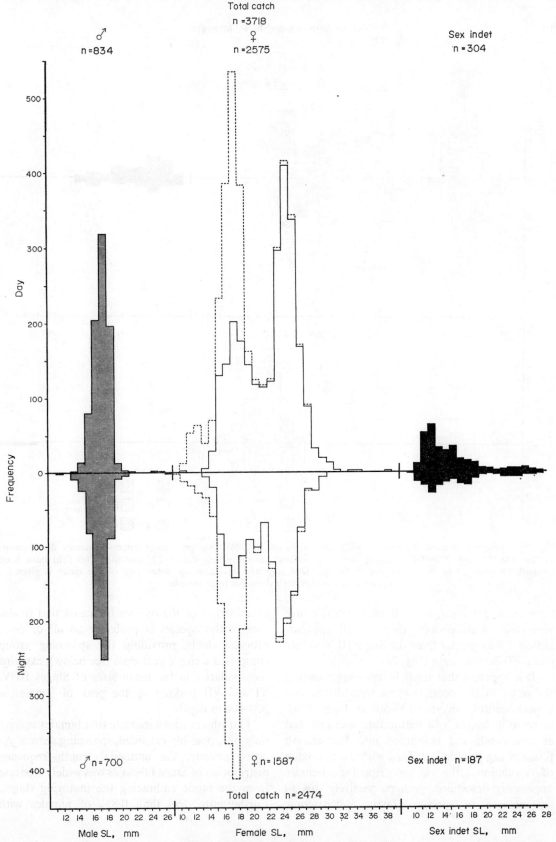

Fig. 6. *Cyclothone braueri*. Length–frequency distribution of the measurable total (pecked line), together with male, female and sexually indeterminate components.

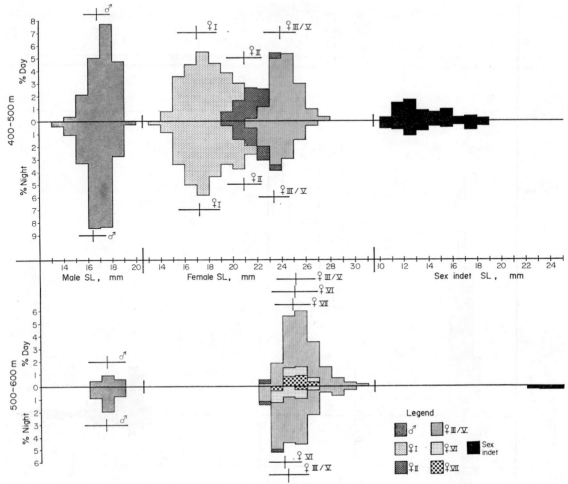

Fig. 7. *Cyclothone braueri*. Day (*n* = 3650) and night (*n* = 2384) total percentage length–frequency distributions in the 400–500 and 500–600 m depth layers. Individual values of less than 0·2% (totalling 266 fish) have been omitted. Ovarian maturity stages are indicated. One standard deviation on either side of the mean is given for each stage, together with those for male samples.

from the upper layer, those from 500–600 m are very close. Furthermore, they are all approximately 1 mm greater than the Stage III/V values from 400–500 m depth (Fig. 7).

It is apparent that small females approaching the onset of the oogenic cycle were distributed almost entirely in the 400–500 m layer. With increase in length, ova maturation was initiated at this depth and continued into the growth phase of egg production. Then with the approach of ovulation, the larger ripening females apparently descended, perhaps passively due to an increase in specific gravity. Microscopic

examination of the ovaries indicated that ovulation in this species is probably an all or none process which, providing the spawning group comprised a single year class (see below), explains the similarity in the mean sizes of Stages III/V, VI and VII making up the peak of females in 500–600 m depth.

Our observations indicate that females survive only one, possibly extended, spawning season. As noted already, the unimodal length–frequency distribution of Stage I females was widely separate from the mode embracing the maturing stages. Furthermore, less than 0·8% of females with

*Table 4. Summary of statistics of gonad maturity stages of* Cyclothone braueri.

| | | | ♂ | ♀I | ♀II | ♀III/V | ♀VI | ♀VII |
|---|---|---|---|---|---|---|---|---|
| 400–500m | DAY #2 | Number | 743 | 865 | 257 | 586 | | |
| | | Length max | 24 | 23 | 28 | 34 | | |
| | | min | 13 | 10 | 17 | 20 | | |
| | | Mean | 16.74 | 17.06 | 20.96 | 23.91 | | |
| | | Standard Deviation | 1.08 | 1.64 | 1.42 | 1.30 | | |
| | NIGHT #8 | Number | 586 | 617 | 122 | 258 | | |
| | | Length max | 20 | 23 | 26 | 28 | | |
| | | min | 11 | 13 | 17 | 20 | | |
| | | Mean | 16.33 | 17.19 | 20.93 | 23.35 | | |
| | | Standard Deviation | 1.14 | 1.73 | 1.38 | 1.27 | | |
| 500–600m | DAY #16 | Number | 86 | | | 593 | 80 | 99 |
| | | Length max | 25 | | | 31 | 33 | 30 |
| | | min | 15 | | | 21 | 22 | 21 |
| | | Mean | 17.42 | | | 25.11 | 25.02 | 24.83 |
| | | Standard Deviation | 1.56 | | | 1.59 | 1.91 | 1.49 |
| | NIGHT #20 | Number | 94 | | | 376 | 75 | |
| | | Length max | 26 | | | 30 | 27 | |
| | | min | 15 | | | 21 | 20 | |
| | | Mean | 17.35 | | | 24.41 | 24.12 | |
| | | Standard Deviation | 1.83 | | | 1.64 | 1.33 | |

immature ovaries were longer than 21 mm (cf. Fig. 7, 400–500 m) and of these only four specimens, each of 25 mm, were found to have recovering spent ovaries from microscopic examination. This contrasts with the expected polymodal length–frequency distribution of fish in an ovarian stage prior to yolk formation in a population in which females re-enter the oogenic cycle. In support of this view also is the fact that sections cut from ovaries close to maturity contained a majority of large eggs of similar size and very few resting oocytes of a subsequent generation. The ovaries of spent fish were found to be almost completely empty sacs. The maturation of the resting oocytes of the early spawners of the current season was possibly evidenced by a small proportion of females of moderate to large size in which were found unusually small Stage III/V ovaries.

JESPERSEN and TÅNING (1926) concluded that,

in the Mediterranean, *C. braueri* spawned in the spring and summer half of the year. GOODYEAR, ZAHURANEC, PUGH and GIBBS (1972) similarly suggested that spawning was occurring at the end of August in the western, if not the eastern, Mediterranean. The gonads examined from the present collections, made largely during April, show that spawning occurs in spring at least, in the eastern North Atlantic. Moreover, the lack of substantial numbers of post-larvae in the RMT 8 in the surface layers, matched by a similar dearth in the RMT 1, indicates that spawning could only recently have begun.

The fecundity of ripe females was found to be similar to the 200–300 eggs per individual found among Mediterranean *C. braueri* by JESPERSEN and TÅNING (1926). Total egg counts were made from the ovary pairs of eight specimens (Table 5) from 30°N, 23°W. The fecundity varied as 218–416 eggs, with like distribution between left and right ovaries. However, both the above results vary considerably from the estimate of about 1000 made by MURRAY and HJORT (1912) for *C. signata* (= *braueri*). No position of capture accompanied this estimate, but our own hitherto unpublished data from more northerly stations show that females of larger size than those found in lower latitudes contained up to nearly 900 eggs. The diameter of ovulated eggs remaining in the ovary was found to be about 0·5 mm. This compares with values of 0·46 mm given by MURRAY and HJORT (1912) and 0·3 mm by JESPERSEN and TÅNING (1926). It is likely that the discrepancy of the latter was due to measurement prior to the uptake of fluid by the ripe egg at ovulation.

On the basis of the bimodality of their length–frequency data, JESPERSEN and TÅNING (1926) suggested that *C. braueri* lives no longer than 2 years. This was based upon the assumption that the second mode was composed of a single age-class, since nothing was known on the growth rate of the species. The growth rate of *C. braueri* still has to be elucidated, but the present data serve to confirm JESPERSEN and TÅNING's (1926) findings (Fig. 6), and show a similar overall situation to that found in the Mediterranean in winter. The modal sizes of the males and Stage I females coincide and could represent a single age-class. However, assuming normal mortality, the samples imply (Fig. 6) that the size-class of larger females was stronger than that of the smaller ones. It is likely that the modal size of the latter peak (16–18 mm), however, is close to the minimum size entirely retained by the meshes of the RMT 8 net. Data on euphausiids suggest that only above about 15 mm total length will all individuals be retained (P. T. James, personal communication). Thus the true peak may easily be larger and occur at a smaller size than our data suggest. Alternatively sex-reversal may operate in this species, as was thought possible for the genus by KAWAGUCHI and MARUMO (1967). Yet evidence in support of this is entirely lacking at present. There is no indication of intermediate stages between male and female gonads from the many microscopic examinations made. Also, the male length–frequency peak entirely overlaps that of

immature female fish. Nevertheless, sex-reversal, which can occur in the related *C. microdon* (see p. 22), cannot be completely ruled out as it could take place among males in a season other than that sampled. Alternatively, males might be the shorter lived sex, as suggested by MARSHALL (1967), in which case gonad maturation would occur earlier in males than in females. If this is the case, the present data (Fig. 5) suggest that spawning males would be one age-class out of phase with their mates. A third possibility is that growth in males may terminate earlier than in females, with the result that the modal group would be comprised of more than one age-class.

MARSHALL (1967) noted that while a diverse variety of bathypelagic fishes have macrosmatic males, *C. braueri* is unusual among mesopelagic species in this respect. Although he mentioned *C. braueri*, he did not describe the form of the olfactory organs, nor observe any other sexually dimorphic character than size within the genus. Little can be added here to the general comparative morphology of the olfactory nerves and forebrain of male or female *Cyclothone* discussed by MARSHALL (1967). Nevertheless, male *C. braueri* from the present collections show a stage of development of the olfactory organs together with a prolongation of the snout beyond that described from *C. microdon* by MARSHALL (1967). A similar form of development in *C. microdon* was also found, suggesting, perhaps, that the features described below are common to all *Cyclothone* species.

The form of the olfactory organ in *C. braueri* and its development are shown in Fig. 8. At 17 mm (Fig. 8) the olfactory organ is proportionally larger in males than females, although neither the anterior nostril nor the lamellae protrude much above the dorsal profile of the snout. The anterior part of the snout is terminated by the premaxillary bone structure. Unabraded specimens of a larger size (*ca.* 19 mm, Fig. 8) have a much enlarged and elaborated olfactory organ. This apparently obscures the fish's forward vision which, judging from the overlying bone and musculature, is unlikely to be acute. The lamellae of the olfactory organ are erect and feathery, protruding above the

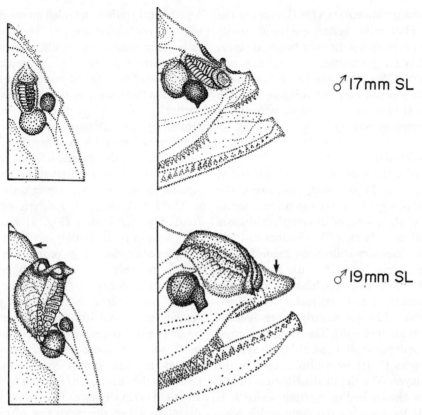

♂ 17 mm SL

♂ 19 mm SL

Fig. 8.  *Cyclothone braueri*. Structural development of the nasal rosette in males. Fleshy prolongation of the snout is arrowed.

dorsal profile of the snout. The organs themselves are V-shaped with the apex lying over the inner margin of the eye. The anterior nostril in larger males protrudes laterally and is almost figure-of-eight shaped. A thin membrane which originates from the dorsal side of the inner half of this nostril spreads backwards in the form of a hood, over most of the inner limb of the olfactory organ to the circumference of the organ above the eye. The olfactory organ extends forwards to be level with the front of the mouth (Fig. 8). However, protruding anteriorly from between the two organs is a fleshy prolongation of the snout (arrowed, Fig. 8), which extends beyond the level of the lower jaw. It is possible that the purpose of this delicate structure is to provide a hydrodynamic fairing to allow a smooth water flow through the nostrils as the fish swims.

It is worthwhile considering the significance of the comparatively late development of the olfactory organs in male *C. braueri* in relation to spawning activity. While no quantitative staging of spermatogenesis was possible as milt occurred in all but the smallest testes, in general the most turgid testes were found in the larger males. Moreover, full development of the olfactory organ and, presumably, the optimum capacity for olfactory homing on females also occurs only in the larger males. Assuming uniform abrasion during capture, a higher proportion of larger males with the delicate olfactory organ and snout developed were found in the 500–600 m sample. Within this layer, from which 25% of the total sample was caught (Fig. 7), the ratio of males to running ripe females was approximately 1 : 1 (day, 89 ♂ : 82 ♀ VI and night, 94 ♂ : 77 ♀ VI). Thus while there was a preponderance of potentially ripe males in the overall population, similar to the situation found by MARSHALL (1967), more of the larger and olfactorily better equipped males occupied the lower part of the distributional range, where the highest proportion of running ripe females occurred. So, while the full development of the olfactory organ may not coincide with the onset of sexual viability, it may well correspond with peak spawning activity.

## Cyclothone microdon

A total of 2710 specimens of *C. microdon*, ranging in size from 11 to 59 mm, was caught. The abundance of this species in the catches was somewhat unexpected, in view of its complete absence in samples taken in 28°N, 14°W (BADCOCK, 1970) and in other *Discovery* collections made further south in the eastern North Atlantic (Institute of Oceanographic Sciences, unpublished). As in the case of *C. braueri*, more *C. microdon* were caught by day (66%), although slightly more females were caught at night (Fig. 9). The species occurred in 500–2000 m depth at all times, although at night a few specimens (four) were also caught in the 400–500 m layer. The depth distribution of the population is shown in Fig. 4, from which it is clear that, at least among the comparable tows (p. 6) in the upper 1000 m, a broad peak in abundance occurred in 500–900 m depth. The marked similarity between the day and night distributions indicates the absence of overall diel vertical migration.

A sample of the smaller specimens indicated that below about 22 mm the only distinguishable gonads were testes, but most fish were too small for accurate sex determination. In each case among this small proportion of recognizable males, the sex could be confirmed by external examination of the olfactory organ (see MARSHALL, 1967, and p. 12). Thus only obvious males were finally separated from the group of smallest specimens. Larger fish (> 22 mm), however, were all dissected and females staged in accordance with Table 2 (see also p. 8). Among those dissected was a specimen of 25 mm (from #4, 800–900 m depth) in which was found an ovotestis. This specimen was macrosmatic and internally the testicular portion of the gonad was considerably the larger. Nevertheless,

histological sections revealed an enveloping region of tissue containing oogonia. This specimen might have been treated as an abnormality, and been overlooked had not 3% of more than 3000 *C. microdon* been found in more or less the same condition from samples taken by day, in a similar vertical series of collections, off Bermuda (Institute of Oceanographic Sciences, unpublished). From this evidence that sex reversal is not unusual in the species, it is not surprising that MARSHALL's (1967) observation that males are smaller than females was confirmed (♂ 16–30 mm; ♀ 23–59 mm; Fig. 9). Analysis of the length–frequency distribution (Fig. 9) revealed similar situations in both the day and night catches. The samples contained a high proportion of small fish, which possibly indicates a bimodal length–frequency distribution (Fig. 9). The greater part of the length range, however, was poorly represented and showed little evidence of modality.

A positive correlation between individual size in *C. microdon* and depth of capture was found, similar to that shown for *C. braueri*, which confirms the conclusions reached by MURRAY and HJORT (1912). Figure 10 shows the standard deviation about the mean size of males, females and specimens of indeterminate sex by depth range of capture. The increase in length with increased depth is greater for females than males, but further emphasizes the absence of diel vertical migration. Females also appeared to extend deeper than males (Fig. 10). Such evidence as there is suggests a steady increase in the relative number of females with depth. Thus, by day the sex ratio in 700–800 m depth was 1 ♂ : 0·1 ♀ with an increase of females to 1 ♂ : 2·3 ♀ in 1250–1500 m depth; likewise by night the sex ratio changed from 1 ♂ : 0·1 ♀ to 1 ♂ : 3·5 ♀ between the same depth ranges. The trend is therefore similar to that indicated for *C. braueri* within its depth of maximum abundance. The overall ratio in the sexed proportion of the *C. microdon* population was 1 ♂ : 0·6 ♀ by day and 1 ♂ : 0·8 ♀ by night.

Despite the paucity of maturing females, it is implicit from Table 6 and Fig. 10 that, at least during early ovarian development, there was a positive correlation between depth and gonad

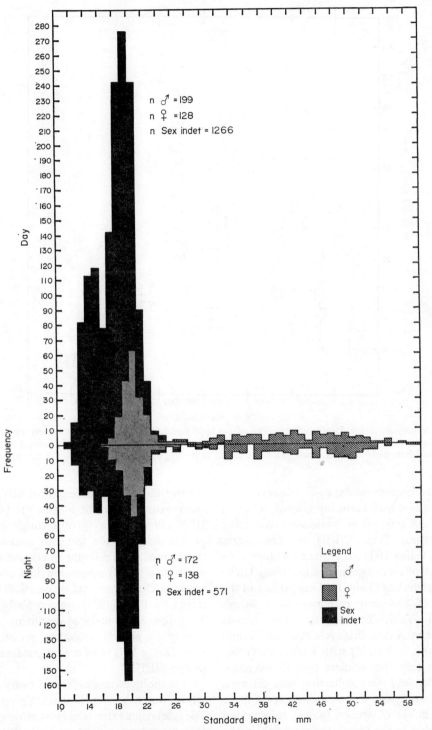

Fig. 9. *Cyclothone microdon*. Length–frequency distribution of day and night measurable totals, with males, females and sexually indeterminate individuals indicated.

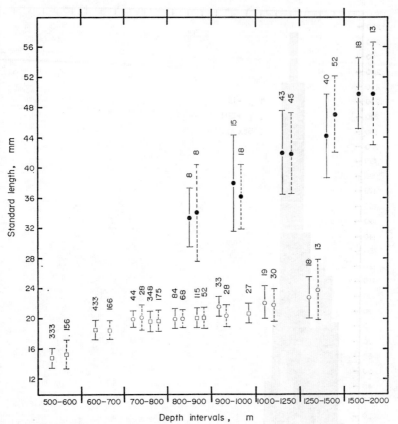

Fig. 10.   *Cyclothone microdon.* Depth–size relations of the total sample indicated by one standard deviation about the means of males, females and sexually indeterminate individuals by depth range of capture. (Solid line, day samples; pecked line, night samples; open circles, ♂; closed circles, ♀; open squares, sexually indeterminate specimens.)

maturity, in a manner similar to *C. braueri* (p. 16). Thus no females with maturing gonads occurred shallower than 800–900 m, while such recovering spent specimens (Stage VII/II) as were caught came from below 1000 m. Females with ovaries in the growth phase of egg maturation (Stage III/V) were most abundant (Table 6). Comparison of the size-range of this and earlier ovarian Stages (I and II—Table 6) suggests that female *C. microdon* enter their first cycle at about 25 mm. This compares favourably with KOEFOED's (1960) observation that the smallest ripe *C. microdon* from the *Michael Sars* collection was 30 mm. Evidence that female *C. microdon* spawn more than once, unlike *C. braueri* (p. 18), is provided by the overlap in the length range of fish in the maturation Stages (II, III/V and VI—Table 6) of

the ovarian cycle. Further, the difference between the maximum sizes of Stages VII (49 mm) and III/V (59 mm) may give a rough indication of growth between the last two spawning seasons represented. It is relevant to comment here upon KOEFOED's (1960) suggestion that the emaciated specimens of *C. microdon*, mentioned and illustrated by PARR (1934), were probably spent fish. Very few such meagre specimens were found among the present collection, yet all were either spent (Stage VII) or in the early stages of recovery (Stage VII/II).

Fecundity estimates from 3 ovary pairs (Stage III/V) from fish of 45–49 mm are given in Table 5. The total counts from the most advanced group of eggs ranged from 2184 to 3301. These are considerably lower than the 10,000 estimated for

Table 5. *Total fecundity counts of* Cyclothone braueri *and* C. microdon.

| # | Depth (m) | SL (mm) | Left ovary | Right ovary | Total |
|---|---|---|---|---|---|
| **C. braueri** | | | | | |
| 2 | 400-500 | 24 | - | - | 299 |
| 2 | 400-500 | 29 | - | - | 281 |
| 16 | 500-600 | 24 | 125 | 196 | 321 |
| 16 | 500-600 | 25 | - | - | 416 |
| 16 | 500-600 | 25 | 127 | 159 | 286 |
| 16 | 500-600 | 26 | 125 | 93 | 218 |
| 20 | 500-600 | 25 | 173 | 170 | 343 |
| 20 | 500-600 | 26 | 172 | 193 | 365 |
| **C. microdon** | | | | | |
| 48 | 1000-1250 | 45 | 1049 | 1135 | 2184 |
| 52 | 1250-1500 | 49 | - | - | 3301 |
| 52 | 1250-1500 | 49 | - | - | 2599 |

Table 6. *Percentage occurrence of ovarian maturity stages from the total day and night catches of* Cyclothone microdon, *together with length range per stage.*

| Ovarian Maturity Stage | Percentage Occurrence | | Length range (mm) |
|---|---|---|---|
| | Day (n=135) | Night (n=138) | |
| I | 3.7 | 2.2 | 23-31 |
| II | 5.2 | 8.0 | 28-51 |
| III/V | 77.8 | 83.5 | 26-59 |
| VI | 10.4 | 2.9 | 32-51 |
| VII | 2.2 | 1.4 | 35-49 |
| VII/II | 0.7 | 2.2 | 41-49 |

*C. microdon* by MURRAY and HJORT (1912). It has already been observed that the fecundity of *C. braueri* is greater in the larger specimens found in the northern parts of its range in the eastern North Atlantic (p. 20), and similarly such latitudinal variation may account for the above discrepancy in the fecundity of *C. microdon*. MURRAY and HJORT (1912) gave the diameter of ripe *C. microdon* eggs as 0·56 mm, which is conspicuously larger than the mean of 0·3 mm diameter found from ripe fish in the present study.

In conclusion, it is notable that the main feature of the *C. microdon* population in 30°N, 23°W was the high percentage of adolescent specimens. This contrasted with the small percentage of ripening fish (Stages II and III/V), a much lesser proportion of running ripe individuals (Stage VI) and an almost complete dearth of *Cyclothone* spp. post-larvae (total catch, 39, including nine *C. braueri*). The low proportion of ripe or ripening females and the scarcity of post-larvae combine to suggest that the spawning season was in its earliest stages, while the abundance of adolescents must have originated from a previous spawning. The data available are insufficient to show whether or not these samples reflect a stable situation in the area. Evaluation of this would be possible if the effects of seasonality, growth rate and proximity to the extremity of the geographical range of this species were investigated.

### Cyclothone pseudopallida

BADCOCK (1970) first recorded *C. pseudopallida* as such in the eastern North Atlantic from 308 specimens collected in 28°N, 14°W. The present collections total 222 juveniles and adults (12–45 mm) caught in approximately equal numbers by day and night. The overall depth distribution found (400–1250 m depth, Fig. 4) was more extended than the more southerly captures indicated (550–800 m depth; BADCOCK, 1970). The depth of peak abundance was, however, 500–800 m depth by both day and night. The similarity of day and night samples from the strata within these limits suggests that no diel vertical migration is made by the species.

To facilitate further analysis of the data, males were separated from females either externally by the size of the olfactory organ (see MARSHALL, 1967), or by internal examination of the gonads. Females were staged in accordance with Table 2. Although the data are not numerous, it is clear from the length–frequency distribution of the total samples (males, staged females and specimens of indeterminate sex—Fig. 11) that the

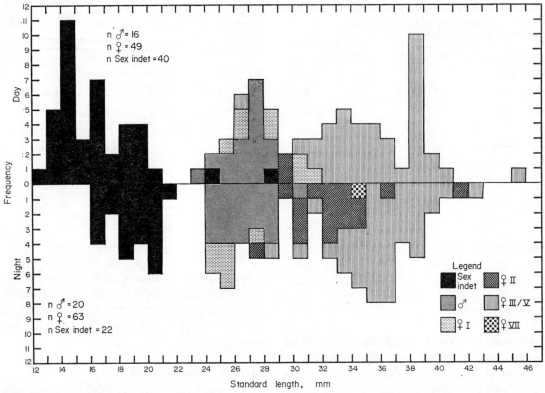

Fig. 11. *Cyclothone pseudopallida.* Length–frequency distribution of day and night measurable totals with males, females and sexually indeterminate individuals indicated. Ovarian maturity stages are also shown.

population was multimodal with broad agreement between day and night catches. The peak of the smallest specimens (12–22 mm) was comprised entirely of fish of indeterminate sex. With increased size, the next peak (23–29 mm) contained predominantly males (range 23–30 mm). The length–frequency distribution of specimens larger than 31 mm was less easy to interpret, but was comprised entirely of females; thus indicating that female *C. pseudopallida* grow larger than males, as in other species of the genus. Females were separable from a size of 24 mm and overlapped the lengths of males over the range 24–30 mm (see above). Microscopic examination of the testes of *C. pseudopallida* gave no evidence of sex-reversal.

As in the case of *C. braueri* and *C. microdon*, the data for *C. pseudopallida* show a positive correlation between size and depth of capture (Fig. 12). In the upper part of the vertical range,

males were most abundant by day and night (500–600 m, $\delta$ : $\varphi$, day, 1 : 0·7, night, 1 : 0·9). They decreased numerically and proportionally with depth (600–700 m, $\delta$ : $\varphi$, day, 1 : 13·5, night, 1 : 7·7; 700–800 m, night only, 1 : 12). No males were caught deeper than 600–700 m depth by day and 700–800 m by night. The overall sex ratio was close to 1 : 3 in both sets of samples. Small females, in Stages I and II, were most abundant from the shallower stratum (500–600 m, Fig. 12). Larger, Stage III/V, females were most numerous among the catches from the central stratum of peak abundance (600–700 m, Fig. 12), and a similar situation was apparent in the next layer deeper (700–800 m). The single spent female occurred in the night sample 600–700 m depth. Thus a similar situation existed to that found for *C. braueri* and *C. microdon*, in which females in the more advanced stages of maturity were caught below males and less advanced females.

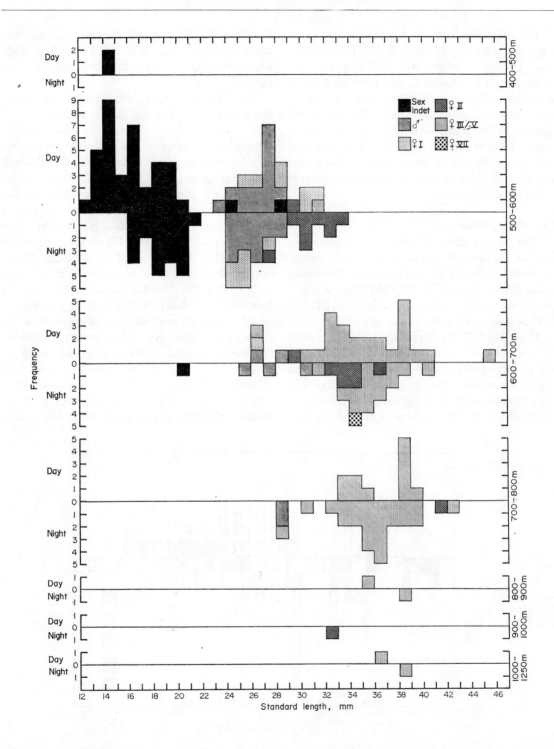

Fig. 12.  *Cyclothone pseudopallida*. Depth–size relation of total measurable day and night samples of males (*n* = 36) staged females (*n* = 112) and indeterminate individuals (*n* = 62) by depth of capture.

It is concluded from the data that first maturity is reached at least by 26 mm (Fig. 12). The distribution of maturity stages (Fig. 11) suggests that, unlike female *C. braueri*, *C. pseudopallida* females spawn more than once. Further, in relation to the apparent length–frequency distribution of staged females, evidence provided by certain individuals (Stage III/V) at 26 mm, Stage VII at 34 mm, and Stage II at 36 and 41 mm) suggests that representatives of three or four age groups are present in the total collection.

## Cyclothone pallida

The present series of collections contained a total of 112 juvenile and adult *C. pallida* (range, 16–63 mm), of which 47 specimens (42%) occurred in the day catches and 65 (58%) in those at night. By day the species was distributed between 600 and 1250 m depth, while the range extended at night to 500–1500 m. The samples showed that at least 80% of the total population was situated in 600–800 m throughout the 24-h period (Fig. 4). The degree of agreement existing between day and night samples from this principal distribution layer indicated that the species is not a diurnal migrant.

BADCOCK (1970) reached the same conclusion based on data from 28°N, 14°W for *C. pallida*, although there the upper distributional limit was about 750 m depth.

Sex determination was feasible in most specimens of 28 mm and larger. Recognition of males was again facilitated by the sexually dimorphic character of the olfactory organ, and females were staged in accordance with Table 2. No clear population structure emerged from analysis of the length–frequency distribution of the total catch (Fig. 13). Little diel variation in the structure of the population is apparent either from the total catch (Fig. 13) or the component samples (Fig. 14). It is clear that although males apparently do not grow as large as females, their length range (28–42 mm) is embraced by that of females (28–63 mm). No evidence of sex-reversal was found from microscopic examination of the testes of *C. pallida*. As with congeners, a positive correlation existed between size and depth of capture (Fig. 14). Thus males occurred with increased size down to 900 m depth, whilst a similar trend in females was apparent over the total depth range of the species (Fig. 14). Such a

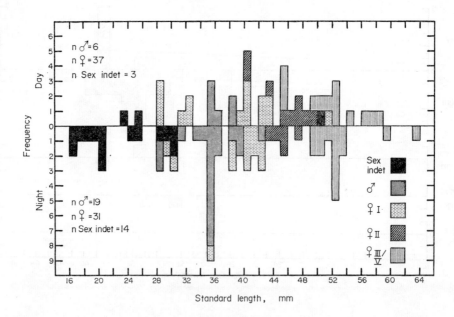

Fig. 13. *Cyclothone pallida*. Length–frequency distribution of day and night measurable totals with males, staged females and sexually indeterminate specimens indicated.

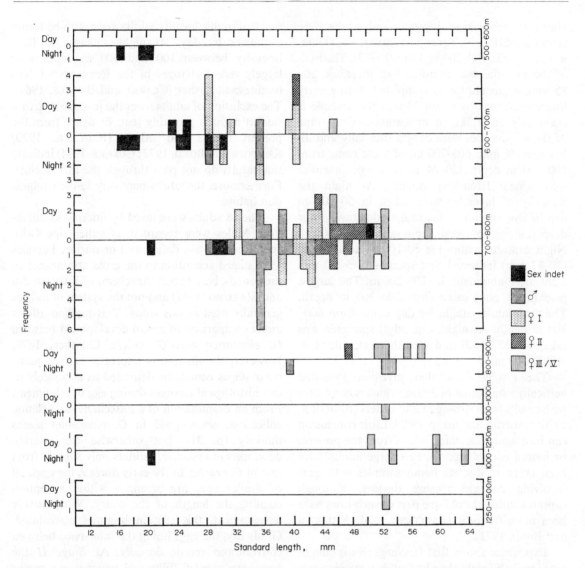

Fig. 14. *Cyclothone pallida*. Depth–size relation of day and night measurable total samples of males (*n* = 25), staged females (*n* = 68) and sexually indeterminate individuals (*n* = 17) by depth range of capture.

discrepancy in depth distribution between the sexes inevitably introduced a stratification of the sex ratio, although no conclusive trend emerged from the few data by day or night. The overall sex ratio shows a considerable variation between day (♂ : ♀, 1 : 6·2) and night (♂ : ♀, 1 : 1·7). Consideration of the length–frequency distribution of the total sample (Fig. 13) in terms of the ovarian maturity stages present suggests that first maturity is attained in females at about 40–45 mm. The

apparent positive correlation between size and maturity stage leads to a depth stratification of gonad maturity (Fig. 14), as in related species sampled during these collections. The evidence is insufficient to indicate whether or not individuals spawn more than once.

*Gonostoma elongatum*

The vertical distribution data are sparse. Of 86 individuals caught, only 13 were taken by day.

However, post-larvae, juveniles and adults were represented (Fig. 3). Three post-larvae, 14–20 mm, were caught in 100–200 m depth (Fig. 3). The bulk of the material was comprised of juveniles, 20–85 mm. Pigmentation is completed during early juvenile life and by about 23 mm the juvenile is essentially adult-like in appearance. Only nine of the 65 juveniles were caught diurnally and all but one (46 mm, 600–700 m) of these came from 500–600 m depth (20–44 mm). Large juveniles were absent from day catches. At night the majority of juveniles were taken in 50–200 m depth. One specimen was caught shallower, four deeper, giving an overall depth range of 25–600 m. Night captures within the 50–200 m depth band (20–67 mm) indicated that specimens > 50 mm were more abundant in 100–200 m. The largest juvenile (85 mm) came from 200–300 m depth. The three adults caught by day came from 600–700 m depth. At night, one adult specimen was taken at 1250–1500 m depth; the remainder (14) at 200–300 m.

The results show that juveniles migrated vertically a minimum of 300 m, which is consistent with results from closing nets of others (BADCOCK, 1970; KRUEGER and BOND, 1972). Little conclusion can be drawn concerning adults from the present, or indeed earlier reports. Few large adults have been taken in discrete depth samples with gear involving cod-end closing devices, although captures within the oblique part of such tows have been more frequent (BADCOCK, 1970; KRUEGER and BOND, 1972).

Experience shows that G. elongatum is caught more easily by night than by day. Net avoidance is possibly reduced by night, and the lower light level nocturnally is presumed important in this respect. From the present series, it is unlikely that juveniles should occupy waters shallower than 500 m depth by day. Possibly adults do not live deeper than juveniles diurnally as indicated (Fig. 3; above) since the increased ambient light intensity may enhance potential avoidance behaviour. On the other hand, the adults of the species are essentially adapted to the deeper mesopelagic environment. The few adult captures made at 600–700 m depth are, therefore, taken as indica-

tive that in this locality adults are not to be found shallower by day. At night, changes in light intensity between 100 and 300 m depth may largely reflect changes in the frequency of bio-luminescent flashes (CLARKE and BACKUS, 1964). The exclusion of adults from the juvenile distribution at night is probably real. Evidence from the present results and others (BADCOCK, 1970; KRUEGER and BOND, 1972; CLARKE, 1974) indicate that adults do not pass through the thermocline. Furthermore, the adults may vary in their migration habits.

The 18 adults were sexed by internal examination. Males were regarded as either ripe (with sperm in the vasa deferentia) or unripe. Females were staged according to the gross appearance of the gonads, based upon the scheme of KAWAGUCHI and MARUMO (1967) and not the system of indices generally used in this study. This method allows direct comparison of gonad development between G. elongatum and G. gracile Günther, 1878. However, the method is imprecise in that particular stages cannot be delimited as accurately as can histological changes during egg development. From an examination of a miscellaneous Atlantic collection, sex-reversal in G. elongatum seems unlikely (p. 31), but otherwise the ovarian development apparently differs only slightly from that of G. gracile. In the early Stage I, the eggs, all of similar size, are arranged within a septum running the length of the ovary. With further development, this septum becomes convoluted. Gradually, the eggs lining the interspace between convolutions recede dorsally. At Stage II the eggs have receded fully, and under a low power binocular microscope, the ovary has well-established, thick, regular cross-striations in it. (These striations, ovarian convolutions, may be seen in Stage I, where they are irregular and half formed in early stages, to evenly but widely spaced in later stages.) In Stage III the septal convolutions are closely packed together. In terms of actual egg development, initially, the eggs are clear; yolk does not form until late ovarian Stage II; and the chorion is formed at about ovarian Stage III.

Fifteen adults of the current series were males

(112–150 mm) and of these only three were apparently unripe (114, 119 and oddly 150 mm). The ripe specimens (112–149 mm) varied in proportions of the generating parts of the testes, with the proportion normally increasing with increased length of fish. The three females (120–126 mm) were all in ovarian *Stage I* condition.

A number of factors may have contributed to the lack of females in the series collection; for example, seasonality, horizontal displacement, net evasion and the lack of repeat trawling. KRUEGER and BOND (1972), in their Bermudan studies, suggested that a "reduced number of large specimens in September" was due to post-spawning death. These authors, however, gave no indication of sex determination and based their adult category upon GREY's (1964) assessment. Further light on these aspects is shed by the 38 extra specimens (111–250 mm) examined from miscellaneous eastern North Atlantic collections. Of these specimens, seven were males (111–143 mm), one of which was unripe (118 mm). In all then, a sample of 56 adults—22 males (111–150 mm) and 34 females (115–250 mm)—was investigated. The number of females per ovarian stage were as follows: *Stage I*, 11, 115–142 mm; *Stage II*, 7, 140–183 mm; *Stage III*, 11, 166–250 mm; *spent*, but in recovering condition, 5, 188–226 mm. The specimens represent a limited and select sample, but it is probably valid to record that females attain a much greater length than do males. Sexual dimorphism in gonostomatids occurs in *Cyclothone* spp. (MARSHALL, 1967, and see p. 20), *Gonostoma gracile* (KAWAGUCHI and MARUMO, 1967) and *Valenciennellus tripunctalatus* (KRUEGER, 1972). The spent females were in a recovering condition, suggesting that in the eastern North Atlantic individuals are able to spawn more than once. Moreover, the extreme size-range of ovarian *Stage III* specimens implies that animals of about 200 mm may be approaching their first spawning; whilst the two fish, 238 and 250 mm, may be in their second cycle. The conclusion reached by KRUEGER and BOND (1972) that *G. elongatum* reaches maturity at *ca.* 150–250 mm (in both sexes) does not seem to fit with the present evidence. GREY (1964) cited mature

specimens of both sexes of 197–250 mm length, but unfortunately did not qualify them further. During the course of this study, it looked as though representatives of the species off Bermuda may attain initial maturity at lengths greater than is found in eastern North Atlantic individuals. Twenty-nine specimens, 56–245 mm, caught in 32°N, 64°W during March 1973 were therefore examined. Recognizable males (6) gave a size-range 106–148 mm, the smallest ripe specimen being 122 mm. Females (8) were of size-range 153–245 mm. Ovarian *Stage I* was represented by a single specimen, 153 mm; *Stage II* by 6, 206–225 mm; and *Stage III* by one individual, 245 mm. Limited though it is, the evidence suggests that initial maturity at a greater size is indeed the case for the species in Bermudan waters. Sexual dimorphism, however, is maintained, and it would seem that KRUEGER and BOND (1972) inadvertently excluded a potentially large proportion of ripe males in their assessment of maturity size. According to their figures (KRUEGER and BOND, 1972, Table 7) fish of the size-range 120–169 mm, were absent from June collections (larger specimens were caught, though). In September this size-range was represented by individuals of 120–149 mm, although fewer larger specimens (maturing females?) were caught. In view of the small number of animals involved overall, however, it would seem wiser to attribute these discrepancies to chance than to any implied biological significance.

KAWAGUCHI and MARUMO (1967) examined the gonads of *G. gracile* and presented evidence showing it to be normally subject to sex-reversal. Sex reversal in *G. elongatum* from the eastern North Atlantic seems unlikely to be a normal occurrence, although the data presented here are limited. The size-range for males totally encompasses that of ovarian *Stage I* females, and partially of *Stage II*. Although males may be normally ripe at about 120 mm (or less), the testes continue developing, and appear best developed in specimens longer than 130 mm. This suggests that, should sex-reversal be the normal mode of development, recognizable *Stage I* females of less than 130 mm would be unexpected. Six of the 11

*Stage I* females were 115–126 mm. On the other hand, CLARKE (1974) has suggested that some *G. elongatum* may be protandrous, and that their frequency of occurrence may depend upon environmental conditions. As he states, however, only a detailed histological survey of the gonads of individuals of all sizes (and from various areas) can finally resolve this problem.

It seems, then, that the lack of larger females in the current collection is not a product of seasonal and/or sex-reversal factors. The collections were made near the northern limits of the species' range in the eastern Altantic (WITZELL, 1973) but the presence of post-larvae and juveniles implies that the area lies within the breeding zone of the species. This, and the occurrence of males nearing spawning condition, further indicates the likely presence of mature females in the area. The capture of mature adults off Bermuda by the RMT 8 also shows that larger specimens are within the scope of capture of the net. Since no repeat series of hauls per depth were made in 30°N, 23°W, absence of mature females from the collection, then, must be regarded as a product of chance.

### Bonapartia pedaliota

Twenty-three post-larval forms were caught at 100–300 m depth by day. At night, post-larvae were restricted principally to 100–300 m depth, whilst juveniles occurred at 200–300 m (Fig. 3). Three adults were caught at 200–500 m depth.

### Valenciennellus tripunctulatus

Six hundred and twenty-nine specimens were captured in 100–500 m depth, of which 173 were taken by day. A further four day and three night captures were made at greater depths, but these have been regarded as contaminants (see p. 10). The vertical distribution is summarized in Fig. 15, with number per sample expressed as a percentage of the total day and night samples, respectively. These animals occupied principally 200–400 m depth. Length–frequencies, by depth of capture, are shown in Fig. 16 for day and night, expressed as a percentage of the total day and night catches, respectively. The animal lengths were measured by dial calipers and placed in 0·5 mm size groups in this case to render them compatible with a separate study on development.

The disparity between day and night catches (173 versus 456) is particularly high. In the case of this species it is possible to ascertain whether or not such disparity, which can be observed in other species also, is a chance anomaly or a stable feature of the species when sampled by the RMT 8. During *Discovery* Cruise 45 (Sta. 7856, 30°N,

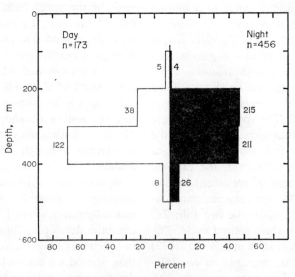

Fig. 15. *Valenciennellus tripunctulatus*. Vertical distribution expressed in percentages of total captures per day and night. Catch numbers superimposed.

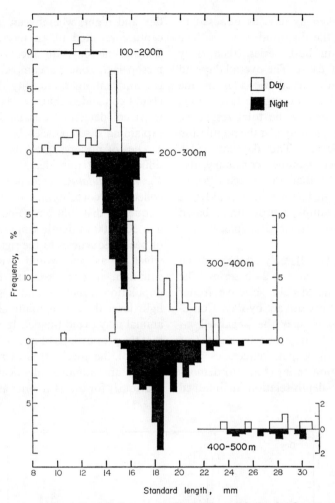

Fig. 16. *Valenciennellus tripunctulatus*. Length–frequency distributions with depth of capture, as percentage of total day (*n* = 168) and night (*n* = 413) measurable captures, respectively.

23°W), a series of repeat tows were made at a single depth. One dawn, one dusk, five daytime and four night tows, each of a fishing duration of 1 h, were made over a single 24-h period using the RMT 1+8 combination net (ROE, 1974) at the selected depth of 250 m (range 230–266 m). The number of *V. tripunctulatus* caught by the RMT 8 varied as 7–18 ($\bar{n}$ = 11·6) by day and as 53–92 ($\bar{n}$ = 71·5) at night. Irrespective of the magnitude of these values, the constancy of the day and night collections suggests a reasonable comparability of sampling. Hence RMT 8 samples of the vertical distribution series may be considered typical on this basis.

It is worth considering the implications of the day–night differences in both sets of results since, if the factors responsible for them can be deduced, some insight into the behaviour of the species is given. The larger night catch may result from a different pattern of behaviour nocturnally and/or physical changes in the environment. It could be interpreted as being a result of short nocturnal migrations, as indeed is implied from the preliminary vertical distribution data (Fig. 15). There is, however, a similarity in the proportions taken by day of the total day and night catches of both the 24-h series and the vertical series. In the 24-h series, 14% of the combined diurnal and nocturnal mean catches was taken by day, as opposed to 27% of the total in the distribution series. The numerical disparity in both series, then, may result from a common cause. The overall day and night length–frequency distributions from the distribution series (Fig. 17) compare favourably, showing that proportionally the total respective samples were each representative of the population to a comparable degree. The daytime catch numbers of the 24-h series, in their constancy, also suggest it is unlikely that the species forms diurnal aggregations—a factor which could give spurious results in a sampling programme based upon a one depth–one sample principle, as was the distribution series.

MERRETT and ROE (1974) analysed the stomach contents of specimens from the 24-h series. The species was, they concluded, a selective feeder which foraged almost exclusively by day. Consequently, *Valenciennellus* may be regarded as relatively more active by day than by night.

The distribution of length–frequencies with depth in the distribution series (Fig. 16) demonstrates an animal size–depth relation, maintained day and night, whereby an increase in depth of capture resulted in an increase in animal size. During the day, an increase in depth in the mesopelagic zone is accompanied by a reduction in the ambient light intensity. PEARCY and LAURS (1966) concluded that avoidance by animals due to visual detection was an important factor in explaining discrepancies between day and night catches of mesopelagic fishes. Although it seems difficult to fit this idea to a species as small as *V. tripunctulatus*, evidence from the present collections corroborates their view. The length–frequency distributions (Fig. 16) in 200–300 m and 300–400 m depth samples show a slight shift of peak frequencies to the right (larger) at night— which is as one would expect in the present situation, if one bears in mind the apparent population structure and assumes that at each light level there is a minimum size at which an animal may avoid the net. It is worth adding here that the proportions of day (15%) and night (85%) catch of the total captures made at 200–300 m depth are almost identical for those cited previously for the 24-h series at 250 m (14%; 86%).

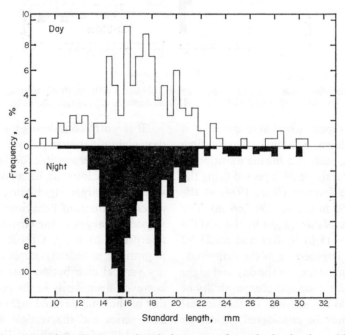

Fig. 17. *Valenciennellus tripunctulatus*. Percentage length–frequency of samples by day (*n* = 168) and night (*n* = 413).

The increase in abundance apparent at 250 m depth in this latter series, then, is likely to be more a result of less effective net evasion nocturnally than one of short migration into this layer.

As the population is indicated as being non-migrant, with or without minor internal vertical displacements, the species is likely to become more susceptible to predation nocturnally from migrating predators. A decline in activity perhaps seems inconsistent under such conditions but it does have possible selective value. By night, feeding activity is diminished (MERRETT and ROE, 1974) and the dispersion of melanophores darkens the body to adapt to the ambient light conditions (BADCOCK, 1969). Thus, while passively avoiding predators at night, *V. tripunctulatus* may be rendered more vulnerable to capture by the RMT 8.

One hundred and thirty-two specimens (14·5–30·5 mm) were arbitrarily selected from different samples for gonad examination. Forty-five individuals (15·5–25·5 mm) were males, indicating a sex ratio of approximately 1 ♂ : 2 ♀. Males of 24·0–25·5 mm were ripe. Female maturity Stages I–VI were represented in the following frequencies: Stage I, 2 (14·5–15·0 mm); Stage II, 20 (15·5–21·0 mm); Stage III/V, 63 (20·0–30·5 mm); Stage VI, 2 (25·5–27·0 mm). During Stage III/V both small and large eggs are present in the ovaries. Counts of the latter, based on the ovary pairs of 35 specimens (Stages III/V and VI), showed the number of large eggs to increase with increased animal size (Fig. 18). A near breeding condition is attained at 25·5+ mm. The length–frequency/depth data (Fig. 17) suggests that the breeding occurs in the lower reaches of the distribution.

Our results agree with KRUEGER (1972) with

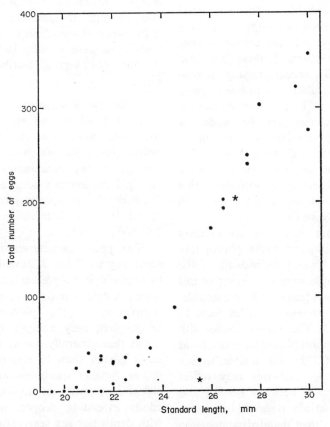

Fig. 18. *Valenciennellus tripunctulatus*. Relation of the number of large eggs per ovary pair (♀ III/V and VI) to standard length. Asterisk denotes Stage VI.

respect to size/depth relations. There are, however, differences apparent between eastern and western Atlantic populations. The photophores and pigmentation are laid down sequentially rather than simultaneously during development (GREY, 1964). BR* are the first photophores initiated (JESPERSEN, 1933) although other groups are present before their completion. As each photophore develops, in general, it attains the ability to occlude itself by superficial pigment dispersion before the development of the ensuing photophore is complete. The potential for occlusion has been noted before (BADCOCK, 1969) but in that it develops early suggests that each photophore may become functional as soon as it is developed. A similar speculation has been made for *Argyripnus atlanticus* Maul, 1952, (BADCOCK and MERRETT, 1972) which also has compound light organs. In the eastern North Atlantic the light organs of *Valenciennellus* are fully developed at 17–19 mm. The dermal melanophores are the principal active agents of the colour change system (BADCOCK, 1969) and these are also developed sequentially, accompanying photophore formation. Pigmentation is, in essence, also complete at 17–19 mm. In post-larvae, then, a partial diel colour change can be made, its effectiveness increasing with further development. KRUEGER (1972) suggested that this ability developed late in the juvenile stage (which he regards as sexually indeterminate specimens with a full photophore complement), but our own observations disagree with this.

The gonads of our specimens are sexually distinguishable at the stage where AC photophore groups are initiated. Twenty individuals of the gonad inspection sample were in varying phases of AC photophore development. All were sexable, with 11 ♂ and 9 ♀ present. Females were in maturity Stages I and II. The smaller females with a full complement of photophores were mostly at Stage II. KRUEGER (1972) was unable to sex animals of these stages and this may reflect developmental differences between eastern and western Atlantic populations. It should be pointed out, however, that a direct detailed comparison between the results of KRUEGER (1972) and our

own cannot be made. For example, Kreuger's sub-adult females include some specimens with ovaries containing a few large eggs, whilst his adults had mostly large eggs (KRUEGER, 1972). In our scheme, all such specimens would be categorized as Stage III/V or VI. In conclusion, differences between the two populations are indicated, i.e. in the initiation of maturity, and animal size (western animals appear to be smaller). Only a direct comparison, however, can substantiate such speculation.

*Family Sternoptychidae*
(*sensu* BAIRD, 1971)

The sternoptychids were represented by 866 specimens in two genera and five species. *Argyropelecus hemigymnus* was the most abundant (609 specimens) followed by *A. aculeatus* (155) and *Sternoptyx diaphana* (29). Four specimens of *A. affinis* group and two of *S. pseudobscura* were also caught. In addition, eight unidentified *Argyropelecus* (non-*affinis* group) and 59 *Sternoptyx* post-larvae were recorded in the RMT 8 catches. A summary of vertical distribution is presented in Fig. 3.

### *Argyropelecus hemigymnus*

Four hundred and six day and 203 night specimens were caught. An overall diurnal depth distribution in 300–600 m was indicated [although four specimens (contaminants) came from deeper]. At night the animals occurred primarily in 200–600 m depth. One specimen came from shallower. At all times the distributional peak occurred in 300–500 m depth.

The specimens were sexed and females staged according to Table 2. Juveniles and post-larvae (with photophores present) were not differentiated since sex determination was possible prior to the completion of the photophore complement. Length–frequency analysis by depth (Fig. 19) shows that generally during the day only individuals greater than 18 mm occurred as deep as 600 m. Smaller specimens were limited to 300–500 m depth, primarily to 400–500 m. Other than adults extending deeper, no size-stratification with depth nor sex segregation in this species is

*Designation after GREY (1964).

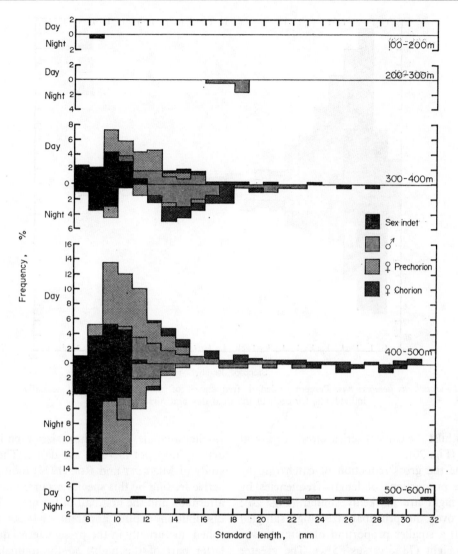

Fig. 19. *Argyropelecus hemigymnus*. Percentage length–frequency distribution, with depth of capture, by day (*n* = 400) and night (*n* = 197), with males, female pre- and post-chorion stages, and sexually indeterminate individuals indicated.

obvious (Fig. 19). The data demonstrate a decrease in the catch proportion of specimens 12–18 mm at 400–500 m depth by night, and a corresponding increase of this size-group in 300–400 m depth, indicating that some migratory activity occurred. The apparent migration was made by males and Stage III/V females from within this size-group (Fig. 19). Off Fuerteventura, *A. hemigymnus* shifted its distribution limits by about 200 m (BADCOCK, 1970), but no such vertical movement was observed here. Off Bermuda, the population is apparently static (GIBBS and ROPER, 1970). BAIRD (1971), on the other hand, considered *A. hemigymnus* as usually a migrant. The behavioural difference apparent between the present and the Fuerteventura population may be due to their respective developmental states. At Fuerteventura, the size–frequency peak lay between 14 and 20 mm (i.e. principally males and Stage III/V females, see Fig. 19) (BADCOCK, 1970, Fig. 7B, 2 mm size

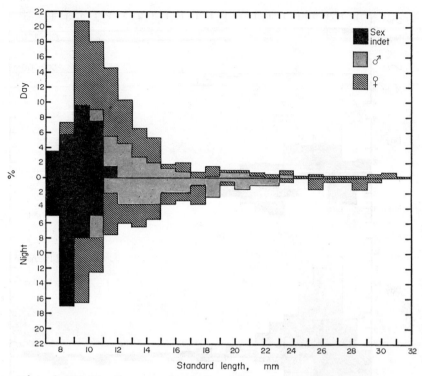

Fig. 20. *Argyropelecus hemigymnus.* Percentage length–frequencies of males, females and sexually indeterminate individuals, for each of the total day and night samples.

groups), whilst the current series shows a peak at 8–12 mm (Fig. 20).

Despite the gross reduction of catch rate by night, the proportions of length–frequencies by day and night are markedly similar (Fig. 20); although over the principal size-range sampled (7–14 mm) a smaller proportion of animals was taken by night (78% versus 85%). The greater day catch could indicate that the animal is more active by night and thus better able to avoid the net. Both the Fuerteventura population and this considered here were similar in that little relation between size of animal and depth of capture was shown (although smaller individuals were vertically less dispersed than larger ones, see above). If increased activity was responsible for the smaller night catches at 30°N, a reduction in the proportion of larger specimens could be expected, as was found in the SOND results (BADCOCK, 1970). This was not apparent, but then the population was comprised mainly of

smaller animals, and at these sizes even increased activity may not increase avoidance. The feeding study of MERRETT and ROE (1974) indicated that active feeding by this species occurred during late afternoon and the first half of the night. The peak distribution depths apparent, 300–500 m, were fished nocturnally in the present series during the latter part of the night, so the animals caught were unlikely to be feeding actively. If one examines the overall sex ratios of ♀ : ♂ : *sex indet.* by day and night, more specific changes in the population sampled may be seen. The ratios compare as approximately 2·5 : 1 : 1·3 (day) and 1·3 : 1 : 1·4 (night). Proportionally there was an apparent loss of females from the night samples; although in terms of actual numbers of animals caught, a nocturnal 'loss' by each sex-class was made. Among the females, the greatest 'loss' was sustained by the Stage I and II groups. BADCOCK (1970) pointed out that the extent of migration for this species was "not necessarily uniform throughout the size range" and cited specimens of 17 mm and less migrating 50–100 m. Migrations of this order would not necessarily be detectable when sampling 100 m depth bands, but could result in a decrease of vertical space occupied by the species within a particular depth band, which may affect catch numbers. The pinpointing of a single migratory group, i.e. 12–18 mm, could be an incidental consequence of this.

In the collections, the least developed individuals which have been positively identified as *A. hemigymnus* are those in stages during which the AN photophore group (designation after BAIRD, 1971) is just forming. However, six individuals of *Argyropelecus* (7–9 mm) in stages prior to photophore pigmentation were also caught by the RMT 8 in depths shallower than 600 m. Thirty-five similar specimens (5–9 mm) occurred in the RMT 1 catches. These probably belong to *A. hemigymnus*. The distributions of these two developmental groups (i.e. prephotophore and early photophore stages) and of slightly more advanced individuals, as indicated by the captures of the one net are reflected by those of the other. No detectable migrations were made by either developmental group, and they were, in effect, segregated vertically. Specimens with photophores developing (5–9 mm) were confined mainly in 300–500 m depth; whilst those individuals without pigmenting light organs (5–9 mm) occurred principally in 100–300 m. It would appear, then, that the onset of photophore pigmentation is accompanied by a migration into deeper waters. The number of prephotophore specimens caught was low, but distinction between those captured in 100–200 m depth and those in 200–300 m could not be made on morphological grounds. The lack of size-stratification with depth as shown by the more advanced members of the population in 300–500 m (p. 36) may prove to be characteristic of the prephotophore stages in shallower waters.

### A. aculeatus

Eighty day and 75 night specimens were collected. Diurnally the species was confined to 300–500 m depth (two specimens came from deeper) (Fig. 3). Overall, the population sampled was essentially in early juvenile stages. Consequently the majority of specimens occurred in one depth, 300–400 m. At night, the distribution encompassed 100–500 m (four individuals came from deeper), with a peak at 200–300 m depth. The data are sufficient to suggest that not all the juvenile population migrates, and, furthermore, that detectable migrations are not made by animals until a certain degree of development has been attained. All animals less than 9 mm remained at depth, and in the 9–10 mm size-group some individuals migrated, others did not. One cannot preclude, however, the possibility that minor, undetectable migrations may be made by apparent non-migrants within the 300–400 m depth band.

The extent of general migration cannot be defined, but migrant animals would appear to rise through the order of 100 m by night.

### Sternoptyx diaphana

Sixteen specimens taken by day came from 500–900 m depth. At night 13 specimens were caught in 600–900 m depth. Adults were in a numerical minority overall.

### Sternoptyx sp. post-larvae

These specimens were in developmental stages prior to supra-anal photophore formation, and consequently could not be identified with confidence. Twenty-one specimens were taken by day in 500–700 m and 900–1000 m depth. At night 38 individuals were caught, in 500–800 m depth. Peak captures by day and night occurred in 500–700 m depth.

### Family Chauliodontidae

Two species, *Chauliodus danae* and *C. sloani*, were represented in the collections. Sixty-nine *C. danae* and 15 *C. sloani* were caught. In addition, 80 post-larvae unidentified to species were taken.

### C. danae

During the day 28 specimens were caught, represented by five post-larvae (with photophores, 28–30 mm) from 600–700 m depth; 18 juveniles, 27–66 mm, from 400–600 m depth; and five adults, 75–114 mm, from 400–700 m (four were taken at 600–700 m depth) (Fig. 3).

At night 41 individuals, 29–103 mm, were caught. The single post-larva (with photophores) came from 500–600 m depth. The 28 juveniles (31–63 mm) taken indicated a disjointed and dispersed nocturnal distribution. Specimens were caught in 100–300 m and 400–900 m depth, with peak numbers taken in 100–200 m and 500–600 m depth. Adults and subadults, 61–103 mm, were also considerably dispersed, and occurred in 200–400 m and 500–700 m depth (Fig. 3).

The apparent migrant and non-migrant elements of the population could not be separated on the basis of feeding behaviour (MERRETT and ROE, 1974), nor size, sex nor sexual maturity, which suggests that individual migration is not of a diel nature. The two fractions were also sampled in approximately equal numbers (21 versus 20). Adults that had migrated at night lay deeper than migrant juveniles (Fig. 3) [HAFFNER (1952) found a similar tendency]. The adult–juvenile depth relations of the non-migrant element matched those of the diurnal population (Fig. 3). Juveniles were principally confined to 400–600 m; adults to 600–700 m depth. Adults and juveniles, when they migrate, appear to travel a similar distance, ca. 300 m.

The total sample of 17 adults and subadults (61–114 mm) comprised 10 ♂ and 7 ♀. Stage I were recognizable at 61–66 mm; Stage II at 72 mm; Stage III/V at 75–114 mm. Males have a proportionally larger eye (REGAN and TREWAVAS, 1929), and, in addition, a smaller postocular photophore, than females. Such distinction, however, could not be ascertained in juveniles.

### C. sloani

By day all specimens (11 juveniles) came from 400–600 m depth. At night, one juvenile was caught in 400–500 m depth. Three post-larvae (bearing photophores), however, were taken in 600–1000 m depth.

### Chauliodus spp. post-larvae

Seventy-three post-larvae without photophores were caught; in a further seven, with photophores, specific identification was precluded by damage. The non-photophore specimens were confined to 50–300 m depth by day, and 25–200 m by night. The photophore-bearing individuals were caught at night in 600–800 m (Fig. 3). The absence of photophore-bearing post-larvae between 300 and 500 m depth in the overall Chauliodus data suggests that a downward post-larval migration may be made to the adult levels of diurnal distribution at the onset of photophore formation.

### Family Idiacanthidae
### Idiacanthus fasciola

Although poorly represented in the collections, I. fasciola is worthy of mention in view of the apparent different migratory habits of males and females. By day, the species was represented by seven individuals caught at 600–2000 m; at night by 14 specimens from 25–200 m and 900–1250 m depth. All females apparently underwent a nocturnal vertical migration, whilst the 10 males caught (three by day) were confined to 900–1250 m depth.

### MYCTOPHOIDEI

The myctophoid fishes were represented by five families, 19 genera and 36 species. A summary of non-myctophid captures is given in Table 7.

### Family Myctophidae

Fifteen genera and 31 species were represented. Catch numbers per species were generally low and, as would be expected, the smaller species were more easily caught (Table 8). In general, more specimens per species were taken by night. The discrepancy between day and night catches was probably due, in the most part, to the more intense fishing effort made in the upper 100 m compared with that at greater depths, although undoubtedly net evasion would also be a contributing factor. Post-larval, juvenile and adult forms were caught. In general, it was found that juveniles could be divided, on the basis of coloration, into two groups within a species. Thus newly transformed, light coloured forms, and

Table 7. *Capture data of non-myctophid myctophoids* (Sta. 7856), *namely: unidentified paralepidids;* Omosudis lowei *Günther, 1887;* Scopelarchus analis (Brauer, 1902); *S.* michaelsarsi *Koefoed, 1955;* Benthalbella infans *Zugmayer, 1911;* Scopelosaurus *sp. Number and length range is given by depth of capture by day and night.*

| Depth range (m) | Unidentified paralepidids D | N | O. lowei D | N | S. analis D | N | S. michaelsarsi D | N | B. infans D | N | Scopelosaurus sp. D | N |
|---|---|---|---|---|---|---|---|---|---|---|---|---|
| 10-25 | | | | | | | | | | | | |
| 25-50 | | | | | | | | | | | | |
| 50-100 | | | | | | 12(20-28) | | | | | | 9(34-54) |
| 100-200 | | | | | 4(17-40) | 17(18-33) | 1(16) | | | | | 4(44-64) |
| 200-300 | | | 1(18) | | 5(18-31) | 1(49) | | | 1(17) | | | 1(55) |
| 300-400 | | | | | 1(41) | | | | 1(31) | | | |
| 400-500 | | | | | 11(25-35) | | | | 1(59) | | | 1(32) |
| 500-600 | | | | | 2(29-30) | | | | | 1(75) | | 1(32) |
| 600-700 | | | | | 2(30-34) | | | | | | 1(46) | |
| 700-800 | | | 4(9-13) | | 1(35) | | | | | | | |
| 800-900 | 1(44) | | 4(11-27) | 4(10-12) | 2(39-58) | | | | | | | |
| 900-1000 | 1(58) | 1(40) | 7(11-34) | 5(10-36) | 1(63) | | | | | | | |
| 1000-1250 | 1(61) | 1(66) | | 1(28) | | | | | | | | |
| 1250-1500 | | | | | 1(72) | | | | | | | |
| 1500-2000 | | | | | | | | | | | | |
| Total no. | 3 | 2 | 16 | 10 | 30 | 30 | 1 | – | 3 | 1 | 1 | 16 |
| Overall length range (mm) | 40-66 | | 9-36 | | 17-72 | | 16 | | 17-75 | | 32-64 | |

more pigmented, dark coloured forms occurred (see Appendix). Where possible larvae were identified to species, and the terms 'larva' and 'transforming larva' (= metamorphosis) are those as used by Moser and Ahlstrom (1972). Sources for larval identification used were Tåning (1918), Pertseva-Ostroumova (1964) and Moser and Ahlstrom (1970, 1972). Qualification, and in some cases further description, of little-known larvae found in this area has been necessary (see Appendix). In view of their importance, larvae from the RMT 1 collections have been examined for additional information. Overall, *Notoscopelus resplendens, Benthosema suborbitale, Diogenichthys atlanticus* and *Hygophum hygomi* were the dominant larval species, in order of diminishing number. *Hygophum reinhardti*, transforming *H. benoiti* and *Electrona rissoi* were represented by small numbers. Of 434 larval and transforming larvae taken by these nets (RMT 1 + 8), 175 could not be identified to species (the material included *Lobianchia* and *Lampanyctus* spp.).

In the account following, only the principal species are considered. A summary of distributions is given in Fig. 21. The data from both RMT 1 and RMT 8 were utilized in assessing larval distributions, but the distributions cited for juveniles and adults are based solely upon RMT 8 catch data.

### Hygophum

*H. hygomi.* Larvae were caught in 50–100 m depth by day and 25–100 m by night. Transforming larvae were taken in 500–600 m depth by day (seven) and night (seven). In addition, three such specimens were caught in 200–300 m depth by night.

No adults were caught. By day juveniles were taken in 400–700 m depth, though principally in 500–600 m. At night a few early juveniles remained at depth, 500–600 m; the rest occurred in 10–100 m depth. Although direct comparison of the upper 100 m samples may not be made, 10–25 m depth appeared as optimal for the species. The size-range principally sampled was 15–20 mm

*Table* 8.    *Myctophid catch data.*

| Species | Juveniles & adults* | | | Larvae † | | | Total |
|---|---|---|---|---|---|---|---|
| | D | N | SL range (mm) | D | N | SL range (mm) | |
| Electrona rissoi (Cocco, 1829) | 2 | 2 | 10–15 | 3 | 0 | 6.2–9.0 | 7 |
| Hygophum benoiti (Cocco, 1838) | 0 | 1 | 16 | 1 | 0 | 13.0 | 2 |
| H. hygomi (Lütken, 1892) | 48 | 121 | 12–43 | 10 | 43 | 3.5–15.0 | 222 |
| H. reinhardti (Lütken, 1892) | 14 | 27 | 13–35 | 5 | 3 | 12.0–15.0 | 49 |
| H. taaningi Bekker, 1965 | 0 | 3 | 16–24 | 0 | 0 | – | 3 |
| Benthosema suborbitale (Gilbert, 1913) | 92 | 284 | 10–30 | 17 | 44 | 4.0–10.8 | 437 |
| Diogenichthys atlanticus (Tåning, 1928) | 46 | 65 | 12–23 | 6 | 52 | 3.0–11.5 | 169 |
| Myctophum selenops Tåning, 1928 | 1 | 0 | 21 | 0 | 0 | – | 1 |
| Gonichthys coccoi (Cocco, 1829) | 1 | 1 | 18–35 | 0 | 0 | – | 2 |
| Lobianchia dofleini (Zugmayer, 1911) | 53 | 30 | 11–38 | 0 | 0 | – | 83 |
| L. gemellari (Cocco, 1838) | 8 | 7 | 14–56 | 0 | 0 | – | 15 |
| Diaphus effulgens (Goode & Bean, 1895) | 0 | 5 | 47–51 | 0 | 0 | – | 5 |
| D. holti Tåning, 1918 | 0 | 1 | 35 | 0 | 0 | – | 1 |
| D. metopoclampus (Cocco, 1829) | 1 | 3 | 20–25 | 0 | 0 | – | 4 |
| D. mollis Tåning, 1928 | 7 | 15 | 29–52 | 0 | 0 | – | 22 |
| D. rafinesquei (Cocco, 1820) | 7 | 23 | 22–78 | 0 | 0 | – | 30 |
| Notolychnus valdiviae (Brauer, 1904) | 42 | 47 | 12–24 | 0 | 0 | – | 89 |
| Taaningichthys bathyphilus (Tåning, 1928) | 7 | 4 | 18–60 | 0 | 0 | – | 11 |
| T. minimus (Tåning, 1928) | 3 | 1 | 28–47 | 0 | 0 | – | 4 |
| Lampanyctus ater Tåning, 1928 | 3 | 1 | 42–93 | 0 | 0 | – | 4 |
| L. cuprarius Tåning, 1928 | 10 | 32 | 35–77 | 0 | 0 | – | 42 |
| L. festivus Tåning, 1928 | 0 | 3 | 40–66 | 0 | 0 | – | 3 |
| L. lineatus Tåning, 1928 | 1 | 2 | 26–76 | 0 | 0 | – | 3 |
| L. photonotus Parr, 1928 | 1 | 1 | 39–52 | 0 | 0 | – | 2 |
| L. pusillus (Johnson, 1890) | 8 | 11 | 15–33 | 0 | 0 | – | 19 |
| Lepidophanes gaussi (Brauer, 1906) | 13 | 5 | 29–46 | 0 | 0 | – | 18 |
| L. güntheri (Goode & Bean, 1895) | 0 | 1 | 65 | 0 | 0 | – | 1 |
| Bolinichthys indicus (Nafpaktitis & Nafpaktitis, 1969) | 15 | 30 | 18–43 | 0 | 0 | – | 45 |
| B. supralateralis (Parr, 1928) | 0 | 1 | 95 | 0 | 0 | – | 1 |
| Ceratoscopelus warmingi (Lütken, 1892) | 21 | 76 | 32–72 | 0 | 0 | – | 97 |
| Notoscopelus resplendens (Richardson, 1844) | 6 | 10 | 22–67 | 23 | 52 | 3.2–22.0 | 91 |
| Totals | 410 | 813 | | 65 | 194 | | 1482 |

*RMT 8 data; †RMT 1 + 8 data.

and neither by day nor night was size-stratification with depth apparent.

*H. reinhardti.* Three larvae were caught at night, 10–100 m depth. Transforming larvae and

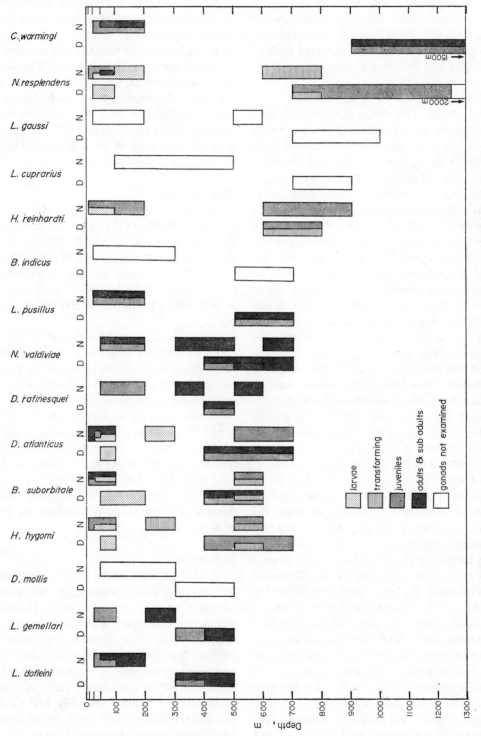

Fig. 21. Vertical distribution of myctophids. (Column width has no quantitative implication.)

light coloured juveniles were taken in 600–800 m depth by day; light juveniles were caught nocturnally in 600–900 m. No adults were captured. Fully pigmented juveniles were confined to 600–800 m by day and 10–200 m by night.

## Benthosema

*B. suborbitale.* The results showed that both depth of occurrence and migratory behaviour were related to stage of development. Larvae were caught at 50–200 m depth by day and 25–100 m by night, which implies a slight nocturnal migration into shallower waters. Transforming larvae and light coloured juveniles (see Appendix) were non-migrants, confined to 500–600 m. Dark juveniles and the adults were vertical migrators. By day they occupied 400–600 m depth, with the majority of specimens coming from 500–600 m. Nocturnally, the species was principally distributed in 10–50 m depth, with a few specimens caught in 50–100 m. Although direct comparison between the samples from the 10–25 and 25–100 m depth range was not possible, the data indicated that individuals of 22 + mm tended to be excluded from 10–25 m, whilst smaller individuals were abundant throughout the 10–50 m depth range. CLARKE (1973) in his Pacific study, found a converse situation where smaller specimens tended to be higher in the water column, with adults more vertically dispersed.

The discrete depth ranges fished below 100 m were too broad to demonstrate whether diurnally any animal size–depth relations amongst the migratory fraction of the population occurred. The vertical segregation of larvae and transforming larvae, on the other hand, does suggest strongly that either just prior to transformation or at its initiation, the larvae migrate down to the lower levels of the diurnal distribution of adults. The non-migratory, light coloured juveniles also occur at these depths. The stage at which juveniles become part of the migrant population cannot be defined, but the present evidence implies that only when pigmentation is fully developed is the nocturnal journey to the upper waters made. CLARKE (1973) also noted that small individuals (10–12 mm) remained at depth at all times, although

he gave no indication of their developmental stage.

Adults were sexed according to caudal gland characteristics. One hundred and ninety-three individuals could be sexed in this manner, giving a ratio of approximately 2 ♂ : 1 ♀. The size-range of these specimens was 17–30 mm and no size difference between males and females was apparent. Females with egg maturity Stage II spanned 17–21 mm; Stage III/V, 21–30 mm. In general, individuals of 20 + mm could be sexed; those < 20 mm could not.

## Diogenichthys

*D. atlanticus.* Larvae (3·0–11·5 mm) were caught in 50–100 m depth by day and 25–100 m by night (optimal depth 50–100 m). A single specimen, 11 mm, was caught in 200–300 m depth (night). Transforming larvae were not taken, but light coloured juveniles (12–13 mm) were caught in 600–700 m by day, and 500–700 m depth by night.

Juveniles and adults occurred in 400–700 m depth by day, the majority of specimens coming from 500–600 m. At night 10–100 m depth was occupied, but juveniles (12–14 mm) were practically limited to 25–50 m.

A sex ratio of approximately 1 ♂ : 1 ♀ was found for 93 specimens, spanning a size-range 12–23 mm. No size difference between males and females was apparent. The majority of females were Stage III/V (15–23 mm), but two specimens were ripe (Stage VI, 17–19 mm). Stage II females (13–15 mm) were confined to 500–600 m depth by day.

## Lobianchia

*L. dofleini.* By day the species inhabited 300–500 m depth (one specimen, a presumed contaminant of 12 mm, was found in the 500–600 m depth sample). Catch numbers for each sample within the distribution range were similar, but specimens from 300–400 m depth comprised principally of juveniles whilst those from 400–500 m were mostly adult. At night the catch number was inexplicably low (Table 8). The species was restricted to 25–200 m depth, and adults were totally absent at 25–50 m.

Juveniles by day appear confined to the shallower parts of the distribution. The relative lack of adults from this level could be a result of net evasion. The maintenance of segregation at night (albeit demonstrated by sparse data) render such a suggestion less likely, however; which implies that adults and juveniles alike make a vertical migration of the order of 300 m. GOODYEAR, ZAHURANEC, PUGH and GIBBS (1972) found that the species was size-stratified with depth, day and night, in the Mediterranean.

*L. gemellari.* By day the species occurred in 300–500 m depth. At night, juveniles (14–22 mm) were caught in 25–100 m, adults (50–56 mm) in 200–300 m depth. CLARKE (1973) also noted that individuals over 40 mm occurred only below 150 m depth at night.

### Diaphus

*D. rafinesquei.* The species was represented in only one day sample, from 400–500 m depth (22–70 mm). At night juveniles were caught in 50–200 m depth. Adults were taken only in 300–600 m depth. The differences in migratory behaviour between juveniles and adults has already been noted (NAFPAKTITIS, 1968; BADCOCK, 1970; GOODYEAR, ZAHURANEC, PUGH and GIBBS, 1972).

*D. mollis.* Day caught specimens came from 300–500 m depth. At night the species was taken from 50–300 m. An increase in animal size with increased depth was apparent.

### Notolychnus

*N. valdiviae.* By day the population was centred between 400 and 700 m, with the majority of specimens coming from 400–500 m depth. Juveniles < 18 mm were confined to the latter band, but adults were caught throughout the depth range. At night the population was considerably dispersed, occupying the depths 25–200, 300–500 and 600–700 m (Fig. 21). The population thus contained migrant, partial migrant and non-migrant elements. The fishing procedure in the upper 100 m precluded a total assessment of these fractions, but the data

indicated that juveniles were full migrators, whilst adults occurred in all three categories. CLARKE (1973) noted a similar phenomenon in the Pacific, and suggested that the proportion of non-migrants or partial migrants was dependent upon season. In September and December, the proportion was high; in March and June it was low (CLARKE, 1973).

The specimens were sexed according to eye and snout characteristics. The sex ratio of 79 specimens was approximately 1 ♂ : 1 ♀. The male size-range was 15–22 mm; female range, 15–24 mm. Specimens that could not be sexed encompassed 13–18 mm. Female maturity Stage II occurred in individuals 15–19 mm; Stage III/V, 19–24 mm. As indeed CLARKE (1973) found, there was no apparent difference in sex ratio or maturity between the shallow and deep components of the night catch.

### Lampanyctus

*L. cuprarius.* The species was confined to 700–900 m depth by day; and 100–500 m, principally 200–400 m depth, by night.

*L. pusillus.* Juveniles and adults occupied 500–700 m depth by day, and 25–200 m by night.

### Lepidophanes

*L. gaussi.* During the day the species apparently lay in 700–1000 m depth. Nocturnally it occurred in 25–200 m depth. A single specimen was also caught in 500–600 m at night.

### Bolinichthys

*B. indicus.* The species was caught in 500–700 m depth by day. At night the distribution was vertically dispersed, lying in 25–300 m.

### Ceratoscopelus

*C. warmingi.* No recognizable larvae were caught. Juveniles and adults occupied 900–1500 m depth by day, and no size-stratification with depth was observed. The whole population migrated to 25–200 m depth at night, occupying principally 50–100 m. CLARKE (1973) found that small juveniles, 15–19 mm, were not regular migrants, but animals of these size-groups were not represented here.

## Notoscopelus

*N. resplendens.* The larvae occupied 25–100 m depth by day; 25–200 m by night. The transforming specimens (20–22 mm) were caught in 700–800 m depth by day and 600–800 m by night. Juveniles and adults were poorly represented. Juveniles (22–25 mm) were caught in 700–1250 m by day and 10–100 m by night. Adults were taken in 1500–2000 m depth by day (one individual) and 50–100 m by night. The species, however, is known to travel to the surface (e.g. BADCOCK, 1970).

### Family Omosudidae
## Omosudis lowei

The 26 individuals caught were all post-larval and juvenile. By day these were distributed in 700–1000 m depth, and a single one came from 200–300 m. No nocturnal migration was noted. The 10 night specimens caught were confined to 800–1250 m depth, but nine of these came from 800–1000 m. The RMT 1 catches (five specimens) concur with these results.

### Family Scopelarchidae

The scopelarchids were represented by two genera and three species. *Benthalbella infans* was represented by four specimens [of these, three are cited by MERRETT, BADCOCK and HERRING (1973)]. *Scopelarchus michaelsarsi* was represented by a single post-larva. Sixty specimens of *S. analis*, however, were caught (Table 7).

## S. analis

No adults were caught. An overall day range of 100–1500 m depth was given. Post-larval forms occurred in 100–300 m depth. The transitional stage between post-larva and juveniles has not been descriptively defined, but specimens approaching, and in, transformation were caught only in 400–500 m depth. Juveniles were caught at all depths, but a tendency for larger specimens to be caught deeper was apparent (Table 7). At night the total population was confined to 50–

300 m depth (18–49 mm). Post-larval and presumed transforming forms occupied 50–200 m depth, whilst juveniles were taken throughout the distribution depth range.

### REMAINING SPECIES

Species, other than stomiatoid and myctophoid fishes, were, with the exception of *Macrorhamphosus gracilis,*\* numerically poorly represented. A summary of indicated vertical distributions is given for some of these in Table 9.

Although no adults were taken, *Macrorhamphosus gracilis* was second in the order of species' abundance caught by the RMT 8. The night catch, however, was considerably larger than the day (Table 9), due primarily to an exceptional sample of the species (3475 individuals) taken in 100–200 m depth. The size of animals represented in this latter catch (7–31 mm), moreover, were generally not found in the day catches (maximum size, 11 mm). As the RMT 1 catches show, however, the layer which animals of this size occupy by day was not sampled by the RMT 8 (Table 10). Several distributional trends are indicated by the RMT 1 data, which are also implied by the less complete data of the RMT 8 catches.

By day the species was confined in 0–50 m depth. The maximum size caught (RMT 1) was 13 mm and although several size-groups were each represented in every stratum down to 50 m, preference for particular depths by specific sizes of animal was indicated (Table 10). Thus, 2–3 mm individuals showed an apparent marked selection for the 10–25 m stratum, whilst still being quite abundant in 25–50 m depth. Four-millimetre specimens similarly showed preference for 10–25 m depth, but 5–7 mm individuals occurred mainly in the upper 25 m (Table 10). Larger specimens were caught in 0–10 m depth.

At night, the species occurred at 0–200 m

---

\*Note added in proof: Since submitting this paper, EHRICH (1975, Investigations on *Macrorhamphosus scolopax* (Linnaeus, 1758) (Pisces, Syngnathiformes) from the subtropical N.E. Atlantic. *International Council for the Exploration of the Sea, Contribution to Statutory Meeting,* C.M. 1975/G:13, pp. 11) has shown that *M. gracilis* is but the pelagic juvenile stage of *M. scolopax.*

*Table 9. Catch data of certain species.*

| Species | Number Day | Number Night | SL range (mm) | Depth range (m) Day | Depth range (m) Night |
|---|---|---|---|---|---|
| Bathylagus greyae Cohen, 1958 | 9 | 40 | 20-65 | 600-900 | 25-500 |
| B. compsus Cohen, 1958 | 2 | 3 | 29-121 | 800-1000 | ( 50-200 ) (900-1000) |
| Bathylagus sp. post larvae | 7 | 3 | 9-24 | (100-300) (600-700) | 50-200 |
| Bathylagus sp. juvs. (brown) | 5 | 9 | 17-22 | 600-800 | 200-500 |
| Bathylaco nigricans Goode & Bean, 1895 | 2 | 2 | 19-55 | 800-1000 | 900-1500 |
| Holtbyrnia schnakenbecki (Krefft, 1953) | 4 | 7 | 16-26 | 400-600 | 300-600 |
| searsid juvs. | 9 | 15 | 10-36 | 700-1500 | 600-2000 |
| Melanonus zugmayeri Norman, 1930 | 21 | 6 | 13-34 | 600-900 | 100-500 |
| Scopeloberyx opisthopterus (Parr, 1933) | 12 | 16 | 25-31 | 800-1500 | 800-1500 |
| S. robustus (Günther, 1887) | 3 | 2 | 13-49 | 1000-2000 | 900-2000 |
| Eurypharynx pelecanoides Vaillant, 1882 | 12 | 6 | 30-559 | 700-2000 | 800-1500 |
| Eurypharynx? leptecephali | 3 | 1 | 28-53 | 50-200 | 100-200 |
| Avocettina infans (Günther, 1878) | 4 | 4 | 357-574 | 1000-2000 | 1000-2000 |
| Cryptopsarus couesei (larval) Gill, 1883 | 5 | 6 | 8-10 | 1250-2000 | 1250-2000 |
| Macrorhamphosus gracilis (Lowe, 1839) | 350 | 3735 | 2-31 | 10-50 | 10-200 |

depth. Unlike the RMT 8 catch, the overall nocturnal RMT 1 catch was smaller than the day (Table 10). This may be attributed to two factors. Firstly, the total time fished in the upper 200 m was $1\frac{1}{2}$ h less than by day; secondly, the species is vertically more dispersed by night. Individuals larger than 7 mm were, however, much better represented in the night collections (Table 10), and this discrepancy may be a product of enhanced vision and more effective net avoidance by the species during daylight.

Recently JOHN (1973) has reported a probable behavioural change among the post-larvae of *Macrorhamphosus* sp. in this region. He found that animals 2-3 mm were more abundant in catches at the surface by night than by day; and a converse situation occurred with larger animals. The reverse migrations of animals larger than 3 mm is clearly indicated in Table 10, and, furthermore, the range of migration is shown as having been roughly proportional to size of animal. In addition, Table 10 demonstrates that the 2-3 mm size-group contained both reverse migratory and non-migratory elements. In view of the lack of repeat hauls, however, the 'normal' migration of the 2-3 mm group as reported by JOHN (1973) cannot be considered as convincingly demonstrated by the present data. Nevertheless, when one considers the difference in the length of towing time for the two 0-10 m samples, an indication that 2 mm individuals may have behaved in such a manner is given in Table 10.

Reverse migrations in the surface waters have been demonstrated for the post-larvae and juveniles of several groups of fishes (PARIN, 1967; JOHN, 1973), and perhaps the most remarkable feature of the *Macrorhamphosus* migration is its relatively extensive range. The abundance of small post-larvae implies the simultaneous presence in the water column of adults in this region, and the absence of the latter in the catches must be attributed to net avoidance and/or chance. Adults of *M. gracilis* live both at the bottom and in midwater (EHRICH and JOHN, 1973). Limited, unpublished data taken from large-net midwater samples collected near Madeira suggest that there adults are in abundance at about 350 m depth by day, and at 140-190 m at night; although noc-

*Table* 10.   Macrorhamphosus gracilis. *Relations of standard length and depth by day and night (RMT 1 data).*

| Depth (m) | Standard Length (mm) | | | | | | | | | |
|---|---|---|---|---|---|---|---|---|---|---|
|  | 2 | 3 | 4 | 5 | 6 | 7 | 8 | 9 | 9+ | Sum |
| **DAY** | | | | | | | | | | |
| 0–10 | 10 | 37 | 62 | 26 | 18 | 23 | 9 | 22 | 13 | 220 |
| 10–25 | 727 | 311 | 102 | 20 | 21 | 8 | 0 | 1 | 0 | 1190 |
| 25–50 | 112 | 271 | 96 | 3 | 2 | 0 | 0 | 0 | 0 | 484 |
| 50–100 | 0 | 0 | 0 | 0 | 0 | 0 | 0 | 0 | 0 | 0 |
| SUM | 849 | 619 | 260 | 49 | 41 | 31 | 9 | 23 | 13 | 1894 |
| **NIGHT** | | | | | | | | | | |
| 0–10 | 42 | 14 | 1 | 0 | 1 | 0 | 0 | 0 | 0 | 58 |
| 10–25 | 154 | 43 | 0 | 1 | 0 | 0 | 0 | 0 | 0 | 198 |
| 25–50 | 277 | 287 | 21 | 0 | 0 | 0 | 0 | 0 | 0 | 585 |
| 50–100 | 63 | 75 | 46 | 7 | 3 | 3 | 1 | 0 | 0 | 198 |
| 100–200 | 0 | 0 | 0 | 0 | 0 | 1 | 15 | 43 | 242 | 301 |
| SUM | 536 | 419 | 68 | 8 | 4 | 4 | 16 | 43 | 242 | 1340 |
| TOTAL | 1385 | 1038 | 328 | 57 | 45 | 35 | 25 | 66 | 255 | 3234 |

turnally they are also taken in smaller numbers in depths down to about 430 m. Another point of interest arises then, in that the habits of juveniles and adults are apparently distinct and at some point in development a further change in behaviour must be presumed. Clearly a detailed investigation of this species could be rewarding.

In addition to the species tabulated in Table 9, the following species occurred in small numbers in the collections: *Platytroctes apus* Günther, 1878; *Searsia koefoedi* Parr, 1937; *Opisthoproctus soleatus* Vaillant, 1888; *O. grimaldii* Zugmayer, 1911; *Serrivomer beani* Gill and Ryder, 1883; *S. parabeani* Bertin, 1940; *S. brevidentatus* Roule and Bertin, 1929; *Nemichthys scolopaceous* Richardson, 1848; *Cyema atrum* Günther, 1878; *Poromitra capito* Goode and Bean, 1883; *Anoplogaster cornuta* Cuvier and Valenciennes, 1883; *Platyberyx opalescens* Zugmayer, 1911; *Chiasmodon niger* Johnson, 1863; *Howella brodiei*

Ogilby, 1889; *Brama brama* (Bonnaterre, 1788); *Brotulotaenia* (*nigra* Parr, 1933?); *Linophryne sexfilis** Bertelsen, 1973; *Ceratias holboelli* Kröyer, 1844; *Melanocetus johnsoni* Günther, 1864; *Caulophryne jordani* Goode and Bean, 1895.

Partly identified specimens of the following also occurred: gempylid post-larvae; *Eutaeniophorus*? post-larvae; *Melamphaes* sp.; *Linophryne*; *Gigantactis*; and indeterminate oneirodids.

### SUMMARY AND CONCLUSIONS

The total collections of midwater fishes taken at the six positions of the meridional section (see p. 4) incorporate about 80,000 specimens and a wide diversity of species. It was decided to consider the results in convenient latitudinal groupings to do justice to the scope these data provided. Thus insight into the dynamic relations of fishes

---

*Type specimen, from 1250–1000 m, *not* 1250–100m, as misprinted in BERTELSEN (1973).

at inter- and intra-specific level has been given only for one locality, 30°N, 23°W. How conclusions from this study relate to the findings at the remaining stations will be discussed when the complete transect has been reported.

The present collection of 17,443 specimens comprises 37 families, 66 genera, with 98 species recognized. The results broadly confirm the diurnal vertical segregation of adapted forms indicative of oceanic faunal zonation described in the past (e.g. MARSHALL, 1960; PEARCY and LAURS, 1966; PAXTON, 1967; BADCOCK, 1970; GIBBS and ROPER, 1970; KRUEGER and BOND, 1972; GOODYEAR, ZAHURANEC, PUGH and GIBBS, 1972; CLARKE, 1973, 1974). Despite the wide variation in apparent abundance of the species represented, the trends in the vertical pattern of distribution suggested by the less numerous species generally conform with one or other of the patterns indicated by the plentiful species.

A minimum of 110 species was present (confirmation of only 98 of these was possible, generally due to damage or developmental state), of which 71 were common to day and night catches, 14 represented solely in day samples, and 25 only in night catches. The minimum number of species present per depth by day and night is shown in Table 1. The figures of the two columns, however, are not directly comparable since not all species included were taken by both day and night, but they are sufficient to show the depth displacement of the peak in species' diversity at night, from a diurnal depth range in 400–900 m to one in 50–300 m. Much of the detail of vertical displacement found in this study, however, is obscured in Table 1, partly due to the different day–night species composition and partly to the changes in migratory behaviour that occur at varying stages of life in many species. Table 11 shows the relation in the water column of typically non-migrant species with the migratory elements of the population. It is based upon the *identified* species from Tables 3, 7, 8 and 9 that were common to day and night catches. For the sake of clarity, the non-migratory sections of an habitually migrant population (e.g. post-larvae, etc.) have been omitted. Columns 1 and 2 show the number

of species per depth by day and night, whilst column 3 gives the net gain or loss of species' numbers by night. The salient features shown (Table 11) are as follows:

(1) In agreement with Table 1, the peak diversities by day (column 1) and night (column 2) occurred in 400–900 and 50–300 m depth, respectively.

(2) The increase in species' number in the upper layers by night was matched by a decrease over 400–900 m depth (column 3).

(3) The diurnal distribution of migrant species' diversity increased with depth to a peak in 400–700 m, with a major decline occurring in depths below 1000 m (column 4).

(4) The migratory elements of any population were not found nocturnally at depths greater than 600 m. Whilst the greatest species' diversity occurred at 50–300 m depth nocturnally, some migration into the 400–600 m band also occurred (columns 4 and 5, not shown by column 3).

(5) The number of habitually non-migrant species per depth showed little variation below 100 m (column 6). Thus, proportionally more species were habitual migrants in the 400–800 m band, whilst at greater depths the converse situation was apparent.

The more abundantly caught species taken per depth by day and night of habitual migratory and non-migratory types, and the respective overall apparent dominant species, are tabulated in Table 12. Among non-migrants, as expected, the apparent abundance relations were generally maintained by day and night, and at depths greater than 200 m they comprised the overall dominant species at each depth. Domination by migrants occurred only at night, in the upper 200 m. It is also worth noting that the majority of migrants listed were common to both day and night columns, despite the vertical compression that has resulted from migration.

Attempts to categorize migratory behaviour have been made (MARSHALL, 1960; BADCOCK, 1970) based upon apparent species' behaviour as exemplified by juveniles and adults. In many species, however, migratory behaviour is dependent upon developmental stage, and among

*Table* 11.   *The relation in the water column of typically non-migrant species with the migratory elements of the total population.*

| Depth (m) | No of species Day | No of species Night | Gain or loss of spp number at night | Migrants Day | Migrants Night | No spp common to day and night catches |
|---|---|---|---|---|---|---|
| Column | 1 | 2 | 3. | 4 | 5 | 6 |
| 10-25 | 2 | 7 | +5 | 1 | 6 | 2 |
| 25-50 | 1 | 18 | +17 | 1 | 15 | 1 |
| 50-100 | 4 | 24 | +20 | 4. | 21 | 3 |
| 100-200 | 7 | 28 | +21 | 4 | 19 | 6 |
| 200-300 | 8 | 20 | +12 | 1 | 13 | 6 |
| 300-400 | 10 | 13 | +3 | 4 | 7 | 4 |
| 400-500 | 21 | 13 | -8 | 12 | 5 | 6 |
| 500-600 | 19 | 12 | -7 | 12 | 1 | 7 |
| 600-700 | 21 | 6 | -15 | 15 | 0 | 6 |
| 700-800 | 17 | 7 | -10 | 9 | 0 | 6 |
| 800-900 | 21 | 10 | -11 | 10 | 0 | 10 |
| 900-1000 | 16 | 11 | -5 | 7 | 0 | 8 |
| 1000-1250 | 10 | 8 | -2 | 3 | 0 | 6 |
| 1250-1500 | 9 | 9 | 0 | 2 | 0 | 7 |
| 1500-2000 | 8 | 6 | -2 | 1 | 0 | 5 |

adults factors such as breeding condition may be causal in its modification. Although not, perhaps, a constant feature throughout its juvenile and adult life, the migratory habitude of a species first develops during the juvenile stages. Among the stomiatoids and myctophids two characteristic modes of post-larval development are apparent. In some species (e.g. *Cyclothone*, *Chauliodus* and *Benthosema*) the post-larvae acquire their photophores more or less simultaneously; a process which coincides with a rapid downwards migration from the surface waters to the juvenile–adult diurnal layers (JESPERSEN and TÅNING, 1926; this study). In some stomiatoid species the post-larvae acquire their photophores more gradually and in more easily definable sequences (e.g. *Valenciennellus*, *Bonapartia*, *Gonostoma* and the hatchet fishes). Whilst the onset of photophore pigmentation in *Argyropelecus* may coincide with a rapid downwards migration to depth, among the gonostomatids (*sensu* GREY, 1964) of this ilk,

the ontogenetic migration is probably a gradual process (GREY, 1964), as exemplified by *Valenciennellus* in this study. The present results demonstrate, however, that the basic mode of post-larval development has no bearing whatsoever upon whether a species is an habitual migrant or non-migrant in adult life. Migratory behaviours are *species* characteristic.

CLARKE (1973) showed that small juveniles of certain habitually migrant myctophid species remained at depth at all times. As judged from the limited data of the present study, juveniles of myctophids do not adopt the normal juvenile–adult migratory characters until a certain degree of pigmentation has been attained.

In general, juveniles and adults of midwater fishes may be characteristically full migrants, shallow or deep partial migrants, or non-migrants (*sensu* MARSHALL, 1960; BADCOCK, 1970). Behaviour varies from species to species, and no correlation between migratory pattern and diurnal

Table 12. The most abundantly caught habitual migratory and non-migratory types, ordered first and second per type, and the overall dominant species, per depth by day and night.

| Depth (m) | DAYTIME Non-Migrants 1st | DAYTIME Non-Migrants 2nd | DAYTIME Migrants 1st | DAYTIME Migrants 2nd | NIGHTTIME Non-Migrants 1st | NIGHTTIME Non-Migrants 2nd | NIGHTTIME Migrants 1st | NIGHTTIME Migrants 2nd |
|---|---|---|---|---|---|---|---|---|
| 10-25 | (Cyclothone p.l.) | - | *Macrorhamphosus ‡ | - | - | - | *B. suborbitale ‡ | H. hygomi ‡ |
| 25-50 | - | - | *(Macrorhamphosus) | - | - | - | *B. suborbitale ‡ | Macrorhamphosus ‡ |
| 50-100 | *(Cyclothone p.l.) | - | (Macrorhamphosus) | - | Cyclothone p.l. | - | *Macrorhamphosus ‡ | C. warmingi ‡ |
| 100-200 | *(B. pedaliota) | (V. tripunctulatus) | (S. analis) | - | (Cyclothone p.l.) | (V. tripunctulatus) | *Macrorhamphosus ‡ | S. analis |
| 200-300 | *V. tripunctulatus ‡ | B. pedaliota | (S. analis) | - | *V. tripunctulatus ‡ | (B. pedaliota) | A. aculeatus | e.g. L. cuprarius |
| 300-400 | *A. hemigymnus ‡ | V. tripunctulatus ‡ | A. aculeatus ‡ | L. dofleini | *V. tripunctulatus ‡ | A. hemigymnus ‡ | L. cuprarius | (N. valdiviae) |
| 400-500 | *C. braueri ‡ | A. hemigymnus ‡ | N. valdiviae | L. dofleini | *C. braueri ‡ | A. hemigymnus ‡ | (B. greyae) | e.g. (M. zugmayeri) |
| 500-600 | *C. braueri ‡ | C. microdon ‡ | B. suborbitale ‡ | H. hygomi | *C. braueri ‡ | C. microdon ‡ | (L. gaussi) | - |
| 600-700 | *C. microdon ‡ | C. pseudopallida | B. indicus | e.g.D. atlanticus | *C. microdon ‡ | C. braueri | - | - |
| 700-800 | *C. microdon ‡ | C. pallida | (L. gaussi) | e.g.(L. cuprarius) | *C. microdon ‡ | C. pallida | - | - |
| 800-900 | *C. microdon ‡ | e.g.(S. opisthopterus) | e.g. (L. gaussi & L. cuprarius) | - | *C. microdon ‡ | e.g. (S. opisthopterus) | - | - |
| 900-1000 | *C. microdon ‡ | (C. braueri) | (C. warmingi) | e.g.(L. gaussi) | *C. microdon ‡ | (C. braueri) | - | - |
| 1000-1250 | *C. microdon ‡ | (C. braueri) | (C. warmingi) | (N. resplendens) | *C. microdon ‡ | (C. braueri) | - | - |
| 1250-1500 | *C. microdon ‡ | e.g.(S. opisthopterus) | (C. warmingi) | (S. analis) | *C. microdon ‡ | (C. braueri) | - | - |
| 1500-2000 | *C. microdon | (C. braueri) | (N. resplendens) | - | *C. microdon | e.g.(C. couesei) | - | - |

* Overall dominant species per depth; ‡ species sample > 49 specimens; parentheses, species sample of < 10 specimens.

depth distribution could be made. Nevertheless, at depths greater than about 800 m the proportion of non-migratory species increases. The juveniles and adults of a species often fall into the same migration class, but there are cases where they do not. For example, juveniles of *Gonostoma elongatum* are full migrators, whilst adults are partial ones (Fig. 3).

Feeding habit has been regarded as a prime motivator of the cyclic migratory behaviour of midwater fishes (e.g. MARSHALL, 1960). This viewpoint has been complemented recently by the study of MERRETT and ROE (1974), which was based upon samples collected at the same position as those used in this report. They show that not only are certain midwater fishes selective feeders but also the vertical distributions of predator and prey coincide, in such cases, only at particular times. No biological factors examined, however, can explain the remaining at depth by elements of the *Notolychnus* adult population at night (p. 45, Fig. 21), noted both in this study and in the Pacific work of CLARKE (1973). Nor indeed does the apparent non-diel migratory behaviour of *Chauliodus danae* at present lend itself to satisfactory interpretation.

Reverse migrations were noted only among the post-larvae and early juveniles of *Macrorhamphosus*, which occurred in the upper layers. Recent work has shown the temporal and migratory relations of fish post-larvae in the upper layers to be complex (see reviews of PARIN, 1968; HEMPEL and WEIKERT, 1972), and the post-larvae of some species are known to undergo reverse migrations. These, however, occur over relatively short distances, and the reverse migration of up to 200 m by *Macrorhamphosus* juveniles seems the more remarkable.

Incorporated in the ontogenetic variation in the vertical distribution of some species is a size–depth relation linked to advancing maturity. This is clearly demonstrated in the case of the *Cyclothone* species from these collections. Also, a similar relation is implied for *V. tripunctulatus*, where the larger specimens bearing the most large ripening eggs were found in the deepest part of the distribution. The abundant *C. braueri* is most obvious in this respect, as its life span apparently covers a single spawning cycle only. It is possible that, in this case at least, the segregation of the ripest males and females at the bottom of the distributional range of the species may be passive. An increase in specific gravity conceivably results from compression of the swimbladder by swollen, ripe gonads. Running ripe fish were, however, but a small proportion of the total population. The preponderance of male *C. braueri* occurred in the same stratum (400–500 m) as females with immature or ripening gonads. Nevertheless, almost all testes examined were milt filled and the sexually dimorphic olfactory organs were moderately developed. Individually smaller than females, males were almost four times more numerous in this shallower layer, both by day and night. In the stratum below (500–600 m) males were out-numbered by all females by nearly 10 : 1; yet approximate parity existed between males and running ripe females. The delicate olfactory organ and associated snout prolongation in many of the males was fully developed. So while mating was most likely in the deeper levels of the distribution, potential male viability and olfactory location occurred over the entire depth range of the adult population. Additional data are expected to show that somewhat similar patterns exist for *C. pseudopallida*, *C. pallida* and *C. microdon* also, although they are likely to be complicated by longer life cycles in these species. Preliminary evidence from collections off Bermuda (Institute of Oceanographic Sciences, unpublished data) indeed confirms the situation in *C. braueri* and clarifies similar patterns, comprising more age groups, in the other three species. Furthermore, it reveals that *C. microdon*, at least, is protandrous while apparently confirming that *C. braueri* does not undergo sex-reversal.

It should be remembered, however, that the above conclusions are based upon observations from one locality, at one time. While each of the stations from the meridional section will provide a similarly limited insight, even among the variously more abundant species, ultimate consideration of the whole should provide a more representative perspective.

*Acknowledgements*—We would like to thank our colleagues in Marine Biology at the Institute of Oceanographic Sciences for their helpful discussions on various aspects of this study. We are particularly indebted to Dr. J. E. CRADDOCK (Woods Hole Oceanographic Institution), Mr. P. M. DAVID (Institute of Oceanographic Sciences), Dr. R. H. GIBBS, JR. (Smithsonian Institution) and Professor N. B. MARSHALL (Queen Mary College, London), all of whom read the manuscript and provided useful discussion and constructive criticism of it. Our thanks also go to Miss R. LARCOMBE (Institute of Oceanographic Sciences) who generally assisted and patiently measured much of the identified material.

## REFERENCES

AHLSTROM E. H. and R. C. COUNTS (1958) Development and distribution of *Vinciguerria lucetia* and related species in the eastern Pacific. *Fishery Bull. Fish Wildl. Serv. U.S.*, **58** (139), 363–416.

ANGEL M. V. (1969) Planktonic ostracods from the Canary Island region; their depth distributions, diurnal migrations, and community organisation. *J. mar. biol. Ass. U.K.*, **49**, 515–533.

ANGEL M. V. and M. J. FASHAM (1973) SOND Cruise 1965: factor and cluster analyses of the plankton results, a general summary. *J. mar. biol. Ass. U.K.*, **53**, 185–231.

ANGEL M. V. and M. J. FASHAM (1975) Analysis of the vertical and geographic distribution of the abundant species of planktonic ostracods in the northeast Atlantic. *J. mar. biol. Ass. U.K.*, **55**, 709–737.

BACKUS R. H., J. E. CRADDOCK, R. L. HAEDRICH and D. L. SHORES (1970) The distribution of mesopelagic fishes in the equatorial and western North Atlantic. *J. mar. Res.*, **28**(2), 179–201.

BACKUS R. H., G. W. MEAD, R. L. HAEDRICH and A. W. EBELING (1965) The mesopelagic fishes collected during Cruise 17 of R.V. *Chain*, with a method for analysing faunal transects. *Bull. Mus. comp. Zool. Harv.*, **134**, 139–158.

BADCOCK J. (1969) Colour variation in two mesopelagic fishes and its correlation with ambient light conditions. *Nature, Lond.*, **221**, 283–285.

BADCOCK J. (1970) The vertical distribution of mesopelagic fishes collected on the SOND Cruise. *J. mar. biol. Ass. U.K.*, **50**, 1001–1044.

BADCOCK J. R. and N. R. MERRETT (1972) On *Argyripnus atlanticus*, Maul 1952 (Pisces, Stomiatoidei), with a description of post-larval forms. *J. Fish. Biol.*, **4**, 277–287.

BAIRD R. C. (1971) The systematics, distribution, and zoogeography of the marine hatchetfishes (family Sternoptychidae). *Bull. Mus. comp. Zool. Harv.*, **142**, 1–128.

BAKER A. DE C. (1970) The vertical distribution of euphausiids near Fuerteventura, Canary Islands (*Discovery* SOND Cruise, 1965). *J. mar. biol. Ass. U.K.*, **50**, 301–342.

BAKER A. DE C., M. R. CLARKE and M. J. HARRIS (1973) The NIO combination net (RMT 1+8) and further developments of rectangular midwater trawls. *J. mar. biol. Ass. U.K.*, **53**, 167–184.

BERTELSEN E. (1973) A new species of deep-sea anglerfish, *Linophryne sexfilis* (Pisces, Ceratioidea). *Steenstrupia*, **3**, 65–69.

BODEN B. P. and E. M. KAMPA (1967) The influence of natural light on the vertical migrations of an animal community in the sea. *Symp. zool. Soc. Lond.* No. 19, 15–26.

CLARKE G. L. and R. H. BACKUS (1964) Interrelations between the vertical migration of deep scattering layers, bioluminescence, and changes in daylight in the sea. *Bull. Inst. océanogr. Monaco.* No. 1318, 1–36.

CLARKE M. R. (1969) Cephalopoda collected on the SOND Cruise. *J. mar. biol. Ass. U.K.*, **49**, 961–976.

CLARKE M. R. and C. C. LU (1974) Vertical distribution of cephalopods at 30°N 23°W in the North Atlantic. *J. mar. biol. Ass. U.K.*, **54**, 969–984.

CLARKE T. A. (1973) Some aspects of the ecology of lanternfishes (Myctophidae) in the Pacific Ocean near Hawaii. *Fishery Bull. Fish Wildl. Serv. U.S.*, **71**(2), 401–434.

CLARKE T. A. (1974) Some aspects of the ecology of stomiatoid fishes in the Pacific Ocean near Hawaii. *Fishery Bull. Fish Wildl. Serv. U.S.*, **72**(2), 337–351.

CLARKE W. D. (1966) Bathyphotometric studies of the light regime of organisms of the deep scattering layers. A.G.C. Res. and Dev. Tr. 66-02. G. M. Defense Res. Lab., Santa Barbara, Calif.

COHEN D. M. (1973) Zoogeography of the fishes of the Indian Ocean. In: *Ecological studies, analysis and synthesis*, Vol. 3, B. ZEITZSCHEL, editor, Springer-Verlag, New York, pp. 451–463.

CURRIE R. I., B. P. BODEN and E. M. KAMPA (1969) An investigation on sonic scattering layers: the RRS *Discovery* SOND Cruise, 1965. *J. mar. biol. Ass. U.K.*, **49**, 489–514.

EBELING A. W. (1962) Melamphaidae. I. Systematics and zoogeography of the species in the bathypelagic genus *Melamphaes* Günther. *Dana Rep.* No. 58, 1–64.

EHRICH S. and H. C. JOHN (1973) Zur Biologie und Ökology der Schnepfenfische (Gattung *Macrorhamphosus*) vor Nordwestafrika und Überlegungen zum Altersaufbau der adulten Bestände der Großen Meteorbank. *'Meteor' Forsch.-Ergebn.* D, No. 14, 87–98.

FOXTON P. (1969) SOND Cruise 1965. Biological sampling methods and procedures. *J. mar. biol. Ass. U.K.*, **49**, 603–620.

FOXTON P. (1970a) The vertical distribution of pelagic decapods (Crustacea, Natantia) collected on the SOND Cruise 1965. I. The Caridea. *J. mar. biol. Ass. U.K.*, **50**, 939–960.

FOXTON P. (1970b) The vertical distribution of pelagic decapods (Crustacea: Natantia) collected on the SOND Cruise 1965. II. The Peneidea and general discussion. *J. mar. biol. Ass. U.K.*, **50**, 961–1000.

FOXTON P. (1971–1972) Observations on the vertical distribution of the genus *Acanthephyra* (Crustacea, Decapoda) in the eastern North Atlantic, with particular reference to species of the '*purpurea*' group. *Proc. R. Soc. Edinb.* (B), **73**(30), 301–313.

GIBBS R. H. and C. F. E. ROPER (1970) Ocean Acre preliminary report on vertical distribution of fishes and cephalopods. *Proceedings of an International symposium on biological sound scattering in the ocean*, 1970, pp. 119–133.

GOODYEAR R. H., B. J. ZAHURANEC, W. L. PUGH and R. H. GIBBS (1972) Ecology and vertical distribution of Mediterranean midwater fishes. *Mediterranean Biological Studies. Final Report*, **1**(3), 91–229.

GREY M. (1964) Family Gonostomatidae. In: *Fishes of the western North Atlantic*, Y. H. OLSEN editor, *Mem. Sear Fdn mar. Res.*, **1**, 78–240.

HAEDRICH R. L. and J. E. CRADDOCK (1968) Distribution and biology of the opisthoproctid fish *Winteria telescopa* Brauer 1901. *Breviora*, No. 294, 1–11.

HAFFNER R. E. (1952) Zoogeography of the bathypelagic fish *Chauliodus. Syst. Zool.*, **1**, 113–133.

HARRISSON C. M. H. (1967) On methods for sampling mesopelagic fishes. *Symp. zool. Soc. Lond. No. 19*, 71–126.

HEMPEL G. and H. WEIKERT (1972) The neuston of the subtropical and boreal north-eastern Atlantic Ocean. A review. *Mar. Biol.* **13**, 70–80.

JESPERSEN P. (1933) *Valenciennellus tripunctulatus. Faune Ichth. Atl. Nord.* No. 14, 80–81.

JESPERSEN P. and A. V. TÅNING (1926) Mediterranean Sternoptychidae. *Rep. Dan. oceanogr. Exped. Mediterr.* II, (Biol.) (A2), 1–59.

JOHN H-C. (1973) Oberflächennahes Ichtyoplankton der Kanarenstrom-Region. '*Meteor*' *Forsch.-Ergebn.* D. No. 15, 36–50.

KAMPA E. M. (1970) Underwater daylight and moonlight measurements in the eastern North Atlantic. *J. mar. biol. Ass. U.K.*, **50**, 397–420.

KAWAGUCHI K. and R. MARUMO (1967) Biology of *Gonostoma gracile* (Gonostomatidae). I. Morphology, life history and sex reversal. *Inf. Bull. Planktol. Japan*, 1967, 53–69.

KOEFOED E. (1960) Isospondyli. 2. Heterophotodermi. 2, from the "Michael Sars" North Atlantic Deepsea Expedition 1910 (with addenda et corrigenda to Isospondyli 1 and 2 : 1). *Rep. scient. Results Michael Sars N. Atlant. deep Sea Exped.*, **4** (part II), No. 8, 1–15.

KRUEGER W. H. (1972) Biological studies of the Bermuda Ocean Acre. IV. Life history, vertical distribution and sound scattering in the gonostomatid fish *Valenciennellus tripunctulatus* (Esmark). Report to U.S. Navy Underwater Systems Center. Contract No. N.00140-72-C-0315, pp. 1–37.

KRUEGER W. H. and G. W. BOND (1972) Biological studies of the Bermuda Ocean Acre. III. Vertical distribution and ecology of the bristlemouth fishes (family Gonostomatidae). Report to U.S. Navy Underwater Systems Center. Contract No. N. 00140-72-C-0315, pp. 1–49.

MARSHALL N. B. (1960) Swimbladder structure of deepsea fishes in relation to their systematics and biology. *Discovery Rep.*, **31**, 1–122.

MARSHALL N. B. (1967) The olfactory organs of bathypelagic fishes. *Symp. zool. Soc. Lond.* No. 19, 57–70.

MERRETT N. R. (1971) Aspects of the biology of billfish (Istiophoridae) from the equatorial western Indian Ocean. *J. Zool. Lond.*, **163**, 351–395.

MERRETT N. R., J. BADCOCK and P. J. HERRING (1973) The status of *Benthalbella infans* (Pisces, Myctophoidei), its development, bioluminescence, general biology and distribution in the eastern North Atlantic. *J. Zool. Lond.*, **170**, 1–48.

MERRETT N. R. and H. S. J. ROE (1974) Patterns and selectivity in the feeding of certain mesopelagic fish. *Mar. Biol.*, **28**, 115–126.

MOSER H. G. and E. H. AHLSTROM (1970) Development of the lanternfishes (family Myctophidae) in the California Current. Part 1. Species with narrow-eyed larvae. *Bull. Los Ang. Cty. Mus. nat. Hist. Sci.* No. 7, 1–145.

MOSER H. G. and E. H. AHLSTROM (1972) Development of the lanternfish, *Scopelopsis multipunctatus* Brauer 1906, with a discussion of its phylogenetic position in the family Myctophidae and its role in a proposed mechanism for the evolution of photophore patterns in lanternfishes. *Fishery Bull. Fish Wildl. Serv. U.S.*, **70**(3), 541–564.

MURRAY J. and J. HJORT (1912) *The depths of the ocean*, Macmillan, 821 pp.

NAFPAKTITIS B. G. (1968) Taxonomy and distribution of the lanternfishes, genera *Lobianchia* and *Diaphus*, in the North Atlantic. *Dana Rep.* No. 73, 1–131.

PARIN N. V. (1967) Diurnal variations in the larval occurrence of some oceanic fishes near the surface. *Oceanology*, **7**(1), 115–121. Translated from Russian by Scripta Technica Inc. for the American Geophysical Union.

PARIN N. V. (1968) *Ichthyofauna of the epipelagic zone*. Translated from Russian by I.P.S.T. Ltd. for U.S. Dept. of the Interior and the National Science Foundation, Washington D.C. 1970, 206 pp.

PARR A. E. (1934) Report on experimental use of a triangular trawl for bathypelagic collecting with an account of the fishes obtained and a revision of the family Cetomimidae. *Bull. Bingham oceanogr. Coll.* **4**(6), 1–59.

PAXTON J. R. (1967) A distributional analysis for the lanternfishes (family Myctophidae) of the San Pedro Basin, California. *Copeia*, 1967, No. 2, 422–440.

PEARCY W. G. and R. M. LAURS (1966) Vertical migration and distribution of mesopelagic fishes off Oregon. *Deep-Sea Res.*, **13**, 153–166.

PERTSEVA-OSTROUMOVA T. A. (1964) Some morphological characteristics of myctophid larvae (Myctophidae, Pisces). In: *Fishes of the Pacific and Indian oceans. biology and distribution*, T. S. RASS, editor. Translated from Russian by I.P.S.T. Ltd. Jerusalem, 1966, pp. 79–97.

PICKFORD G. E. (1946) *Vampyroteuthis infernalis* Chun. An archaic dibranchiate cephalopod. I. Natural history and distribution. *Dana Rep.* No. 29, 1–40.

Pugh P. R. (1974) The vertical distribution of the Siphonophores collected during the SOND Cruise, 1965. *J. mar. biol. Ass. U.K.*, **54**, 25–90.

Regan C. T. and E. Trewavas (1929) Fishes of the family Astronesthidae and Chauliodontidae. *Oceanogrl. Rep. 'Dana' Exped.* 1920–22, No. 5, 1–39.

Robison B. H. (1972) Distribution of the midwater fishes of the Gulf of California. *Copeia*, 1972, No. 3, 448–461.

Roe H. S. J. (1972a) The vertical distributions and diurnal migrations of calanoid copepods collected on the SOND Cruise, 1965. I. The total population and general discussion. *J. mar. biol. Ass. U.K.*, **52**, 277–314.

Roe H. S. J. (1972b) The vertical distributions and diurnal migrations of calanoid copepods collected on the SOND Cruise, 1965. II. Systematic account: families Calanidae up to and including the Aetideidae. *J. mar. biol. Ass. U.K.*, **52**, 315–343.

Roe H. S. J. (1972c) The vertical distributions and diurnal migrations of calanoid copepods collected on the SOND Cruise, 1965. III. Systematic account: families Euchaetidae up to and including the Metridiidae. *J. mar. biol. Ass. U.K.*, **52**, 525–552.

Roe H. S. J. (1972d) The vertical distributions and diurnal migrations of calanoid copepods collected on the SOND Cruise, 1965. IV. Systematic account of families Lucicutidae to Candaciidae. The relative abundance of the numerically most important genera. *J. mar. biol. Ass. U.K.*, **52**, 1021–1044.

Roe H. S. J. (1974) Observations on the diurnal vertical migrations of an oceanic animal community. *Mar. Biol.*, **28**, 99–113.

Steedman H. F. (1974) Laboratory methods in the study of marine zooplankton. A summary report on the results of Joint Working Group 23 of SCOR and UNESCO, 1968–1972. *J. Cons. perm. int. Explor. Mer*, **35**, 351–358.

Sverdrup H. U., M. W. Johnson and R. H. Fleming (1942) *The oceans. Their physics, chemistry ana general biology*, Prentice-Hall, 1087 pp.

Tåning A. V. (1918) Mediterranean Scopelidae (*Saurus, Aulopus, Chlorophthalmus* and *Myctophum.*) *Rep. Dan. oceanogr. Exped. Mediterr.*, **2** (Biol.), (A7), 1–154.

Witzell W. N. (1973) Gonostomatidae. In: *Clofnam* I. *Checklist of the fishes of the north-eastern Atlantic and of the Mediterranean*, No. 37, J. C. Hureau and Th. Monod, editors, UNESCO, pp. 114–122.

### APPENDIX

*Myctophid larvae*

The larval developments of *N. resplendens*, *B. suborbitale* and *H. hygomi* are traced below. In early stage larvae, prior to notochord flexion, body length (BL) was taken from the tip of the snout to that of the notochord. In post-notochord flexion specimens standard length (SL) was taken. All measurements were made to the nearest 0·1 mm using dial calipers. The terminology of Naffaktitis (1968) has generally been used where reference to particular

photophores is made. The accounts are based upon 75 specimens of *N. resplendens*, 61 *B. suborbitale*, and 53 *H. hygomi*. The material thus provides only for an outline description of larval and transforming characteristics.

(a) *Notoscopelus resplendens*. Both *N. japonicus* (Tanaka, 1908) and *N. resplendens* have been figured in accounts characterizing the larvae at generic level (Pertseva-Ostroumova, 1964, Fig. 10; Moser and Ahlstrom, 1972, Fig. 6, respectively). *Notoscopelus elongatum* (Costa, 1844) is the only species where the larvae have been described in detail (Tåning, 1918). The following brief description of *N. resplendens* larvae is based on 72 specimens 3·2 mm BL–19·7 mm SL. Three transforming specimens 20–22 mm SL were caught, but badly damaged.

The larger larvae are distinguished from other myctophid larvae in the area by virtue of their round eye, the dorsal fin exceeding the length of the anal fin, and their characteristic pigment pattern (Fig. 22d). The degree of pigmentation, however, is dependent upon stage of development. In the earliest forms examined, 3·2–3·8 mm BL, prior to notochord flexion, pigment occurs as single spots at the jaw symphysis; on the snout; embedded in the midline of the frontal area; and on the posterior edge of the anal papilla (Fig. 22a). In addition, two spots occur in the gill cavity, and a further spot postero-ventrally at the base of the brain. The peritoneum of the swimbladder is also pigmented in the dorsal aspect (Fig. 22a). This basic pigment pattern is retained throughout the larval period (although it is not necessarily always the most distinct present), and further pigment modifications are attained purely by the addition of melanophores. In the above-mentioned early forms, fin bases are indiscernible, apart from the pectoral lobes. Just prior to notochord flexion, however, differentiation of the dorsal and anal fin bases is initiated. Furthermore, modifications in the head and body shape occur, and the embedded pigment of the lateral line series begins to develop anteriorly (Fig. 22b). Up to this stage photophores are absent. $Br_2$ forms during notochord flexion; additionally, the initial melanophores of the dorsal and ventral series appear (Fig. 22c). The caudal fin rays begin differentiation in this period, and hereafter the dorsal and anal fin rays develop more or less concurrently. Pigmentation is also furthered. By 7·0 mm SL (Fig. 22d) the body is relatively robust, the dorsal and anal fins are practically complete, the pectoral fin nears completion of fin rays, and the first rays of the pelvic fin are distinct. The pigmentation is more advanced, and the second photophore, $PO_5$, has appeared. In *Notoscopelus* larvae, Vn is the third photophore to appear (Moser and Ahlstrom, 1972). In the smallest larva of the current series on which it was observed, 9·8 mm SL, fin development was complete. At about 11 mm SL, pigment aligned vertically at the caudal base starts forming. In the most advanced specimens the dorsal pigment series runs from the nape to the procurrent rays. Further secondary lines of pigment run parallel between the dorsal series and the lateral line. The embedded lateral line pigment itself is overlain by a row of more superficial melanophores. The pigmentation of the brain region also becomes more complicated during these later stages. After notochord flexion the melanophore pattern in this region is simple. It is comprised almost invariably of an olfactory spot, two

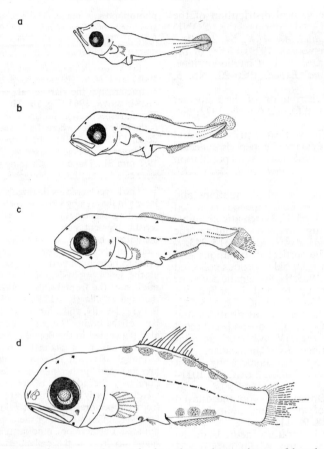

Fig. 22. Larval development of *Notoscopelus resplendens* shown by specimens of lengths (a) 3·8 mm; (b) 4·7 mm; (c) 5·4 mm; (d) 7·0 mm.

pairs of more superficially disposed occipital spots, and a deep-seated spinal cord pigment (Fig. 22c and d). With further development pigment is added in the occipital region. TÅNING (1918), in his description of the larval forms of *Lampanyctus* (= *Notoscopelus*) *elongatus*, notes "occipital pigment, generally paired melanophores, which may, however, partially fuse". A similar situation was found in the present specimens, but it was noted that the occipital melanophores were dispersed, i.e. 'partially fused', in day specimens only. These melanophores were aggregated in night specimens (Fig. 23). The specimens were not in ideal condition, and it could not be ascertained whether or not the melanophores of the dorsal and ventral series were more dispersed by day.

(b) *Benthosema suborbitale*. The larvae of *B. glaciale* (Reinhardt, 1837) and *B. panamense* (Tåning, 1932) are relatively well known (e.g. TÅNING, 1918; MOSER and AHLSTROM, 1970), but of *B. suborbitale* only a brief description (PERTSEVA-OSTROUMOVA, 1964) could be found. Forty-five larvae and 16 transforming larvae were represented in the present collections (4·0 mm BL–11·0 mm SL). The distinction between late transforming and early juvenile forms is slight, and has been based upon the nature of the eyes. Individuals with round eyes have been regarded here as juveniles.

The larva of *B. suborbitale* is typically narrow-eyed (*sensu* MOSER and AHLSTROM, 1970), with the ventral choroid tissue slightly developed. Pigmentation is limited, confined principally to the head region and characteristically arranged. It is situated as illustrated (Figs. 24 and 25) and thus associated with the snout, jaw symphysis, Br photophores, the isthmus and the pectoral fins (in the last case, even during fin ray development). The opercular pigment spot lies on the inner lining in a position where the superior OP arises. In specimens of 7·1 mm SL and greater, indications of a further (external) upper opercular spot can be seen (Fig. 24c), but this is not a constant feature. Pigment occurs in the posterior lining of the gill cavity as two distinct patches: one, more externally disposed at the level of the pectoral fin base; the other, slightly more ventral and effectively placed medially (thus a medial pair anterior to the stomach is apparent). Further, there is pigment located just posterior to the brain, a spot on each side of the spinal cord. This pigment, however, becomes occluded

Night                    Day

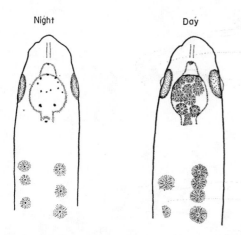

Fig. 23. Comparison of occipital and meningeal melanophores of larval *Notoscopelus resplendens* by night and day.

a

b

c

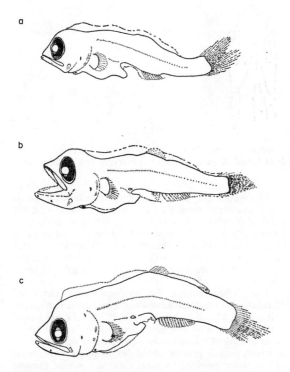

Fig. 24. Larval development of *Benthosema suborbitale* shown by specimens of lengths (a) 6·0 mm; (b) 6·4 mm; (c) 7·1 mm.

in specimens of 7·1 mm SL and greater by a gradual thickening of epaxial muscle. No pigment was observed along the ventral midline of the tail. In specimens in which notochord flexion has not occurred (4·0–4·9 mm BL) the

pigment spots are fewer. The gill cavity melanophores are undeveloped in the smaller individuals, and the superior one develops just prior to notochord flexion. Neither the opercula nor the spinal cord spots appear until after notochord flexion. The isthmus pigment appears as two distinct melanophores prior to notochord flexion, but afterwards they fuse to form a single pigment patch.

$Br_2$ photophores are present on all specimens, but only in the larger specimens is there further photophore development. $PO_1$ and $PO_2$ are developing at 8·5 mm SL; and at 8·6 mm, in addition, $Br_1$.

Notochord flexion is complete at about 5·0 mm SL, and both caudal and pectoral fins are developing. The principal caudal fin rays are complete at 5·5 mm, but the procurrent rays are undeveloped. At 6·0 mm the ventral buds are discernible and the anal fin rays beginning to develop. The pectoral fin rays are complete at 6·3 mm and the procurrent rays differentiating. Both anal and procurrent rays are complete at 7·1 mm, at about which stage the dorsal fin starts developing. By 8·2 mm all fins, with the exception of the pelvics, are differentiated.

During transformation, modifications in eye shape and dentition occur, the pelvics are completed, and the dorsal fin fold recedes posteriorly leaving the dorsal adipose fin distinct. The transforming specimens (8·8–10·8 mm SL) are in relatively poor condition and the sequence of photophore development presented is tentative. In the least advanced specimen Br and PO are numerically complete, and the superior OP and $AOa_{1-2}$ are present. Hereafter, PLO, PVO, inferior OP and VO photophores develop, although $VO_2$ does not form until after the $AO_P$ series is initiated. $AO_{P1}$ and $Prc_1$ appear simultaneously, and $Prc_2$ before $AO_P$ is numerically complete. $SAO_{1-2}$ and VLO appear with $AO_{P1}$ and $SAO_3$ is present before $Prc_2$, as is Pol. The most advanced specimen had all photophores developed with the exception of Dn and So. The latter two light organs are present in all juveniles.

Essentially, pigmentation characteristic of the larva is retained during transformation. The central opercular melanophore becomes less distinct due to the overlay by the superior OP. The gill cavity and posterior isthmus pigment groups are generally distinct in our specimens, but other typically larval pigment is only variously present (presence or absence, however, is related more to the degree of abrasion of the animal than to the developmental state). In late transforming individuals faint cranial and stomach pigmentation is apparent.

Pigmentation is continued through the early juvenile stages so that both light and dark coloured juveniles occurred in the collections. The larval pigment of the gill cavity, jaw symphysis and posterior isthmus is variously distinct in light coloured juveniles (10–11 mm), whilst in dark juveniles (10+ mm) only the jaw symphysis spot is clearly apparent. Light and dark juveniles were found in other species, e.g. *Diogenichthys atlanticus*, *Hygophum hygomi*, and behavioural differences between the two juvenile forms were noted.

(c) *Hygophum hygomi*. Larvae were traced back from juvenile and transforming individuals to a stage prior to notochord flexion, thus eliminating the potential identity confusion between larvae of *H. hygomi* and *H. benoiti* (see Tåning, 1918). The pigmentation appears characteristic

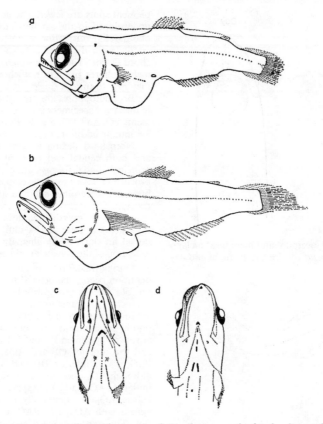

Fig. 25. (a) and (b), continued larval development of *Benthosema suborbitale* shown by specimens of lengths (a) 8·2 mm; (b) 8·6 mm; (c) and (d), ventral aspect of the heads of larval *B. suborbitale* (c) and *Hygophum hygomi* (d).

and is essentially similar in both larval and transforming forms. It is ventrally disposed and associated with the lower jaw symphysis, the isthmus, a position just posterior to the $PO_1$ site (Fig. 25d) and either side of the anal papilla (see also TÅNING, 1918, Fig. 17). Pigment is also to be located as a pair of discrete spots on each side of the gill cavity. Other pigmentation is generally absent from the specimens, but occasionally an odd abdominal spot was observed. In the stages where notochord flexion has not occurred (3·5–5·7 mm BL), an additional pair of melanophores occurs either side of the presumptive anal fin, just prior to its origin. Notochord flexion is completed by 7·3 mm SL. The overall larval size-range was 3·5 mm BL–12·7 mm SL, and $Br_2$ was present only in post-notochord flexion specimens. In the largest specimens, additional pigment speckles the pectoral fin rays, but no further photophore development is made, and the pelvics are undeveloped.

The sequence of photophore development could not be clearly elucidated from the transforming specimens caught (12·0–15·0 mm). $PO_{1-4}$ and $VO_{1-4}$ are developed early, followed by $Br_3$ and the superior PVO. $Br_1$ and the superior OP are initiated before AOa. $AOa_1$ and $PO_5$ appear at about the same time, along with PLO and $SAO_2$. $Prc_{1-2}$ and the posterior $AO_P$ are amongst the last photophores to be initiated.

The larvae cannot be confused with those of *H. reinhardti*, which are more slender and differ in pigment pattern (see MOSER and AHLSTROM, 1970). Of other larvae in the area, those of *H. hygomi* are most similar to those of *B. suborbitale*. Distinction between the two can be made, however, even in pre-notochord flexion stages on the basis of the better developed ventral choroid tissue of the eye in *Hygophum* (see MOSER and AHLSTROM, 1970, Fig. 1), and the more extensive pigmentation of the isthmus in this species (Fig. 25).

*Prog. Oceanog.*, 1976, Vol. 7, pp. 59–90. Pergamon Press. Printed in Great Britain.

# The mixing and spreading phases of MEDOC. I

PETER D. KILLWORTH*

(*Received* 12 *May* 1975; *in revised form* 22 *August* 1975; *accepted* 29 *September* 1975)

Abstract—The processes which produce bottom water in the Mediterranean are studied. Observations show three phases. The first, called 'preconditioning', is not studied. The second 'violent mixing' phase occurs when the cold Mistral begins. Cooling at the surface leads to intense vertical mixing in a narrow chimney-like region. When the wind stops, the 'sinking and spreading' phase begins. Interleaving of water masses occurs; horizontal eddying on scales of 15 km is observed; and 600 m columns of water can be lifted up to 500 m.

A two-dimensional model is used to explain the mixing phase. Non-penetrative vertical convection explains the observations well. Advection of water from outside the column is small, and slows down the descent of the outer parts of the column as observed. The column does not break up even when it reaches the bottom, or if the strong winds cease.

The most efficient mechanism for the spreading phase is baroclinic instability. In conditions of vertical overturning there is a large amount of potential energy available to drive the instability. For an eddy viscosity of 50 $m^2$ $s^{-1}$, after 10 days sinking the growth time is 3–4 days and the eddies are mainly concentrated at the surface with a major axis length of about 15 km. These tend to stabilize the top few hundred meters of the column. Later, finite amplitude effects will produce slower perturbations in the rest of the column.

## CONTENTS

*Department of Applied Mathematics and Theoretical Physics, University of Cambridge, Silver Street, Cambridge.

59

## 1. INTRODUCTION

AMONG the questions which have interested oceanographers for centuries is "why is the sea cold at great depths?" At the beginning of the nineteenth century HUMBOLDT (1814) deduced that cold deep water must have been advected from polar regions as there seemed to be no other heat sinks. Since the ocean could only lose or gain heat at the atmospheric–ocean interface, it followed that the majority of the cold deep water in the oceans had once been at the surface. Indeed, the surface radiative layer, about 10 m thick, is the only place where extrema of temperature or salinity can be established. (There is a bottom layer, but the effects are confined to weak geothermal heating and small intrusions of salts.) However, the regions where sinking of dense surface waters takes place are both surprisingly narrow (STOMMEL, 1962) and few: the main sinking region is in Antarctica, in the Weddell Sea (DEACON, 1937). There are also regions such as the Norwegian Sea (WORTHINGTON, 1970) and the Gulf of Lions (MEDOC GROUP, 1970), which are of less significance to the volume of bottom water but which contribute to its heat and salinity budget. Since the sea-surface temperature has a vital effect on the global climate we need to understand not only the way the dense waters affect the upper circulation of the oceans but also how dense water is formed, sinks, and spreads out under other water masses.

The observing conditions in the regions of dense water formation, however, vary strongly from region to region. The Antarctic is effectively inaccessible in winter, when most of the heat transfer is believed to occur. The outflow from the Norwegian Sea is sporadic (WORTHINGTON, 1969) and strongly influenced by the North Atlantic Polar Front, itself very variable. Consequently it is not surprising that the Gulf of Lions, certainly the least important of the regions, has been frequently chosen for study. Its advantages are several: accessibility, rather better weather, confined location, and reasonable regularity in bottom water production. Its main disadvantage is that in a mild winter there may be no bottom water formed.

Two multi-ship campaigns took place in the Gulf of Lions in 1969 and 1970. The details of the operations have been given by SANKEY (1973) and MEDOC GROUP (1970). Bottom water appears to be produced during three consecutive phases, as follows. During the early part of winter, the 'pre-conditioning' phase takes place. A doming of the isopycnals, associated with a cyclonic circulation, is set up in the Gulf in the vicinity of 42°N, 5°E. In the centre of the dome, the static stability is much reduced from the values even 50 km away. The positioning of the dome makes prediction of the approximate site of dense water formation relatively straightforward. This pre-conditioned phase appears to be relatively stable in time; in 1969 little change occurred between Jan 1 and Jan 24, for example (MEDOC GROUP, 1970).

The second, 'violent mixing', phase begins quite abruptly. Its onset in 1969 and 1970 coincided with the beginning of the Mistral, and, as we shall see, appears to be caused by it. However, note that there are usually other storms before this phase which have little effect. Vertical mixing takes place, apparently confined to a narrow region (estimates of its width vary, but 30 km is a reasonable figure). A 'chimney' of mixed, dense water appears, extending from the surface down to depths which steadily increase to about 2000 m (the local depth is 2500 m). The chimney extends about 20 km N–S and 50 km E–W. We shall at various times model it either to be roughly rectangular or roughly circular, as is convenient. Large ($0.1$ m s$^{-1}$) vertical currents are observed (VOORHIS and WEBB, 1970; GASCARD, 1973). Smaller, 'local' columns can appear about 10 km from the main column, with diameters about 10 km (LACOMBE, 1974).

The onset of the third, 'sinking and spreading' phase roughly coincides with the end of the Mistral and also with the time that the mixed column would have reached the bottom. Less dense water has appeared above the column even 3 days after the Mistral has ended (STOMMEL, 1972), presumably advected from the north or south. During the succeeding days a fragmentation and interleaving of the column occurs. This appears to be somewhat irregular in nature: there is evidence (SANKEY, 1973) that whole columns, 600 m in height, can be lifted 500 m or more, implying a rather energetic process; parts of the column, tagged by their oxygen content (MEDOC GROUP, 1970), seem to slide out into horizontal layers; there is general surface eddying with length scales and speeds of order 15–20 km and $0.15$ m s$^{-1}$ respectively; and a general southward drift of tagged water of about $0.03$ m s$^{-1}$ (SWALLOW and CASTON, 1973). In 1972 the drift was to the northwest, however (LACOMBE, 1974).

These observations raise many theoretical problems:

(a) what causes the preconditioning? Is it a topographical effect, is it produced by the whole Mediterranean circulation, or is it produced by surface cooling in the early winter?

(b) is the pre-conditioned phase vital for bottom water formation?

(c) what is the nature of the violent mixing— is it due to salinity input at the surface (evaporation–precipitation), or heat loss at the surface, both sensible and latent; these both increase the surface density and cause static instability. Or is it due to mechanical (wind-driven) stirring?

(d) why is the column so narrow, when the surface forcings are wider?

(e) what determines when the column begins to collapse—is it the presence of the bottom, the turning-off of the Mistral, or what?

(f) what determines the time scale of the columnar collapse?

(g) what determines the final destination of most of the mixed water? and

(h) can the whole process account for the observed Mediterranean outflow at Gibraltar?

Some of these questions will be discussed here, and the others examined in the body of this paper.

The pre-conditioning phase has been discussed elsewhere both observationally (SWALLOW and CASTON, 1973) and theoretically (HOGG, 1973). The suggestion in both cases is that the cause of

the doming is the cone-like topography of the Rhône fan. Hogg notes that there is the likelihood that the cyclonic gyre as produced by his model is unstable to a quasigeostrophic baroclinic perturbation. This is indeed the case, but the $e$-folding time for such a disturbance is long (at least 100 days) and is of the same order as that for the mid-ocean calculation of GILL, GREEN and SIMMONS (1974). The alternatives to topography as the main cause of the pre-conditioning are several (SANKEY, 1973). He argues that (a) the difference between a station in the centre of the gyre and one 50 km away are consistent with cooling and evaporation for about 6 weeks produced by the prevailing weather, (b) the large-scale geostrophic circulation tends to be *around* the dense patch, thus reducing the horizontal mixing, and (c) the weather system is effectively localized in the Gulf of Lions, due to channeling of winds by land topography and the reasonably short fetch involved. Despite this, however, it seems that the pre-conditioned area still needs a source of denser water at depth (SWALLOW and CASTON, 1973), even if meteorological conditions can account for the surface water properties. The divergence effect (SAINT-GUILY, 1963) and the 'virtual shelf' effect (LACOMBE and TCHERNIA, 1971) may be relevant here. Thus the full explanation of the pre-conditioning is still an open question, which we shall neglect henceforth.

Another unsolved problem is whether the bottom water formation can account for the outflow at Gibraltar. This has been fully discussed by SANKEY (1973), who concludes that there seems no way for the outflow to possess its observed properties; also noted by LACOMBE (1974). This problem also will not be discussed further.

The rest of this paper will address itself to those phenomena which can realistically be termed local: the violent mixing and sinking and spreading phase. (The pre-conditioning may or may not be a local effect.) The two phases are discussed in turn. Section 2 discusses various theories of the violent mixing phase. Section 3 is devoted to a two-dimensional model of this phase, in which it is shown that a one-dimensional (vertical) non-penetrative convection model, coupled with geostrophic adjustment, is quite adequate for modelling this phase. Section 4 examines why the sinking and spreading phase should ever occur, in terms of an instability process. Section 5 presents the appropriate quasi-geostrophic equations for an overturned fluid, together with the problems inherent in such a system. It also discusses the effect of small ($\lesssim$ 2 km) scale disturbances on larger scale flow in terms of their transfer of buoyancy and momentum. Section 6 makes estimates of the growth rate of the columnar collapse. Section 7 details the model used to investigate the stability of the column. Finally, section 8 gives the results of computations in such a region, from which it is concluded that the column is baroclinically unstable from the moment it begins to form, but that the e-folding growth times decrease drastically as the column descends. This produces an instability which appears only after the column has reached great depths.

## 2. THE VIOLENT MIXING PHASE— DISCUSSION

The onset of the violent mixing phase, as we have seen, coincided in 1969 and 1970 with the beginning of the Mistral. Anati and Stommel have examined the violent mixing in a series of papers, ANATI (1970), ANATI and STOMMEL (1970) and STOMMEL (1972), based on $\sigma_t$ data at representative points within the column. Their conclusions were as follows. ANATI (1970) examined penetrative versus non-penetrative vertical convection of the kind discussed by KRAUS and TURNER (1967), as possible explanations for the violent mixing. He notes that the 'slab' models are designed for mixed layers of depths of order 100 m rather than 2000 m as observed here. However, the fit of a one-dimensional non-penetrative convection model to the observed data is excellent, save that the rate of descent of the column is not predicted quite accurately. ANATI and STOMMEL (1970) observe that the vertical integral of salinity appears to be constant during the mixing process, suggesting that large horizontal advection of fresher water (e.g. within the Ekman layer) does

not occur. Their best-fit solution to the constituents of the column indicate that only 30 m of surface water ($2.4$ m$^2$ s$^{-1}$) has been advected into the column. Finally, STOMMEL (1972), under the assumption of small horizontal movement, calculated that the loss of buoyancy along a section one degree east of the column during the mixing was roughly independent of latitude. It appears from float tracks (CASTON and SWALLOW, 1970) and moored current meter records (GOULD and LARBY, 1971) that the horizontal motions were not well correlated during this time, suggesting possible cellular convection on scales of order 6 km (SANKEY, 1973).

Thus the evidence from observations suggests uniformly that the primary forcing for the violent mixing is the surface cooling by the wind—indeed, STOMMEL (1972) estimates the loss of heat within a column as $4.37 \times 10^8$ J m$^{-2}$ (10,400 cal cm$^{-2}$). Further, if the effects of horizontal advection and/or turbulent mixing are neglected, then non-penetrative convection fits the observations best. Curiously, although potential temperature is thus the main cause of the convection, salinity is the better marker of water masses. We shall assume for modelling purposes that density is a function of potential temperature only. Hence the remaining questions are: are there other effects of the wind, as yet unconsidered, which are important; and what effect have horizontal processes on the observed density structure?

During the mixing, the winds are more normal to the long axis (50 km) of the column than parallel to it (STOMMEL, 1972). This indicates that the largest fluxes due to the wind (the Ekman fluxes) will tend to be along the long axis of the column, which tends to reduce any advective properties the wind may have. For example, a wind stress of 1 N m$^{-2}$ would give a flux of 10 m$^2$ s$^{-1}$; spread over a mixed layer of 100 m this gives a mean velocity of $0.1$ m s$^{-1}$, which means that a fluid particle would take about 6 days to go from one end of the chimney to the other (about half the time of the whole phase). So the effects of wind-induced fluxes are unlikely to be strong in this situation. Such fluxes can play little part, too, in the final composition of the chimney: a flux of $10$ m$^2$ s$^{-1}$ takes 46 days to fill an area of chimney 2 km deep by 20 km wide. For later use, we might also note that it takes this long to empty such an area.

It is less certain that there will be no dynamical effect due to the wind stress. The currents in a 2 km deep, one-layer model would be of order $5 \times 10^{-3}$ m s$^{-1}$ with the above wind stress, parallel to the strip. Such velocities are larger than those near the bottom calculated in the next section, but have little or no effect since they advect quantities along, and not across, the strip except near the edges. Since the largest velocities are normally along the strip, we shall choose to neglect the wind in what follows. This results in a convenient simplification in section 3.

The only problem remaining is the effect of motions induced by the density field. The fact that the chimney remains essentially stationary suggests that there is little motion across the strip (the direction in which the density changes most rapidly). This is what would be expected: the geostrophically-induced velocity field (at most $0.1$ m s$^{-1}$; SWALLOW and CASTON, 1973) is parallel to the strip, and so any motions normal to the strip are either ageostrophic or driven by density differences along the strip. Observationally, the latter are small during the violent mixing. So any tendency of the chimney to break up or sink during the mixing can only be produced by ageostrophic motions. The next section will show that these are *small* ($3 \times 10^{-3}$ m s$^{-1}$) in an idealized two-dimensional model. In other words, a dense water column surrounded by less dense water will not tend to disperse in a rotating fluid, unless one appeals to some instability process (cf. section 4). Even if there was a tendency for the dense column to move outwards (e.g. a pressure disturbance, internal waves, etc.) it is likely to move only a distance of the order of the Rossby radius of deformation (BLUMEN, 1972). The first-mode baroclinic radius for the area, away from the column, is about 5 km, and the *local* radius at a given depth is smaller than this. This distance is small compared with the width of the column, so little motion away from the column can be expected under approximate geostrophic conditions in two dimensions.

## 3. THE VIOLENT MIXING PHASE—A TWO-DIMENSIONAL MODEL

(a) *Equations*

To test the ideas of the previous section, a two-dimensional model of the violent mixing phase was constructed. The two dimensions are the vertical ($z$ relative to the surface), and a horizontal direction ($x$). Essentially we would wish the chimney to be thought of as possessing cylindrical symmetry, with no variation of properties in the azimuthal ($y$) direction, but for the purposes of the model $x$ and $y$ are assumed rectilinear. This is done for convenience only; the justification may be found in HART (1974). We can also interpret the assumption of no $y$-variation as the modelling of a rectangular column which is infinitely long in the $y$ direction.

The equations used, and the assumptions made, are:

$$u_t + \nabla \cdot (\mathbf{u}\, u) - f\, v = -\frac{p_x}{\rho_0} + A_H u_{xx} + A_V u_{zz},$$

(1)

$$v_t + \nabla \cdot (\mathbf{u}\, v) + f\, u = A_H v_{xx} + A_V v_{zz}, \quad (2)$$

$$0 = -p_z - g\rho, \quad (3)$$

$$\rho_t + \nabla \cdot (\mathbf{u}\, \rho) = \frac{K_V}{\gamma} \rho_{zz} + K_H \rho_{xx}, \quad (4)$$

$$\nabla \cdot \mathbf{u} = 0, \quad (5)$$

where the velocity relative to ($x$, $y$, $z$) is ($u$, $v$, $w$); suffices denote differentiation; $t$ is time; $\rho$ the density relative to the uniform $\rho_0$; $p$ the pressure; $g$ the acceleration due to gravity (10 m s$^{-2}$); and

$$\nabla \cdot (\mathbf{u}\, \varphi) \equiv (u\varphi)_x + (w\varphi)_z$$

is the two-dimensional divergence. $A_H, A_V, K_H, K_V$, and $\gamma$ are related to eddy effects and will be discussed below. The length scales involved in the violent mixing and sinking and spreading are

< 200 km, so that the coriolis vector $f$ can be taken as constant in what follows. Its value is $9.73 \times 10^{-5}$ s$^{-1}$, corresponding to a latitude of 42°N. The hydrostatic assumption is made, and the horizontal component of the coriolis vector is neglected (also discussed below). Finally, the fluid is assumed Boussinesq and incompressible, with density a linear function of temperature only.

The assumption of hydrostatic balance is valid providing that the vertical mixing process is parameterized in some way. Section 5 will deal with this problem in greater detail. Here, we simply choose the mixing model of BRYAN and Cox (1968): if the fluid is locally statically stable, put $\gamma$ in (4) equal to 1; if the fluid is unstable, put $\gamma$ to zero, corresponding to infinite vertical mixing. The latter is equivalent to averaging $\rho$ with depth over the mixing region.

The justification for using such a formulation in this model is as follows. (a) It appears to give an excellent representation of the violent mixing phase. In some ways this is surprising. The wave spectrum of motions during this phase show a pronounced peak at periods of 12 h, which is also predicted from theory (GASCARD, 1973). The time step used numerically, however, is half an hour. This assumes, then, that the main mixing motions have time scales shorter than this. So there are at least some spectral components of the motion which are receiving cavalier treatment. (b) The formulation gives essentially the non-penetrative convection which, as we have seen, appears to dominate the mixing. (c) Calculations were made where momentum was mixed in the same manner as density, giving a slab-like motion more properly applicable to mixed layers, but there was little change on the density field produced. It is likely that any mixing formulation which eventually produced vertical homogeneity would serve equally well, providing only that density was conserved. (d) Large vertical velocities are removed from the solution. This means that the hydrostatic assumption becomes valid—a great simplification—and also the neglect of the horizontal component of the coriolis vector in (2) is legitimate.

However, vertical homogeneity of density does not imply horizontal homogeneity. A fluid which is stratified horizontally, but not vertically, is unstable to small-scale slanting cellular motions (OOYAMA, 1966) and other three-dimensional motions (see section 4). These are seldom seen in large-scale circulation models due to the large eddy coefficients used. For the small length scales in this problem, rather lower coefficients are in order. Since we shall see later that the growth times for disturbances are long during the initial descent of the column, it is necessary to choose a coefficient which is sufficiently large to damp out the two-grid length instabilities which would otherwise be generated numerically by such instabilities. Also, too small a coefficient would not model the horizontal mixing which must presumably occur at least within the chimney. A value for $K_H$ and $A_H$ of 50 m$^2$ s$^{-1}$ was used for this calculation. Typical values for the large-scale circulation are about $5 \times 10^4$ m$^2$ s$^{-1}$. WOODS (1973) shows that as time and space scales for motions decrease, so does the value of $A_H$. The value used here corresponds roughly to figures quoted by Woods for motions in the seasonal thermocline on the time scale of a month. Sections 5 and 8 will show that there is some justification for a value of this order.

Values for the vertical eddy coefficient in the literature lie in the range $10^{-4}$ to $10^{-2}$ m$^2$ s$^{-1}$ for the open ocean. The calculation used $10^{-4}$ m$^2$ s$^{-1}$ for $A_V$ and $K_V$.

These values of the coefficients have little effect on the solution on time scales of order 10 days. Typical diffusive distances are 6 km horizontally and 9 m vertically. (It is only 90 m if $A_V$ is taken to have the larger value $10^{-2}$ m$^2$ s$^{-1}$.) This implies that the only large-scale movements of density are by vertical overturning. In fact there is no reason to suppose that virtually the same results would not be seen if a non-hydrostatic, dissipationless (but much more costly) model had been run instead.

(b) *Initial and boundary conditions*

The boundary conditions used are

at    $z = 0, -H$:    $u_z = v_z = w = 0$,        (6)

corresponding to no stress at surface (wind stress neglected) or at bottom, together with a rigid lid to filter out surface gravity waves. Little motion occurs at depth, so that the lack of a frictional boundary layer at $z = -H$ ($H$ is the depth) has little effect. Also

at $$z = 0: \quad A_V \rho_z = Q,$$

at $$z = -H: \quad \rho_z = 0, \tag{7}$$

where the input of density $Q$ at the surface is caused by the intense cooling due to the Mistral. $Q$ is easily estimated from STOMMEL's (1972) estimate of the heat loss of the chimney as $L = 4.37 \times 10^8$ J m$^{-2}$. If we average this heat loss over a period taken somewhat arbitrarily as 12 days (which allows for periods of low winds) and assume that no heat is advected or diffused into the column from outside, then $Q$ is obtained as

$$\frac{L \alpha}{c_p(12 \text{ days})} = 2.4 \times 10^{-5} \text{kg m}^{-2}\text{s}^{-1},$$

where $\alpha$, the thermal expansion coefficient of seawater, is taken as $2.4 \times 10^{-4}$ °C$^{-1}$, and $c_p$ is the specific heat of seawater. Even if a significant amount of buoyancy is advected into the chimney from outside, the only result of taking an incorrect $Q$ is to speed up or slow down the advance of the chimney. Note that, following STOMMEL (1972), $Q$ is independent of $x$.

Conditions on vertical boundaries are taken as cyclic in $u$, $v$, $w$ and $\rho$. Providing the length of the box used for the model is much greater than the width of the chimney there is no interaction between one chimney and the next.

The initial condtions are $\rho$ specified, $v$ in geostrophic balance, and $u$, $w$ zero. The distribution of $\rho$ was chosen to resemble the classic picture of the pre-conditioning (e.g. SWALLOW and CASTON, 1973, Fig. 6). The doming, low stability in the central region, and rapid decrease of density at the surface once the central core is left, have all been retained. The formula used is a little complicated, and was chosen after considerable experimentation:

$$\rho_0^{-1}\rho(x, z) = \tfrac{1}{2}\{\rho_1 + \rho_2 + (\rho_2 - \rho_1) \tanh$$
$$[(\chi - |x|)d^{-1}]\} + 20 \times 10^{-3}, \tag{8}$$

where $|x|$ is the distance in kilometers from the centre of the pre-conditioned area,

$$d = 10 \exp(-z'/400),$$

$$\chi = \min\{30 + w' - z'/5, 50 + w'\},$$

$$\rho_1 = 9.11 \times 10^{-3}\{1 - b \exp[-0.11(-z')^{0.536}]\},$$

$$\rho_2 = \begin{cases} 9.11 \times 10^{-3}\{1 - b' \exp[-0.0339 \\ (-z')^{0.63}]\}, & z' < -150 \\ \\ 9.05 \times 10^{-3}, & z' > -150 \end{cases}$$

$$b = 1 - 8.6/9.11,$$

$$b' = 1 - 9.02/9.11,$$

and $z'$ is the modified depth, equal to min (0, $z + 25$), where $z$ is the actual (negative) depth in meters. The parameter $w'$ which occurs in the definition of $\chi$ is related to a crude measure $w''$ of the width of the pre-conditioned area (*not* the column), by

$$w' = \tfrac{1}{2}(w'' - 60),$$

where $w''$ is in kilometers. For this model, $w''$ was taken as 60 km. This is a deliberately large value so that good resolution could be obtained within the column. In fact the width appears to be irrelevant to the dynamics of the violent mixing phase.

The density is given effectively by a weighted average between two extreme functions $\rho_1(z)$ and $\rho_2(z)$ shown in Fig. 1. The weighting varies with both $x$ and $z$ to reproduce the doming, correct positioning of several of the isopycnals, etc., as can be seen from Fig. 3.

The initial velocity $v$ is then calculated from the thermal wind relation

$$v_z = \frac{-g\rho_x}{f\rho_0}, \tag{9}$$

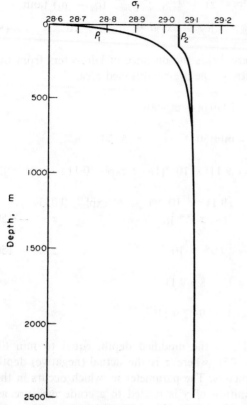

FIG. 1. The weighting functions $\rho_1(z)$, $\rho_2(z)$ used in the initial conditions for the two-dimensional model.

and the condition

$$\int_{-H}^{0} v \, dz = 0. \qquad (10)$$

Some calculations were run with the alternative initial condition

$$v = 0, \quad z = -H, \qquad (11)$$

with little appreciable difference.

    Since $\rho$ is symmetric about the centre of the pre-conditioned region, it follows that $v$ is anti-symmetric. Equations (1)–(5), and the boundary conditions then imply that $\rho$ and $w$ remain symmetric, $u$ and $v$ antisymmetric. Thus the region of interest can be narrowed to $0 \leq x \leq L$, with the centre of the column at $L$, and

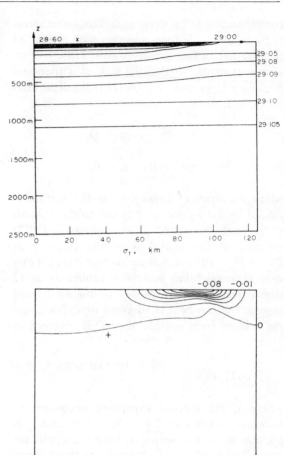

FIG. 2. The initial conditions for the two-dimensional model. The centre of the column is at the right, with cyclic symmetry at both sides. The contour interval for $\sigma_t$ is 0·05, except that additional contours of 29·08 and 29·09, and 29·105 have been added. The contour interval for $v$ is 0·01 m s$^{-1}$.

$$\left. \begin{array}{l} u = v = 0 \\ w_x = \rho_x = 0 \end{array} \right\} \quad x = 0, L. \qquad (12)$$

The initial conditions are shown in Fig. 2.

    The equations were integrated on a $52 \times 51$ grid, with $\Delta x = 2\cdot5$ km, $\Delta z = 50$ m (corresponding to $H = 2500$ m, $L = 125$ km) with a time step of 0·5 h. The local Rossby radius is 5 km, and it was felt that phenomena (e.g. fronts) on this scale should be at least partially resolved. Effects on smaller scales are subsumed in $A_H$.

The method of integration follows KILLWORTH (1973). Equation (5) shows the existence of a stream function $\psi$ for $u$ and $w$, which shows that the vertically-integrated flux in the $x$-direction is a function of time only. Then (1), (2) can be integrated to show, for the initial and boundary conditions used here, that the integrated flux in the $x$-direction is identically zero. The problem then reduces to that of KILLWORTH (1973). On an IBM 370 double precision must be used for the density field to avoid a spurious instability due to rounding errors.

(c)  *Results*

The model was integrated for times up to 20 days, although attention is concentrated on the first 10 days. The centre of the column reaches the bottom at a time of about 13 days after the onset of the Mistral, and by 20 days the area of complete overturning occupies a half-width of 25 km. At no time before the column reaches the bottom does the meridional velocity $u$ exceed $3 \times 10^{-3}\,\mathrm{m\,s^{-1}}$, or the upwelling $w$ exceed $10^{-4}\,\mathrm{m\,s^{-1}}$; after 13 days these figures are increased by at most 150%. The values are much smaller than this except near the boundary between overturning and non-overturning regions.

Figures 3 and 4 show the density field, together with $v$ and $\psi$, at 5 and 10 days respectively. The density field clearly shows the columnar structure and the slanting boundary between overturning and non-overturning regimes, and agrees remarkably with observations (e.g. ANATI and STOMMEL, 1970, Figs. 9, 10). By 10 days there is overturning at every surface grid point, although outside the column this extends over only one grid interval.

The only appreciable difference between the model density and the observed structure is that the frontal structure near the surface slopes away from the column in the model and does not in the observations cf. ANATI and STOMMEL (1970, Fig. 11c). There are, however, two points to note. First, the spacing between the stations used to construct an observational picture of the chimney

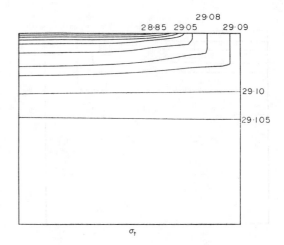

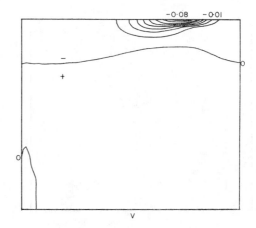

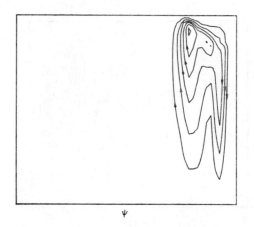

FIG. 3. The solution after 5 days. Contour values for $\sigma_t$ and $v$ are as in Fig. 2. Contour interval for the stream function $\psi$ is $0.04\ \mathrm{m^2\,s^{-1}}$.

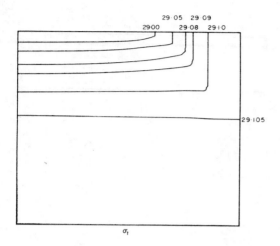

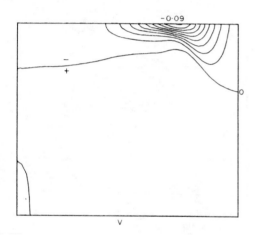

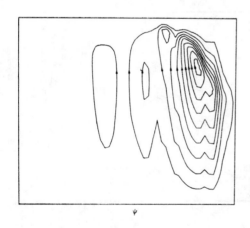

FIG. 4    The solution after 10 days. Contour intervals as in
Fig. 3.

is of the order of 13 km and is much too wide to resolve the front. Secondly, LACOMBE (1974) draws attention to other, 'mini-columns' of small vertical extent within O(10 km) of the main column. Now the further away from the centre of the column, the stronger the original surface stratification and hence, of course, the slower the descent of overturning. However, surface waves provide quite efficient lateral mixing in the top 50 m and would tend to make the production of mini-columns more difficult. So it is suggested that the appearance of such mini-columns is an observational indication that overturning is taking place to a smaller extent away from the centre of the chimney.

The northward ($v$) velocity remains geostrophic to within 1%. Its deviations from geostrophy occur at the front between overturning and non-overturning regimes, since the time scale of density variation differs between the regimes. Within a time of order $f^{-1}$ of overturning reaching a given point, however, geostrophy has returned. The meridional circulation is driven by deviations from geostrophy in the direction parallel to the front; for example, Figs. 3 and 4 shows that the maximum meridional stream function values occur at or near the front, together with weaker motions far from the column.

These latter motions are dominated almost exclusively by inertial oscillations, as one would expect from the above comments. The residual, longer period, motion can simply be explained by considering linearized, dissipation-free versions of (1)–(5), with a basic stratification assumed. Neglecting $\partial^2/\partial t^2$ against $f^2$, these can be written

$$f^2\psi_{zz} + N^2\psi_{xx} = 0, \qquad (13)$$

where $N(z)$ is the basic Brunt–Väisälä frequency. Equation (13) then shows, from its similarity to Laplace's equation, that extrema of $\psi$ occur only on boundaries to the region of validity of (13). These can only be at or near the front, since $\psi$ vanishes on all other boundaries to the stably stratified régime. Hence the residual weak motions are driven purely by ageostrophic effects at the front between overturning and non-overturning. The vertical scale will be of the order of the depth, as observed, since, from (13),

$$z \sim \frac{f x}{N} \quad \text{or} \quad \frac{z}{H} \sim \frac{x}{r} \gtrsim 0(1),$$

where $r$ is the local radius of deformation $NHf^{-1}$.

We can estimate the strength of these motions by using the idea of cross-front geostrophy (HOSKINS and BRETHERTON, 1972). Take as primary balances from (1), (2):

$$f v \sim p_x/\rho_0, \qquad (14)$$

$$v_t + f u \sim 0. \qquad (15)$$

The nonlinear terms in (15) may be important but do not matter for estimation purposes. Then (14), (15) combine with (3) to give

$$u_z \sim \frac{g \rho_{xt}}{f^2 \rho_0}. \qquad (16)$$

If the ageostrophic motions are weak, then $\rho$ is approximately undisturbed away from the column; within the column, however, $\rho_t \sim Q h^{-1}$, where $h$ is the (local) column depth. Since the front will have a width of order the radius of deformation $r$,

$$\rho_{xt} \sim \frac{Q}{hr},$$

and thus

$$u \sim \frac{g Q}{f^2 r \rho_0}. \qquad (17)$$

If $r$ is taken as 5 km (the overall value—local surface values will be bigger), then

$$u \lesssim 5 \times 10^{-3} \text{ m s}^{-1},$$

agreeing well with the numerical values obtained. Equation (16) shows that $u$ will be positive near the surface within the front, and (13) shows that $u$ will decay away from the front in the non-overturning regime; this is confirmed in Figs. 3 and 4.

This meridional motion must advect less dense surface water into the column, thus modifying the rate of descent of the column from its purely non-penetrative value. The modification will be strongest near the edges of the column (where $u$ is greater) and weakest at the centre. We can estimate this effect also. Consider a fluid column, depth $H$, width $l$, lying between $x = b, b + l$ (this is the finite-difference equivalent of an $x$-derivative). Then (4) implies

$$\int_b^{b+l} \int_{-H}^0 \rho_t \, dx \, dz = \int_{-H}^0 u\rho \, dz \bigg|_{x=b}$$

$$- \int_{-H}^0 u\rho \, dz \bigg|_{x=b+l} + Ql, \qquad (18)$$

if horizontal diffusion is neglected. Put

$$u(b + l) = u(b) + \Delta u,$$

$$\rho(b + l) = \rho(b) + \Delta \rho.$$

Then

$$\int_b^{b+l} \int_{-H}^0 \rho_t \, dx \, dz = - \int_{-H}^0 dz \{\Delta u \rho(b)$$

$$+ \Delta \rho u(b)\} + Ql - \int_{-H}^0 dz \, \Delta u \, \Delta \rho. \qquad (19)$$

The last term is small for $l = 2.5$ km (the grid spacing used) and can be neglected. The first terms on the R.H.S. of (19) can be estimated if we assume a crude two-layer structure in the vertical, so that

$$(u, \rho) = (u_1, \rho_1) \quad z > -h_1$$

$$= (u_2, \rho_2) \quad z < -h_1.$$

Thus

$$\int_{-H}^{0} \rho_t \, dx \, dz \sim \Delta\psi(\rho_1 - \rho_2) + \psi(\Delta\rho_1 - \Delta\rho_2) + Ql.$$

$$(20)$$

But $\Delta\rho_2 \ll \Delta\rho_1$; further, $\Delta\psi \sim 0.1$, $\psi \sim 0.3$ m$^2$ s$^{-1}$, $\rho_1 - \rho_2 \sim -6 \times 10^{-2}$ kg m$^{-3}$, and $\Delta\rho_1 \lesssim 10^{-2}$ kg m$^{-3}$. The advective terms thus contribute

$$-6 \times 10^{-3} - 3 \times 10^{-3} \sim -9 \times 10^{-3} \, \text{kg m}^{-1} \text{s}^{-1},$$

while the buoyancy input gives

$$6 \times 10^{-2} \, \text{kg m}^{-1} \, \text{s}^{-1}.$$

So the input of less dense water into a fluid column by advection is greatest where the front is near the surface (i.e. at the edge of the column) and is equivalent to a decrease of the effective cooling of about 15%.

ANATI (1970) examined this effect in the data, and measured it by a ratio

$$t^* = \frac{\text{time for overturning to reach a given level due to convection}}{\text{time for overturning to reach a given level as observed}}$$

He shows that the edges of the column (about 15 km from the centre) deepen slower than expected ($t^* \sim 0.8$). The centre deepens initially faster than expected ($t^* > 1$) but in later stages, at exactly the rate estimated ($t^* = 1$). There are interpretation problems with the southern stations. This compares very well with the numerical results. At 25 km from the centre (about equivalent to 15 km in the observations: recall that the numerical column was made wider than that observed, deliberately), the $t^*$'s range between 0.81 and 0.86, whereas near the centre $t^*$ lies between 0.97 and 0.98. Considering that the errors due to the use of differing stations to estimate $t^*$ will be greatest near the centre of the column (due to its faster sinking rate) the fit between model and observations is excellent.

This can simply be summarized: the density field and the position of the chimney can be predicted quite accurately as a function of $t$ by taking the pre-conditioned density field, pouring in a density input $Qt$ at all points of the surface, and overturning the field downwards until neutral stability is reached. The width and shape of the chimney are thus controlled by what occurs during the pre-conditioning.

### (d) Extensions

Several other calculations were made with the same model. First, noting that the depth-averaged component of $v$ only attained $10^{-4}$ m s$^{-1}$, a calculation was made with $v = 0$ on the bottom. No discernible alteration was made to any of the results. This suggests that, providing the barotropic component is not too large, little loss of accuracy is made if we assume the flow to be baroclinic.

Secondly, the calculation was extended to 20 days to see if any instability might develop which would cause the onset of the sinking and spreading phase. This did not occur: the meridional velocity field remained small, but increasing, and the overturning of the density field continued steadily downward.

Lastly, the buoyancy forcing was turned off after 12 days to see if this might precipitate an instability. However, the velocity field adjusted rapidly to geostrophy, after which there was a very slow dissipation due to viscous effects. There was no tendency for a breakup of the column.

## 4. THE SINKING AND SPREADING PHASE—DISCUSSION

The previous section showed that no columnar breakup is caused in the two-dimensional model when the column hits the floor or when the Mistral (i.e. the cooling) ceases. Therefore, to model the sinking and spreading phase, other effects must be included.

Now a large volume of water must be moved: SANKEY (1973) estimates $5 \times 10^{12}$ m³ in 1969 and $3 \times 10^{12}$ m³ in 1970. A time scale for the breakup can be seen from STOMMEL (1972, Fig. 3) to be about two weeks. Movements on such a scale imply volume fluxes of the order of at least $10^6$ m³ s⁻¹, probably as much as $3 \times 10^6$ m³ s⁻¹. (For comparison, the amount of bottom water feeding the Mediterranean outflow is about $10^6$ m³ s⁻¹ also.) What could provide such fluxes? The following possible reasons why the two-dimensional did not predict the breakup will be discussed in turn.

1. Too large an eddy viscosity was used in the model.

2. The breakup is caused by a small scale process which was not parameterized in the model.

3. The breakup occurs because of three-dimensional effects which were precluded in the model.

It is possible that too large an eddy viscosity was used in the model. Finite-difference solutions to the Navier–Stokes equations usually need some viscous dissipation for stability. As has been noted, running the model of section 3 with no viscosity would lead to a two-grid-point growing oscillation which is the numerical equivalent of a symmetric disturbance of wavelength less than or equal to $2\Delta x$. All, therefore, hinges on the growth rates of such disturbances and what values of $A_H$ are necessary to damp them out. These can easily be obtained by perturbing (1)–(5) about a steady state $\rho(x)$, $V(x,z)$, $u = w = 0$. It can be shown that the most unstable state will be one with $\rho_z = 0$. Seeking normal mode solutions proportional to $\exp\{\sigma t + i(kx + lz)\}$ one obtains for hydrostatic perturbations that there is a critical value of

$$\gamma = \left| \frac{g\rho_x}{f^2\rho_0} - \frac{V_z}{f} \right| = 2\left| \frac{g\rho_x}{f^2\rho_0} \right|,$$

$\gamma_c$, say, below which there can be no instability, given by

$$\gamma_c = \frac{8\pi}{27^{\frac{1}{4}}H}\left(\frac{A_H}{f}\right)^{\frac{1}{4}}.$$

This value assumes $A_H \gg A_V$. For nonhydrostatic perturbations and realistic ocean values, $\gamma_c$ is almost identical to the hydrostatic value. Typical values are: for the column, $\rho_x/\rho_0 \sim 2 \times 10^{-6}$ in 5 km, and for the rest of the fluid, $10^{-5}$ in 5 km, with effective depths of 2 km and 300 m respectively. This gives critical values for $A_H$ of

2, 1 m² s⁻¹

respectively. If we account for the continual overturning of the density field then $\gamma$ is halved, and the values of $A_H$ become

0·5, 0·2 m² s⁻¹.

With the possible exception of WOODS (1973), all values of $A_H$ in the literature are much larger than is required to damp out such instabilities; the success of such values in other oceanographic applications forces the conclusion that too large an eddy viscosity has not been used.

However, there remains the possibility that other small-scale motions, not parameterized in the model, could have a rectified effect which drives the instability. There are certainly wave motions of very large amplitude (1 km) and strength (0·1 m s⁻¹) occurring during the mixing phase (VOORHIS and WEBB, 1970). Much of the energy of these waves lies in a band which can be interpreted (GASCARD, 1973) as inertia-gravity oscillations with periods of 12 h, since there is approximate equipartition between vertical and horizontal kinetic energy (GASCARD, 1974). The amplitudes of such waves can be estimated (SAINT-GUILY, 1972): for an impulsive velocity $V$ of order 0·1 m s⁻¹, amplitudes of $V (2\Omega)^{-1}$ $\sim$ 1 km are produced, where $\Omega$ is the radian frequency of the earth's rotation. This agrees with Voorhis and Webb's observations. A crude estimate of the rectified effect of such waves is given by (amplitude × velocity ~ eddy viscosity), which gives an eddy coefficient of $10^2$ m$^2$ s$^{-1}$, only twice what was used. So little large-scale effect of such waves is expected.

Another class of internal waves is the Rossby waves (cf. LIGHTHILL, 1969, for a discussion of their properties). The baroclinic modes have phase speeds in the preconditioned area of $4 \times 10^{-4}$ m s$^{-1}$, and thus have too long a time scale to affect the column. The barotropic Rossby wave has a maximum long-wave phase speed of 20 m s$^{-1}$, which crosses the column in a fraction of a day and has too short a time scale. Shorter waves, with wave-lengths of about 200 km, have propagation speeds of about 0·02 m s$^{-1}$, and are slow enough to modify the flow considerably. However, if the effects of horizontal density gradients are included, these waves become baroclinically unstable and are treated in the rest of this paper.

Since the energy spectrum may well have a broad peak at 12 h (VOORHIS and WEBB, 1970), nonlinear wave-wave interactions could produce highly energetic motions at longer time scales. Insufficient is known about this phenomenon to make a clear judgement, so we shall neglect it in what follows.

Nonetheless, small-scale motions *must* be important to the dynamics of the sinking phase. We can see this very simply. Note that potential vorticity

$$P = (2\,\Omega + \zeta) \cdot \nabla\rho$$

is conserved in motion without frictional or diabatic effects, where $\Omega$ is the rotation vector of the earth and $\zeta$ the relative vorticity of the flow. It can be shown (HOSKINS, 1974) that motions which are unstable to symmetric disturbances have $P$ positive; a special case of this is a convectively overturned fluid such as the chimney. Hoskins notes that a stable state of weak or no motion has $P$ negative, which implies that frictional and/or heating effects are needed to generate symmetric instability in a previously

stable situation. (This heating is of course provided by the Mistral.) However, Hoskins' argument can be inverted: an unstable situation can never reach a state of stability without invoking friction or heating. Hence, either there must be a significant heat flux through the ocean surface after the Mistral has ended, or else small-scale motions must provide a significant dissipative effect, in order for the flow ever to reach a state of stability once more. Measurements of temperature structure (STOMMEL, 1972) show a small gain at 6°E after the end of the Mistral, but distinctly smaller than the change during the Mistral. This implies that small-scale instability processes, and in particular convective overturning, must be important during the sinking and spreading phase.

There remains, then, the third possibility: that fully three-dimensional motions act to break up the column. Now large-scale advection can be ruled out. We have seen already (section 2) that wind-induced motions are unlikely to disturb the column. Besides, during the sinking phase the wind is effectively nil. Now the large-scale cyclonic circulation at least has fluxes of the right order. SWALLOW and CASTON (1973) find $10^6$ m$^3$ s$^{-1}$ relative to 1800 dbar before the mixing phase, and LACOMBE and TCHERNIA (1971) find about $4 \times 10^6$ m$^3$ s$^{-1}$ relative to 1500 dbar at 6°E in 1963. However, this large-scale flow is directed *around* the column, so any spreading tendencies conceivably associated with it are likely to be small.

Finally, there is the possibility of three-dimensional motions on scales comparable with the lateral scales of the chimney. These appear to be what is observed: SANKEY (1973) interprets the *Discovery* float and current measurements as a rather confused field of "slowly moving eddies with a diameter of about 20 km and speeds of about 0.15 m s$^{-1}$ at the surface". SWALLOW (private communication) finds zero-crossings of transverse and longitudinal correlation coefficients for the eddies of about 16 and 25 km respectively. There is an obvious mechanism for the production of such eddies: baroclinic instability. Such a mechanism has already been hinted at by several authors, e.g. HOGG (1973), HART (1974) and

SAUNDERS (1973), although the latter's experiments suggested the column should be stable to small perturbations. However, as Saunders himself has noted, the horizontal density variations were only crudely represented in his model; we shall see that these produce vertical shears which an instability can feed on.

## 5. QUASIGEOSTROPHIC MOTION IN AN OVERTURNED MEDIUM

The discussion of the previous section leads us to believe that the chimney may be baroclinically unstable. The degree of instability depends on how easily the ocean can give up its large potential energy to feed the eddies which typically result. In the atmosphere, rotational constraints have little effect, and virtually all flows are unstable (CHARNEY, 1947; EADY, 1949). In the ocean, the transformation from mean potential to eddy kinetic energy can also take place but with more difficulty (GILL, GREEN and SIMMONS, 1974), and the growth rates are strongly dependent on the detailed density profile.

It is clear from a consideration of the energetics of such disturbances that the smaller the static stability in the region of the instability or the greater the available potential energy, the faster will the instability usually grow. Hence it appears likely that the fastest growing instabilities will be found either in neutrally (statically) stable situations such as the chimney, or in strong boundary currents where the vertical shear provides a large source of energy (cf. ORLANSKI and COX, 1973).

One can study such instabilities in two ways. ORLANSKI and COX (1973) used BRYAN's (1969) three-dimensional circulation program to integrate the full equations of motion numerically. A recent revision of the program (SEMTNER, 1973) has made it accessible to medium-sized computers, and a companion paper (KILLWORTH, in preparation) presents a fully three-dimensional simulation of the mixing and sinking phases. However, the usual method of study uses the quasigeostrophic approximation (cf. CHARNEY and STERN, 1962; PEDLOSKY, 1964 for a detailed derivation and a discussion of the advantages of this approximation).

The quasigeostrophic approximation rests heavily on various assumptions, e.g. small Rossby number, static stability ($NH/fL \sim 0(1)$, where $N$ is a measure of the Brunt–Väisälä frequency), etc. The latter assumption fails in the case of a neutrally statically stable fluid. A rescaling can be performed when $(NH/fL)^2$ is of order the Rossby number (cf. ROBINSON and GADGIL, 1970) or else formally let $N \to 0$ in the usual theory, to give

$$u = -\psi_y, \tag{21}$$

$$v = \psi_x, \tag{22}$$

$$p = \rho_0 f \psi, \tag{23}$$

$$\rho = -\rho_0 f \psi_z / g, \tag{24}$$

$$\frac{D\psi_z}{Dt} = 0 \left[ + A_H \left( \frac{\partial^2}{\partial x^2} + \frac{\partial^2}{\partial y^2} \right) \psi_z \right], \tag{25}$$

$$\frac{D}{Dt} (\psi_{xx} + \psi_{yy} + f) = f w_z$$

$$\times \left[ + A_H \left( \frac{\partial^2}{\partial x^2} + \frac{\partial^2}{\partial y^2} \right) (\psi_{xx} + \psi_{yy}) \right], \tag{26}$$

where

$$\frac{D}{Dt} \equiv \frac{\partial}{\partial t} - \psi_y \frac{\partial}{\partial x} + \psi_x \frac{\partial}{\partial y},$$

and the bracketed terms allow for a horizontal eddy coefficient $A_H$ for both buoyancy and momentum, discussed below. Variations of $f$ with latitude will again be neglected.

Equations (21)–(26) without dissipation ($A_H = 0$) are very simple. However, they possess various unpleasant features which prevent their use. These are discussed in the appendix, where it is shown that the large potential energy available induces (a) a faster increase in kinetic energy for smaller length scales and (b) a cascade of energy into smaller and smaller length scales as time increases. The result of this is that one of the assumptions of the quasigeostrophic approximation (typically that of small Rossby number) breaks down eventually.

This occurs at length scales of order 1 km, at which point other dynamics come into play. This length scale marks an approximate boundary between what WOODS (1973) has termed the "rotational subrange" (e.g. quasigeostrophic motions with horizontal length scales of order 10 km and more) and the "buoyancy subrange" (length scales less than 1 km, say). There is significant energy transfer into length scales less than 1 km, which act to smear out the features of larger-scale flows. As far as these larger motions are concerned, then, the small-scale mixing acts as a sink for kinetic energy.

This is a different role for horizontal mixing than that in section 3. There it was effectively a *numerical* ploy to allow a hydrostatic model. Indeed, even in normal quasigeostrophic perturbation theory the effects of dissipation are small (e.g. BROWN, 1969) because the natural length scales are large compared with the mixing length scales. In an overturned fluid this is not the case, and the mixing must be parameterized reasonably accurately. This means that order-of-magnitude expressions for Reynold's stresses and eddy fluxes must be computed.

Consider first the flux of buoyancy $\overline{u'\rho'}$ in the usual notation. Since $\rho$ is a conserved quantity, the mixing length argument can be applied (GREEN, 1970; WELANDER, 1973) to yield

$$\overline{u'\rho'} \simeq - A_H \, \overline{\rho}_x,$$

relating the eddy flux to the mean field. Now $\overline{u'\rho'}$ can be estimated by $V\Delta\rho$, where $V$ is a typical mixing velocity and $\Delta\rho$ a typical density variation across the region of mixing. Theoretically, $\Delta\rho$ is known, and $V$ must be estimated from energy arguments (i.e. balancing release of potential energy to increase in kinetic energy). In an overturning situation there are two quite distinct releases of P.E. occurring simultaneously: (i) a (mainly vertical) redistribution of buoyancy input at the surface until the density is independent of $z$; and (ii) a tendency for *horizontally* stratified fluid

to adjust to a state of minimum A.P.E. (cf. GREEN, 1970, fig. 2). We shall estimate the $V$'s corresponding to both processes.

In the overturning, P.E. is being fed in at the surface at a rate $gHQ$ per unit area and is rapidly redistributed via vertical mixing until $\rho_z$ is zero; this means that a column of unit area and depth $H$ gains P.E. at a rate $\frac{1}{2}gHQ$. Hence the remainder is converted to K.E.; crudely estimating the K.E. of a column, we obtain

$$(\tfrac{1}{2}\,\rho_0 V^2 H)_t \sim \frac{gHQ}{2},$$

or, if $t$ is taken as a (median) value of 5 days,

$$V \sim 0\cdot32 \text{ m s}^{-1}.$$

If this value is equally divided between $u$, $v$ and $w$, one has

$$w \sim 0.18 \text{ m s}^{-1}$$

which, although a crude estimate, agrees quite well with VOORHIS and WEBB's (1970) measurements of $w$.

In the adjustment process, matching the potential energy release between a state of uniform horizontal stratification $\rho_x =$ constant and uniform vertical stratification (minimum A.P.E.) given by (A6) to the K.E. gives

$$V^2 \sim \frac{gH\Delta\rho}{6\rho_0}.$$

This value involves lateral movements across the whole of the region to redistribute the P.E. and in general this will be difficult to achieve. Thus only a fraction of this value of $V$ will be available for small-scale mixing. Inside the chimney, $H \sim 2000$ m, $\Delta\rho/\rho_0 \sim 2 \times 10^{-5}$ and $V \sim 0\cdot26$ m s$^{-1}$.

The similarity between these estimates of $V$ shows—at least for these values—an approximate equipartition between the two sources of K.E., and $V \sim 0\cdot3$ m s$^{-1}$ will be taken as representative of both values. Then

$$A_H \sim \alpha VL,$$

where $\alpha$ $(< 1)$ is a scaling constant, as yet un-determined. $L$ is also not precisely known; the value used so far assumes adjustment over the entire width of the box. Assuming the horizontal and vertical length scales are not too dissimilar gives $L \lesssim 5$ km for a depth of 2·5 km. Putting $L = 5$ km gives $A_H \sim 1500\,\alpha$ m² s⁻¹.

The constant $\alpha$ must now be determined. There are no observations of eddy fluxes in the Mediterranean, so instead the estimate of GREEN (1970) will be used. Green found a value of $\alpha$ of 0·019 (a factor of 12$^{\frac{1}{2}}$ exists between Green's and the current values) to fit meridional transport of heat in the atmosphere. The applicability of this to the oceans is naturally somewhat suspect, the more so because the estimate applies only to the adjustment process in a stratified environment. Further, the existence of overturning *forces* mixing to take place during the Mistral, so that this value of $\alpha$ may be a slight underestimate. Direct sub-stitution gives $A_H \sim 30$ m² s⁻¹, so that the final value used, 50 m² s⁻¹, is quite reasonable. Such a value must be an order of magnitude at best, lacking better observations.

The mixing-length arguments cannot be used for momentum mixing since momentum is not a conserved quantity. One also cannnot employ the arguments connecting potential vorticity with buoyancy (GREEN, 1970; WELANDER, 1973), as can be seen from an examination of the terms involved. Hence we *assume* that momentum mixing can be described by the same eddy co-efficient as used for buoyancy.

We have said nothing about mixing outside the overturned region. Arguing after GREEN (1970, eqn. (4)) one obtains $V < 0.003$ m s⁻¹, i.e. the mixing is much less. However, this would imply an $A_H$ varying abruptly between regions of overturning and non-overturning. This presents difficulties in defining the conditions on the boundary between such regions, so instead $A_H$ will be treated as a constant over the entire region.

The vertical mixing is much simpler. In regions of overturning there is infinite vertical mixing of density as in section 3; there is no other vertical mixing. This is justified because vertical eddy coefficients of the order of $10^{-3}$ m² s⁻¹ were found to have virtually no effect in section 3. This formulation is convenient also because no other parameters are involved.

The test of the eddy coefficient used here comes in a comparison of growth rates and length scales for perturbations to the two-dimensional flow. $A_H$ can be chosen to make one of these two quantities correspond to what is observed, but not both unless the physics we have assumed to hold is correct. In section 8 we shall see that the agreement is indeed good.

## 6. QUASIGEOSTROPHIC MOTION IN A PARTLY OVERTURNED MEDIUM

Before proceeding to the most general model, it is worth estimating the growth rates which should occur in the Mediterranean situation by considering a much simpler model without any horizontal variation of velocity. We can only expect its results to be indicative of those for the full problem, since results tend to be sensitive to the details of spatial variation of the basic fields (GILL, GREEN and SIMMONS, 1974).

Consider two superposed horizontal layers of fluid between rigid planes $z = 0, -H$. The upper layer, depth $h$, has $N = 0$, while in the lower layer $N$ has a constant non-zero value. There is a uniform shear $\bar{V}_z$ in the top layer, and a weaker shear $\gamma \bar{V}_z$ $(0 \le \gamma \le 1)$ in the bottom layer. $V$ is zero at $z = 0$. There are no vertical boundaries. This is a crude model of a sheared, overturned mass of water above a less sheared, stratified water mass, and is shown in Fig. 5.

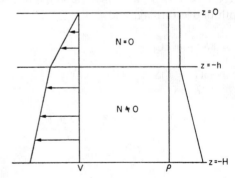

FIG. 5. A crude model of a two-layer system which is overturned in the top layer.

Seeking (temporarily inviscid) perturbations proportional to $\exp\{il(y - ct)\}$, the solutions can be written

$$\psi = \bar{V}_z (z - H \hat{c}), \quad z > -h$$

$$= \bar{V}_z \left\{ - (h + H\hat{c}) \cosh \frac{Nl}{f} (z + h) \right.$$

$$\left. + B \sinh \frac{Nl}{f} (z + h) \right\}, \quad z < -h$$

where $c = H\bar{V}_z\hat{c}$ and $B$ is unknown; the continuity of $\psi$ has already been used. The remaining conditions are the continuity of $\psi$ and $w$ at $z = -h$, and that $w$ shall vanish on $z = 0, -H$. Using (21) to obtain $w(-h)$, these can be written

$$\frac{\psi_z}{\psi} = \frac{\gamma \bar{V}_z}{V - c}, \quad z = -H, \tag{27}$$

$$l^2 \int_{-h}^{0} (V - c)\psi \, dz = \frac{f^2}{N^2} (\gamma \bar{V}_z \, \psi \tag{28}$$

$$- (V - c)\psi_z)\Big|_{z=-h}.$$

These conditions yield a cubic equation for the complex wave velocity $c$:

$$A_1\bar{c}^3 + A_2\bar{c}^2 + A_3\bar{c} + A_4 = 0, \tag{29}$$

where $\bar{c} = \hat{c} + \lambda$, $\lambda = h/H$, a nondimensional depth of overturning, and

$$A_1 = \sigma S + \sigma^2 \lambda C,$$

$$A_2 = \gamma\sigma(1 - \lambda)S + \gamma\sigma^2\lambda(1 - \lambda)C - \sigma^2\lambda^2 C$$
$$- \lambda\gamma\sigma S,$$

$$A_3 = \gamma^2(1 - \lambda)C - \gamma\sigma^2\lambda^2(1 - \lambda)C \tag{30}$$

$$+ \tfrac{1}{3}\sigma^2\lambda^3 C - \frac{\gamma^2 S}{\sigma} + \gamma\sigma\lambda^2 S,$$

$$A_4 = \tfrac{1}{3}\gamma\sigma^2\lambda^3(1 - \lambda)C - \tfrac{1}{3}\gamma\sigma\lambda^3 S.$$

The following symbols have been introduced:

$$\sigma = \frac{NHl}{f}, \text{ a nondimensional wavenumber}$$

and

$$\begin{Bmatrix} S \\ C \end{Bmatrix} = \begin{Bmatrix} \sinh \\ \cosh \end{Bmatrix} \sigma(1 - \lambda).$$

Equation (29) either possesses three real solutions, or else has a root with positive imaginary part, with corresponds to an unstable mode.

We may first examine some limiting cases. If $\lambda = 0$, the fluid is stably stratified everywhere, and

$$H\bar{V}_z \, \bar{c}_i = H\bar{V}_z \, \text{Im}(\bar{c}) = \frac{\pm \gamma H\bar{V}_z}{\sigma}$$

$$\times \left\{ (\coth \sigma/2 - \sigma/2)(\sigma/2 - \tanh \sigma/2) \right\}^{1/2},$$

which is the EADY (1949) result as expected. Conversely, if $\lambda = 1$, the fluid is completely overturned, and

$$H \bar{V}_z \, \bar{c}_i = \frac{H\bar{V}_z}{\sqrt{12}},$$

as found by FJØRTOFT (1950).

Intermediate values of $\lambda$ are more complicated. For simplicity we consider only the limiting ranges $\gamma = 1$ (uniform shear) and $\gamma = 0$ (no shear in the bottom layer). The growth rates for the case of uniform shear are shown in Fig. 6. Note that, if $\omega$ is the dimensional growth rate,

$$\sigma \bar{c}_i = \frac{N}{f\bar{V}_z} \, \omega$$

can be interpreted as a nondimensional growth rate. For $\sigma$ greater than 2·4 there is a stable region near $\lambda = 0$, corresponding to the EADY (1949) high wavenumber cutoff. For $\lambda$ near 1 the growth

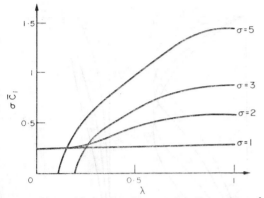

FIG. 6. Growth rates for the case of uniform shear. $\lambda$ is the fractional amount of overturning, and $\sigma\bar{c}_i$ is the non-dimensional growth rate.

rate has a linear dependence on $l$ and is bounded by the FJØRTOFT (1950) result above.

The case of no bottom shear ($\gamma = 0$) is shown in Fig. 7. Curiously, the high wavenumber cutoff for small $\lambda$ disappears, showing that the existence of vertical shear is not a sufficient condition for instability. $\bar{c}_i$ is approximately independent of $\sigma$ in the range $0 \leq \sigma \leq 5$, and its value *approximately* resembles the case $\sigma = 5$, $\gamma = 1$. This can be interpreted simply as the fact that for $\sigma > 2\cdot4$ unstable waves mainly feed on the A.P.E. in the upper régime, and can make little use of the A.P.E. in $z < -h$. When, further, there is no shear in $z < -h$, then waves can *only* feed on the upper A.P.E.

This is shown pictorially in Figs. 8 and 9, which show $\psi$ and $w$ as functions of $z$ for $\sigma = 5$.

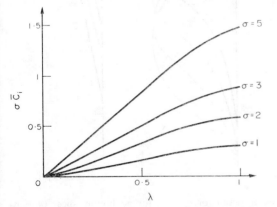

FIG. 7. Growth rates for the case of no bottom shear.

For small $\sigma$, $|\psi|$ and $|w|$ are indistinguishable from the Eady result (cf. McINTYRE, 1970, for example). The cases of shear and no shear both give stream functions which are of order 1 in the overturned region and tend to decay with depth in the stratified region with a length scale of $\sigma^{-1}$. The main difference between $\gamma = 1$ and $\gamma = 0$ lies in the phase of $\psi$, which tends to be larger relative to the bottom in the case of no bottom shear.

The indications from this model are, then, that the larger the vertical shear, the larger the growth rate. For $\sigma > 1$, the main disturbance occurs in the overturned region with a decay with depth below; for $\sigma < 1$, it is spread evenly between the two regions. Since we shall see that, typically, $\sigma > 1$, it follows that at all stages of the breakup, the largest perturbations will be found in the overturned region. Thus influxes of stably stratified water above the column (STOMMEL, 1972) a few days after the end of the Mistral will tend to 'lock' later instabilities on to the region of overturning and exclude them from the surface.

The parameter $\sigma$ can be estimated easily. For $N/f \sim 10$, $H \sim 2$ km, $l \sim 3 \times 10^{-4}$ m$^{-1}$ (corresponding to an eddy diameter of 10 km),

$$\sigma \sim 6.$$

The growth rate $\omega$ can also be estimated. Numerical experimentation (section 8) shows that the amplitude of the stream function is largest where the vertical shear is largest; near the end of the Mistral the top 20% of the fluid is overturned at that location. Figs. 6 and 7 show that the growth rate is about

$$\frac{1}{6}\frac{H\bar{V}_z}{\sqrt{12}}l;$$

with dissipation, this becomes

$$\omega \sim \frac{H\bar{V}_z}{6\sqrt{12}}l - A_H l^2.$$

Structure in the $x$-direction will reduce $\omega$ still further. Then the optimal growth rate and wavenumber are given by

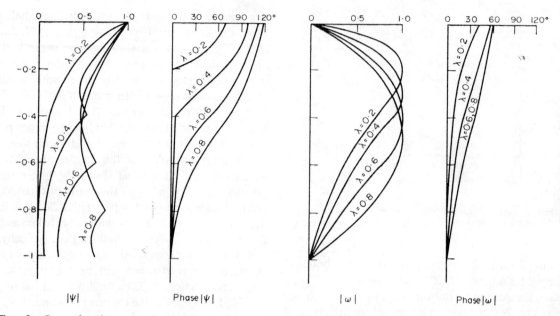

FIG. 8. Streamfunction ψ and vertical velocity w as functions of depth for σ = 5, for uniform shear (γ = 1). The phases are shown relative to the bottom, and amplitudes scaled to unity.

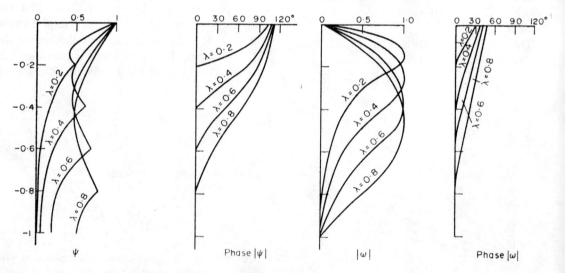

FIG. 9. Streamfunction ψ and vertical velocity w as functions of depth for σ = 5, for no bottom shear (γ = 0). The phases are shown relative to the bottom, and amplitudes scaled to unity.

$$\omega_{max} \sim \frac{H^2 \bar{V}_z^2}{1728 A_H},$$

$$l_{max} \sim \frac{H \bar{V}_z}{12 A_H \sqrt{12}}.$$

Putting $\bar{V}_z \sim 3 \times 10^{-4}$ s$^{-1}$, $A_H \sim 50$ m$^2$ s$^{-1}$, we obtain

$$\omega_{max} \sim 4 \times 10^{-6} \text{ s}^{-1},$$

$$l_{max} \sim 2 \cdot 8 \times 10^{-4} \text{ m}^{-1},$$

i.e. an e-folding time of 2·8 days, and an eddy diameter of 11 km. If $A_H$ is halved, these become 1·4 days, 5·6 km; if $A_H$ is doubled, 5·8 days and 22 km respectively. Thus the choice of $A_H$ is seen to give reasonable estimates of both $\omega_{max}$ and $l_{max}$, as desired, and we shall see from the full model that these estimates are not too much in error.

## 7.  THE SINKING AND SPREADING PHASE —A QUASIGEOSTROPHIC MODEL

To obtain an estimate of the appearance of the perturbations, we shall assume the mean motion at any time to be quasi-steady. This assumption is reasonable despite the similarity between eddy and mean time scales, since it will appear later that the predominant eddies are likely to appear at the end of the Mistral, when the mean motion is genuinely steady. However, the growth times obtained can only be taken as indicative of reality.

The construction and model of solution of the perturbations to (21)–(26) is similar to BROWN (1969). The equations are solved in polar coordinates $(r, \theta, z)$ with $r = 0$ corresponding to the centre of the column, and $r$ replacing $x$ in the foregoing. The use of cartesian coordinates makes it difficult to interpret length scales for the disturbance in the $y$ (now $\theta$) direction, due to the large curvature near $r = 0$. The unperturbed velocity field is

$$(0, V(r, z), 0),$$

and satisfies the thermal wind relation

$$V_z = -g\rho_r(r, z)/f\rho_0,$$

and

$$\int_{-H}^{0} V\, dz = 0.$$

The basic state is found by using the results of section 3. A time $T$ after the onset of the Mistral is specified, and the density field is overturned using a one-dimensional non-penetrative model and a total density input $QT$. The differences between this basic state and the prediction of section 3 are qualitative and are ignored. The width $W''$ of the pre-conditioned area was again 60 km. The horizontal spacing was 5 km (16 points) and the vertical 125 m (20 points). Then, denoting the perturbation stream function by $\psi(r, z, t)e^{il\theta}$, where the azimuthal wavenumber $l$ is an integer, $\psi$ satisfies

$$\nabla^2 \psi_t - \frac{il}{r} \psi \left\{ \frac{V}{r} + V_r \right\}_r + \frac{ilV}{r} \nabla^2 \psi$$
$$= f w_z + A_H \nabla^4 \psi, \quad (31)$$

$$\psi_{zt} - \frac{il}{r} \{ \psi V_z - V\psi_z \} = \frac{-N^2}{f} w$$
$$+ A_H \nabla^2 \psi_z, \quad (32)$$

where

$$\nabla^2 \equiv \frac{1}{r} \frac{\partial}{\partial r} r \frac{\partial}{\partial r} - \frac{l^2}{r^2}.$$

Infinite vertical mixing of $\rho$ (i.e. $\psi_z$) occurs where $N = 0$. Normal quasigeostrophy, relevant for the non-overturned region, assumes $N$ varies only slowly with $r$, so

$$N(r, z) \equiv N(a, z) \quad \text{in the non-overturned region,}$$

where $r = a$ denotes the outer boundary, far from

the column; this gives errors smaller than 5%. Equations (31) and (32) have as boundary conditions

$$\psi = \triangledown^2\psi = 0, \qquad r = 0, a$$
$$w = 0, \qquad z = 0, -H$$

corresponding to motion in a rigid cylindrical box with zero vorticity at the walls and centre.

Equation (31) is solved by centred time-stepping and "running down the diagonal" of the resulting tridiagonal matrix. (It can be shown that the discontinuities of $\psi_r$ at jumps in $N$ are handled correctly by this procedure.) A diagnostic equation for $w$ is obtained by eliminating $t$ derivatives:

$$w_{zz} + \triangledown^2\left(\frac{N^2}{f^2}w\right) = \frac{2il}{r}\left\{V_{rz}\left(\frac{\psi}{r}\right)_r - \psi_{rz}\left(\frac{V}{r}\right)_r\right.$$

$$\left. + \frac{V_z}{r}\left(\psi_{rr} - \frac{l^2-1}{r}\psi\right) - \frac{\psi_z}{r}V_{rr}\right\} \tag{33}$$

in which $A_H$ does not appear. An extra boundary condition for $w$, on $r = 0, a$, is needed if $N \neq 0$. This is obtained from (32), and gives

$$N^2w = 0 \quad \text{on} \quad r = 0, a.$$

Then $w$ is found by over-relaxation. There are regions near the front where this process can in theory fail; however, no trouble was experienced.

The fastest-growing mode was obtained by stepping an initially random $\psi$ field in time until the amplitude of all the slower-growing modes had decayed relative to the desired mode. The quasi-steady mean motion does not necessarily mean that this mode will be observed, since perturbations at earlier values of $T$ may already be growing. However, the mode should give indications of the shape of the eddies at this time.

The phase velocity $c$ is then obtained by assuming a behaviour proportional to $\exp\{il(\theta - ct)\}$ from which

$$\psi_t\psi^{-1} = -ilc.$$

The real part of $c$, $c_r$ say, is the azimuthal phase velocity, and $l$ times the imaginary part ($c_i$) gives the growth rate. Thus $c_r$ and $c_i$ are found as functions of $T$ and $l$.

When tested on the problem in section 6, the model gave very accurate results. However, in that problem the depth of overturning could be chosen to be an integral number of grid points. In the MEDOC situation, the overturned depth depends on $T$ and $r$ and cannot, in general, be an integral number of gridpoints. The resulting 'rounding down' of the overturned depth induces inaccuracies in the predicted growth rate (typically underestimates). These inaccuracies are worst, predictably, when the overturned depth is just less than an integral number of gridpoints: fortunately, however, since computation limits the number of gridpoints, it appears from experimentation that the errors are worst on the modes which are not growing the fastest. Hence, the results in section 8 can be expected to be a reasonable approximation to the continuous case.

## 8. THE SINKING AND SPREADING PHASE—RESULTS

Calculations were performed for perturbations to the two-dimensional solution of section 3 at times $T = 0, 2, 4, 6, 8, 10$ and 14 days, together with some intermediate values at differing resolutions for checking purposes. The growth rate $\omega$, wavenumber and phase speed of the fastest growing mode are shown as function of $T$ in Figs. 10 to 12. For $T$ less than 4 days, the growth rates are slow, with $e$-folding times of the order of 100 days. By $T = 6$ days, there is sufficient overturning in the region of maximum vertical shear (which is, as we have seen, where the instability is centred) to decrease the $e$-folding times to the order of 9–10 days. As $T$ increases, the growth rates rise rapidly, with $e$-folding times reaching about $3\frac{1}{2}$ days for $T = 8$ and 10 days. These times are longer than those estimated in section 6, but still of the same order. Although a $T$ of 10–12 days corresponds most closely with the effective length of the Mistral, calculations were also performed for $T = 14$ days, and the same trend continued: the maximum $\omega$ was $4 \cdot 3 \times 10^{-6}$ s$^{-1}$, or an $e$-folding time of $2 \cdot 7$ days.

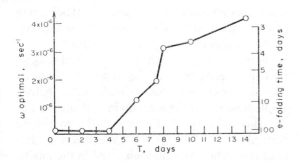

FIG. 10.   Growth rate of the fastest growing mode after overturning for $T$ days. Values of $T$ for which calculations were performed are circled.

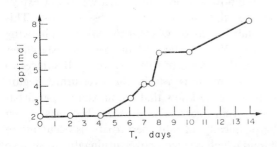

FIG. 11.   Azimuthal wavenumber $l$ of the fastest growing mode after overturning for $T$ days. Wave number $l$ corresponds to an azimuthal eddy diameter of $\pi l^{-1} \times 20\,\mathrm{km}$ at a radius of 20 km.

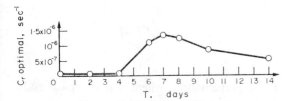

FIG. 12.   Azimuthal phase velocity $c_r$ of the fastest growing mode after overturning for $T$ days. A velocity of $10^{-6}\,\mathrm{s}^{-1}$ corresponds to a velocity of 0.02 m s$^{-1}$ at a radius of 20 km.

The wavenumber associated with the fastest growing instability increases monotonically with $T$. This wavenumber is small (2–3) for $T < 6$ days, but rises rapidly to 6 at $T = 8$ and 10 days. We shall see below that the maximum amplitude of $\psi$

occurs at a radius corresponding to the largest vertical shear (about 20 km). Hence, wavenumber 6 corresponds to an azimuthal eddy diameter of

$$\frac{\pi}{6} \times 20 \text{ km}$$

i.e. 10·5 km. As $T$ increases past 10 days, the wavenumber increases to 8.

In contrast, the phase speed (i.e. angular velocity) of the optimal mode is not monotonic in $T$, and reaches a maximum of $1\cdot3 \times 10^{-6}\,\mathrm{s}^{-1}$ at $T = 7$ days, before decreasing to $8\cdot4 \times 10^{-7}\,\mathrm{s}^{-1}$ at $T = 10$ days. These phase speeds correspond to velocities at a radius of 20 km of 2·6 and 1·6 $\times$ $10^{-2}$ m s$^{-1}$ respectively, with a cyclonic circulation (i.e. the same sign as the mean flow at the surface).

These figures should only be taken as indicative, due both to the broad peak in growth rate if plotted as a function of $l$ (Fig. 13) and to the quasi-steady nature of the mean flow. For $T = 8$ and 10 days, wavenumbers 5, 6 and 7 all produce growth rates greater than $3 \times 10^{-6}\,\mathrm{s}^{-1}$. There was no indication that secondary maxima could occur for large wavenumbers; such maxima would be unlikely *a priori* due to the large viscous damping $A_H l^2$. The angular velocity changes only slightly near the point of maximum growth rate; it increases monotonically with wavenumber in a near-linear fashion (Fig. 14), and can even become anticyclonic for $T$ large and $l$ small.

The form of the eddies is shown in Fig. 15, for $T = 10$ days, and wavenumber 6. The eddies are centred on the region of maximum vertical shear, as anticipated. (Computations of the rate of conversion from mean A.P.E. to eddy K.E. show that its value is two orders of magnitude stronger in the region of strong shear than elsewhere.) Because of this, the eddy strength is mainly confined to the top 500 m, with weaker effects extending to 1000 m. For smaller $T$ or $l$ a weak barotropic field extends to the bottom. The eddies have an azimuthal diameter of 10·5 km, and a radial diameter of order 20 km. These two length scales bracket the length given by SANKEY (1973) closely. The density perturbation is given, from (21), as

---

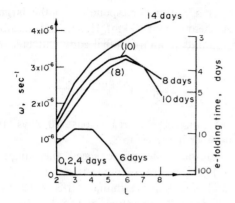

FIG. 13. Growth rate as a function of wavenumber and $T$.

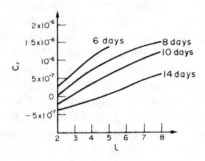

FIG. 14. Azimuthal phase velocity $c_r$ as a function of wavenumber and $T$. Conversion as for Fig. 12.

$$\rho = -\frac{\rho_0 f}{g} \psi_z$$

and is also confined to the surface, where it attains its maximum value. If the maximum eddy velocity is $0.1$ m s$^{-1}$ (the observed size), the value of $\rho$ corresponds to $0.01$ $\sigma_t$, which is enough to overturn most of the remaining stable fluid. This $\rho$ perturbation is positive within cyclonic surface eddies, so the effects of the instability will be seen more strongly in these; conversely within anti-cyclonic eddies, $\rho$ is negative, which tends to stabilize the mean flow in such regions. Whether this can happen depends on the local depth of overturning, as we shall see below.

At depths of over 500 m, there is a small intrusion of water with $\rho$ of the opposite sign to the surface water (seen more clearly in Fig. 16).

It is suggested that this is the (linearized) beginning of the layered intrusions observed in the breakup of the column (MEDOC GROUP, 1970).

Finally in Fig. 15 is shown the vertical velocity $w$. This extends most of the way to the bottom, with downwelling within cyclonic eddies (which would tend to aid the breakup in such areas). For an eddy velocity of $0.1$ m s$^{-1}$, the maximum vertical velocity is $0.003$ m s$^{-1}$ which is much less than the vertical velocities observed in the mixing process (VOORHIS AND WEBB, 1970).

*(a) Comparison with observations*

We now ask how and when we would expect to see this instability in observations. This depends largely on whether the Mistral is blowing or not. We consider first the case of instabilities during the period of surface forcing. It suffices to consider motions within the overturned region only, since we know that the maximum perturbations occur there. The problem becomes: given baroclinically unstable perturbations occurring in a fluid which is steadily overturning due to surface forcing, which effect changes the density field more efficiently: the perturbation, or the surface buoyancy input? We can estimate these as follows. The linearized density equation (32) yields

$$\rho_t \sim \frac{f\rho_0}{g} u V_z$$

for the perturbation, where $u$ is a typical eddy velocity, and conservation of mass gives

$$\rho_t \sim \frac{Q}{h}$$

for the effect due to buoyancy input. Hence

$$\frac{\rho_t \ (\text{perturbation})}{\rho_t \ (\text{buoyancy input})} \sim \frac{f\rho_0 u V_z h}{gQ}$$

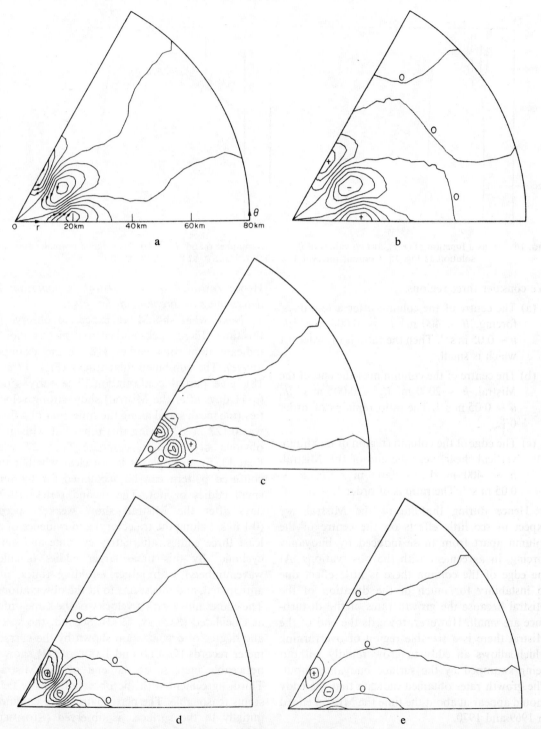

FIG. 15. The form of the eddies predicted, for wave-number 6 and $T_i^{'} = 10$ days. The stream-function has been normalized so that its maximum value is unity. (a) Stream-function at the surface, contour interval 0.2, (b) $\psi_z$ (proportional to density) at the surface contour interval $8 \times 10^{-4}$ m$^{-1}$, (c) $\psi_z$ at a depth of 562 m, contour interval $4 \times 10^{-4}$ m$^{-1}$. (d) Vertical velocity $w$ at a depth of 125 m. There is downwelling within cyclonic eddies. Contour interval $10^{-10}$ m s$^{-1}$. (e) Vertical velocity at a depth of 1250 m, contour interval $10^{-10}$m s$^{-1}$.

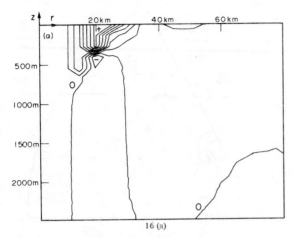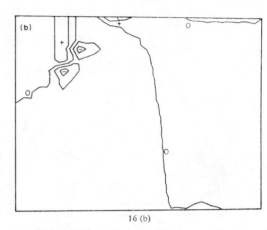

FIG. 16. $\psi_z$ as a function of $(r, z)$, at two values of $\theta$, for wavenumber 6 and $T = 10$ days. Same normalization and solution as Fig. 15. Contour interval $4 \times 10^{-4}$ m$^{-1}$. (a) $\psi_z$ at $\theta = 0$, (b) $\psi_z$ at $\theta = \pi/6$.

We consider three regions.

(a) The centre of the column after a few days' forcing. $h \sim 400$ m, $V_z \sim 0.005$ m s$^{-1}/h$, $u \sim 0.05$ m s$^{-1}$. Then the ratio is of order $0.1$ which is small.

(b) The centre of the column near the end of the Mistral. $h \sim 2000$ m, $V_z \sim 0.005$ m s$^{-1}/h$, $u \sim 0.05$ m s$^{-1}$. The ratio remains of order $0.1$.

(c) The edge of the column (region of maximum vertical shear) near the end of the Mistral. $h \sim 400$ m, $V_z \sim 0.05$ m s$^{-1}/h$, $u \sim 0.05$ m s$^{-1}$. The ratio is of order 1.

Hence during the life of the Mistral, we expect to see little effects on the centre of the column apart from those induced by buoyancy forcing, in agreement with the observations. At the edge of the column there is little effect due to instability for much of the duration of the Mistral because the growth rates of the disturbance are small. However, towards the end of the Mistral there is a sizeable region of overturning which allows an eddy to grow rapidly without being swamped by the surface buoyancy input. The growth rates obtained indicate that the eddy should appear at about the time the Mistral ended in 1969 and 1970.

At the end of the Mistral, the surface buoyancy input ends, and eddies of all sizes are free to grow.

Hence *cessation of the Mistral is important in determining the breakup of the column.*

Next, what should we expect to observe at this time? Float tracks and current meters should indicate numerous eddies (12 in the example above). The combined float tracks (Figs. 17 and 18), give partial confirmation. The early series (5–11 days after the Mistral) show strong velocities (the mean speed along the trajectory of a float was 0·127 m s$^{-1}$ during this time) but with little obvious organization—compare float 27 with float 13, for example. It is not clear whether the confused pattern can be accounted for by very small eddies or not. The second series (19–25 days after the Mistral) show weaker speeds (0·1 m s$^{-1}$ along the trajectory) and evidence of at least three eddies, alternately cyclonic and anticyclonic. By this time, larger eddies (smaller wavenumbers) with larger $e$-folding times, are anticipated, and this seems to fit the observations. The eddies move with a velocity of the same order as calculated (SANKEY, 1973). Second, the speed and degree of organization shown by the current meter records (GOULD and LARBY, 1971) show a noticeable increase at the end of the Mistral. Third, agreement with the observed density field seems reasonable. The perturbations are confined initially to the surface, as observed (STOMMEL, 1972). The beginnings of an interleaving of water layers is also predicted.

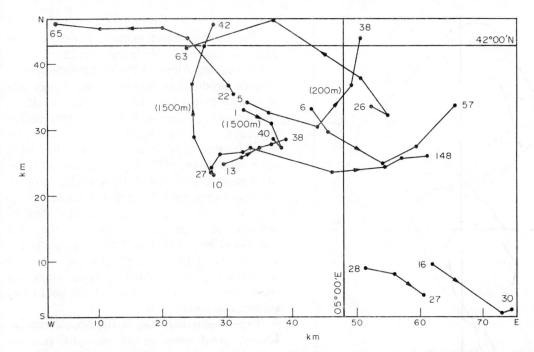

FIG. 17. Float tracks from floats released in February, 1969. Fixes indicated by circles. Base diagram taken from CASTON and SWALLOW (1970). Unless otherwise indicated, the floats have a nominal depth of 500 m. Two figures are given by each track: at the beginning, the float designation, and at the end, the number of hours elapsed between first and last fixes, to give an indication of the speed of the float. The times in days after the end of the Mistral (17 February) for the floats are:

Floats 10, 26, 6 (5 days); 16, 28 (6 days); 1, 27, 13, 5 (11 days).

However, to be able to explain the breakup of the chimney, the baroclinic eddies must be able to carry away a flux of at least $10^6$ m³ s⁻¹ (SANKEY, 1973). This is extremely likely: an eddy velocity of 0·05 m s⁻¹ acting through an azimuthal diameter of 10 km and a depth of 300 m gives a flux of $0·15 \times 10^6$ m³ s⁻¹. There are six such eddies, giving a total flux of $6 \times 0·15 \times 10^6 = 0·9 \times 10^6$ m³ s⁻¹. When we consider that later, non-linear, eddies will be larger, fluxes as high as $3 \times 10^6$ m³ s⁻¹ are not impossible.

(b)  *The solution for large times*

The above description is based on the results of linearized perturbation analysis, and is not strictly valid as the eddies grow in intensity. Despite recent theoretical work on finite-amplitude perturbations (e.g. PEDLOSKY, 1970), the problem remains suitable only for numerical solution of the full three-dimensional problem (in

preparation). In general, however, the initial eddies will tend to bring less dense fluid into the surface layers of the column and thus essentially stabilize the surface. This will create new regions of maximum A.P.E. which are located at depth, resulting in further larger-scale instabilities which are confined to the intermediate and lower parts of the column. Whether the long-term breakup of the column consists of a successive eating away of the column from outside, or whether finite amplitude effects are sufficient to destroy the column within a few days, is an unsolved question.

There remains the problem of the final destination of the water in the column. For time scales of the order of 1 month, the variation of $f$ with latitude can no longer be neglected. If we assume that the mixed water sinks to the bottom, and that its motions becomes steadily slower, then we may apply conservation of potential vorticity, equation (26), to yield

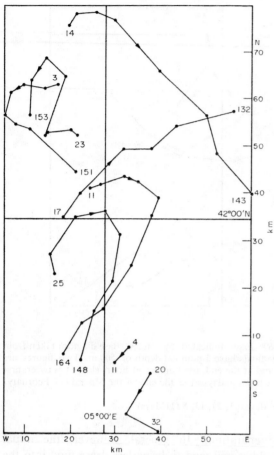

FIG. 18.   Float tracks, from floats released in March, 1969. Legend as Fig. 17. The times in days after the end of the Mistral are: 11, 23, 3, 14 (19 days); 25, 17 (20 days); and 4, 20 (25 days).

$$\frac{D}{Dt}(f + \nabla^2\psi) \sim f w_z < 0.$$

If $\psi$ is by this time sufficiently small that $\nabla^2\psi \ll f$, then this gives

$$\beta\,\psi_x < 0,$$

i.e. a general southward drift. This occurred in 1969 and 1970, but in 1972 the drift was to the northwest (LACOMBE, 1974), for which no explanation (save topographical influences, variation in mean flow, etc.) is given.

## 9. DISCUSSION

The formation and motion of bottom water plays an important role in the circulation of the world's oceans. Although the Mediterranean is the smallest producer of bottom water, it may well serve as a model for other, larger areas such as the Antarctic, where it is possible that similar processes occur on the small scale. It is known, for example, that rapid freezing occurs when leads open up (FOSTER, 1972). Upwelling near the edge of Antarctic continental shelves (KILLWORTH, 1974) can drastically lower the stratification, so that the resulting input of salinity might itself be sufficient to trigger an instability of the kind considered here. This would be a highly efficient way of creating both a supply of dense water and the observed stratification, judging from the speed with which all trace of the MEDOC column disappears.

Experiments continue in the Mediterranean. Ideally, good coverage by neutrally buoyant floats is needed so that the structure of the eddies during the breakup can be studied. Since the position of the column can be predicted with some accuracy from STD casts in the preconditioning phase, suitable positioning for the floats (say, closely spaced around the edge of the column) can easily be chosen. Unfortunately the placement for the floats would coincide with the last few days of the Mistral, when conditions are difficult.

In addition, it is suggested that closely spaced STD casts be made around the edge of the column; not only because this is the likely site for the first instability, but also so that the frontal zone between column and open sea can be resolved as well as possible.

APPENDIX.   ENERGETICS AND PERTURBATION THEORY IN A WEAKLY STRATIFIED MEDIUM

At first sight there seems to be little difference between equations (21)–(26) (the U theory) and those of the normal vertically stratified theory (cf. PEDLOSKY, 1964), hereafter called the S theory. In both theories the vorticity of a fluid element can be altered solely by vortex stretching.

The only difference in the two theories would appear to be that in the U theory, density is advected purely horizontally.

Examination of the energy of the U system shows this view to be erroneous. The total energy of a fluid parcel consists of its kinetic energy (K.E.) and its available potential energy (A.P.E.). The latter is the potential energy relative to a state of minimum P.E. This state is only known when the entire density field is given, so perhaps one should think of A.P.E. only as a global property. Nonetheless it is convenient to consider the contribution to A.P.E. from a fluid element. In the S theory one has

$$\text{A.P.E.} = \frac{-g\rho'^2}{2\rho_{1z}} \qquad (A1)$$

where $\rho_{1z}$ is the gradient of the basic density field. Then

$$\frac{d}{dt} \int (\text{K.E.} + \text{A.P.E.})dV = 0,$$

$$\frac{d}{dt} \int \text{K.E. } dV = - \int w\rho \, dV$$

$$= - \frac{d}{dt} \int \text{A.P.E. } dV, \qquad (A2)$$

where the integration is over a container with rigid walls.

This shows (a) the S system is energetically consistent and (b) the K.E. can be increased if $w$ and $\rho$ are negatively correlated (light fluid up, heavy fluid down). Such motions act to reduce the A.P.E. of the system.

The energetics of the U system can easily be obtained. The formula (A1) for A.P.E. is no longer applicable as the deviations from a state of minimum P.E. are not small. However, MCINTYRE (unpublished lecture notes) gives an expression for A.P.E. which involves no approximations:

$$\text{A.P.E.} = \int dV \, \Pi \, (\rho_1(z), \rho - \rho_1(z)). \qquad (A3)$$

In this formula $\rho_1(z)$ is the (unique) density field which possesses minimum A.P.E. and which can be attained from the field $\rho(x, y, z)$ by mass-conserving motions. (This can be explicitly calculated (DUTTON and JOHNSON, 1967) via a Stieltjes integral, as, indeed, can the A.P.E.) $\Pi$ is defined by

$$\Pi(\xi, \eta) = - \int_0^\eta \{\Sigma(\xi + \eta') - \Sigma\,(\xi)\} \, d\eta',$$

where $\Sigma$ is the inverse of $\rho_1(z)$:

$$\Sigma\,(\rho_1(z)) = gz.$$

Expanding $\Sigma$ in a Taylor series for $\eta'$, we obtain eventually

$$\Pi = \sum_{i=0}^\infty \rho^i g_i(z), \qquad (A4)$$

where $g_i$ are known functions of $z$ alone. Note that truncating the series at $i = 1$ gives (A1). Multiplying (25) by $p_{0z}^{i-1} g_i(z)$, integrating over the box, and summing over $i$ gives

$$\frac{d}{dt} (\text{A.P.E.}) = 0, \qquad (A5)$$

i.e. the net A.P.E. cannot be altered by horizontal non-divergent motions. If one remembers that change in A.P.E. involves vertical motion (here very small) this becomes intelligible.

One might then expect from energy conservation that $d/dt$ (K.E.) is zero also. This is not so; in fact it is given by the first two terms of (A2) just as in the S theory. This apparent contradiction is easily resolved if we estimate the magnitudes of the K.E. and A.P.E. If the dimensions of the box are $L$ horizontally and $H$ vertically, and $\Delta\rho$ is a typical horizontal density variation, then, following GREEN (1970),

$$\text{A.P.E.} \sim \frac{gH^2L^2\Delta\rho}{12}, \tag{A6}$$

$$\text{K.E.} \sim \rho_0 \frac{U^2HL^2}{2}.$$

Using (26) to estimate $\Delta\rho$,

$$\frac{\text{K.E.}}{\text{A.P.E.}} \sim \frac{U}{fL} = Ro \ll 1. \tag{A7}$$

Thus to first order in $Ro$, and *for quasi-geostrophic motions*, conservation of energy reduces simply to conservation of A.P.E., and the K.E. has effectively an 'infinite' supply of A.P.E. upon which to draw. This is in contrast with the S theory, where

$$\frac{\text{K.E.}}{\text{A.P.E.}} \sim \left(\frac{NH}{fL}\right)^2 \sim 1 \gg Ro, \quad \text{by assumption.}$$

For small $N$ the energy is dominated by the A.P.E. In fact, for small Rossby number, the A.P.E. dominates providing $\Delta\rho$ is not much smaller than the vertical variation of $\rho$.

It follows from (A2) that the K.E. is likely to increase in magnitude until one of the quasi-geostrophic assumptions is violated (either $Ro$ ceases to be small, or the time scale decreases). We need to know the rate of increase to know how long equations (21)-(26) will apply. An indication can be obtained by considering an infinitesimal perturbation to a mean flow. For later use, we derive the equations for such a perturbation to a mean field $(0, V(x, z), 0)$. Then thermal wind implies

$$\rho_0 f V_{zz} = -g\rho_{xz} = 0,$$

so that $V = \bar{V}(x) + \hat{z}\hat{V}(x)$, where $\hat{z} = (z + \tfrac{1}{2}H)$. It can be shown that the fastest growing terms for the perturbation stream function $\psi (= \rho_0 fp)$ have the same $z$ dependence, i.e. $\psi = \bar{\psi}(x,y,t) + \hat{z}\hat{\psi}(x,y,t)$. If we seek an $e^{ily}$ dependence, and linearize about the basic state, (25) and (26) become

$$\hat{\psi}_t + il(\bar{V}\hat{\psi} - \hat{V}\bar{\psi}) = 0, \tag{A8}$$

$$(D^2 - l^2)\,\bar{\psi}_t + il\bar{V}\,(D^2 - l^2)\bar{\psi} + il\,\frac{H^2}{12}\,\hat{V}$$
$$\times (D^2-l^2)\hat{\psi} - il\bar{V}_{xx}\,\bar{\psi} - il\cdot\frac{H^2}{12}\,\hat{V}_{xx}\,\hat{\psi} = 0 \tag{A9}$$

where $D \equiv \partial/\partial x$.

If $\hat{V} = 0$ these equations reduce to those for barotropic instability (Lin, 1953), since $\hat{V} = 0$ implies $\rho_x = 0$ (i.e. a homogeneous fluid).

The following can be derived:

(a) It can be shown that the rate of increase of K.E. for $\bar{V} = $ constant lies between

$$\frac{lH|\hat{V}|_{\min}}{3^{\frac{1}{2}}} \quad \text{and} \quad \frac{lH|\hat{V}|_{\max}}{3^{\frac{1}{2}}},$$

where $|\hat{V}|_{\min}$ and $|\hat{V}|_{\max}$ are the largest and smallest values of the vertical mean shear. This corresponds to growth rates of one-half of these values. If $\bar{V} \neq $ constant then elements of barotropic instability are also present.

(b) These values are attainable. The simplest case is $\hat{V} = $ constant, solved somewhat differently by Fjørtoft (1950). He showed that the growth rate was precisely

$$\frac{lH|V_z|}{12^{\frac{1}{2}}}$$

providing that this is smaller than $f$ (otherwise the quasigeostrophic assumption breaks down). He states that for growth rates larger than $f$ there is an area of stability, but this seems unlikely due to other forms of instability (e.g. Ooyama, 1966).

(c) If $\hat{V}$ is not constant then it can be shown that there are no continuous normal mode solutions of the form $\psi(x) \exp i\{ly + wt\}$ which are either confined to a box or are cyclic in $x$. There is a discontinuous solution, namely a delta-function located at the point of maximum vertical shear. If, instead, one examines the temporal development of a single Fourier component in $x$, one finds that there is a cascade of energy into higher and higher wavenumbers as $t$ increases. However, the rate of transfer of energy to higher wavenumbers decreases gradually with wavenumber.

In other words there is a double tendency toward high wavenumber: on one hand the higher the wavenumber the higher the growth rate; on the other hand, even if only low wavenumbers are present, energy will cascade into high wavenumbers for large $t$. Again, one can understand this from extrapolating the S theory. There, the largest growth rate has $NH/fL \sim 1$. So as $N$ decreases, one expects $L$ to decrease also (indeed, in the EADY, 1949, solution, growth rate and length scale are both proportional to $N^{-1}$).

Thus the region of validity of the quasi-geostrophic approximation depends strongly on whether or not large ($\gtrsim f$) growth rates and/or small length scales can be achieved. Both phenomena will normally be damped by most small-scale mixing phenomena. Section 5 shows that such effects can be crudely represented by eddy coefficients. If these are chosen at a 'reasonable' value, then motions with optimal growth rates and length scales are well approximated by (25) and (26).

*Acknowledgements*—This work was carried out under a grant from the Natural Environment Research Council of Great Britain. Thanks are due to the referees, whose comments materially improved both content and presentation.

ANATI D. (1970) On the mechanism of the deep mixed layer formation during Medoc '69. *Cahiers Océanographiques*, **22**, 427–443.

ANATI D. and H. STOMMEL (1970) The initial phase of deep-water formation in the Northwest Mediterranean, during Medoc '69. *Cahiers Océanographiques*, **22**, 347–351.

BLUMEN W. (1972) Geostrophic adjustment. *Reviews of Geophysics and Space Physics*, **10**, 485–528.

BROWN J. A. (1969) A numerical investigation of hydrodynamic instability and energy conversions in the quasi-geostrophic atmosphere: Parts I and II. *Journal of the Atmospheric Sciences*, **26**, 352–375.

BRYAN K. (1969) A numerical method for the study of the circulation of the world ocean. *Journal of Computational Physics*, **4**, 347–376.

BRYAN K. and M. D. COX (1968) A nonlinear model of an ocean driven by wind and differential heating. Parts I and II. *Journal of Atmospheric Sciences*, **25**, 945–978.

CASTON G. F. and J. C. SWALLOW (1970) Neutrally buoyant floats, serial numbers 209–227. N.I.O. Internal Report No. D.3, Institute of Oceanographic Sciences, Wormley, Surrey.

CHARNEY J. G. (1947) The dynamics of long waves in a baroclinic westerly current. *Journal of Meteorology*, **4**, 135–163.

CHARNEY J. G. and M. E. STERN (1962) On the stability of internal baroclinic jets in a rotating atmosphere. *Journal of the Atmospheric Sciences*, **19**, 159–172.

DEACON G. E. R. (1937) The hydrology of the Southern Ocean. *Discovery Report*, **15**, Cambridge University Press, 124 pp.

DUTTON J. A. and D. R. JOHNSON (1967) The theory of available potential energy and a variational approach to atmospheric energetics. *Advances in Geophysics*, **12**, Academic Press, New York, 333–436.

EADY E. T. (1949) Long waves and cyclone waves. *Tellus*, **1**, 33–52.

FJØRTOFT R. (1950) Application of integral theorems in deriving criteria of stability for laminar flows and for the baroclinic circular vortex. *Geofysiske Publikasjoner*, **17**, 1–52.

FOSTER T. D. (1972) Haline convection in polynyas and leads. *Journal of Physical Oceanography*, **2**, 462–469.

GASCARD J.-C. (1973) Vertical motions in a region of deep water formation. *Deep-Sea Research*, **20**, 1011–1028.

GASCARD J.-C. (1974) Deep convection: deep water formation in the north-west Mediterranean Sea. *Procès-Verbaux No. 13, IAPSO Special Assembly, Melbourne 1974.*

GILL A. E., J. S. A. GREEN and A. J. SIMMONS (1974) Energy partition in the large-scale ocean circulation and the production of mid-ocean eddies. *Deep-Sea Research*, **21**, 499–528.

GOULD W. J. and M. J. LARBY (1971) Moored current meter records: N.I.O. moorings 028–037. N.I.O. Internal Report No. A47, Institute of Oceanographic Sciences, Wormley, Surrey.

GREEN J. S. A. (1970) Transfer properties of the large-scale eddies and the general circulation of the atmosphere. *Quarterly Journal of the Royal Meteorological Society*, **96**, 157–185.

HART J. E. (1974) On the mixed stability problem for quasi-geostrophic ocean currents. *Journal of Physical Oceanography*, **4**, 349–356.

HOGG N. G. (1973) The preconditioning phase of MEDOC 1969—II. Topographic effects. *Deep-Sea Research*, **20**, 449–459.

HOSKINS B. J. (1974) The role of potential vorticity

in symmetric stability and instability. *Quarterly Journal of the Royal Meteorological Society*, **100**, 480–481.

HOSKINS B. J. and F. P. BRETHERTON (1972) Atmospheric frontogenesis models: mathematical formulation and solution. *Journal of the Atmospheric Sciences*, **29**, 11–37.

HUMBOLDT A. V. (1814) Reisen in die Aquatorial-Gegenden des neuen Kontinents. Ausgabe, v.H. Hauff, Stuttgart 1861, **1**, p. 22.

KILLWORTH P. D. (1973) A two-dimensional model for the formation of Antarctic bottom water. *Deep Sea Research*, **20**, 941–971.

KILLWORTH P. D. (1974) A baroclinic model of motions on Antarctic continental shelves. *Deep-Sea-Research*, **21**, 815–837.

KRAUS E. B. and J. S. TURNER (1967) A one-dimensional model of the seasonal thermocline II. The general theory and its consequences. *Tellus*, **19**, 98–105.

LACOMBE H. (1974) Deep effects of energy transfers across the sea surface. (Presidential address) *Procès-Verbaux No. 13, IAPSO Special Assembly, Melbourne 1974.*

LACOMBE H. and P. TCHERNIA (1971) Le problème de la formation des eaux marines profondes. *Annales de L'Institut Océanographique, Monaco*, **48**, 75–110.

LIGHTHILL M. J. (1969) Dynamic response of the Indian Ocean to the onset of the Southwest Monsoon. *Philosophical Transactions of the Royal Society*, A, **265**, 45–92.

LIN C. C. (1953) *The Theory of Hydrodynamic Instability*. Cambridge University Press, 155 pp.

MCINTYRE M. E. (1970) On the non-separable baroclinic parallel flow instability problem. *Journal of Fluid Mechanics*, **40**, 273–306.

MEDOC GROUP (1970) Observation of formation of deep water in the Mediterranean Sea. *Nature, London*, **227**, 1037–1040.

OOYAMA K. (1966) On the stability of the baroclinic circular vortex: A sufficient criterion for instability. *Journal of the Atmospheric Sciences*, **23**, 43–53.

ORLANSKI I. and M. D. COX (1973) Baroclinic instability in ocean currents. *Geophysical Fluid Dynamics*, **4**, 297–332.

PEDLOSKY J. (1964) The stability of currents in the atmosphere and the ocean. Part I. *Journal of the Atmospheric Sciences*, **21**, 201–219.

PEDLOSKY J. (1970) Finite-amplitude baroclinic waves. *Journal of Meteorology*, **27**, 15–30.

ROBINSON A. R. and S. GADGIL (1970) Time-dependent topographic meandering. *Geophysical Fluid Dynamics*, **1**, 411–438.

SAINT-GUILY B. (1963) Remarques sur le mécanisme de formation des eaux profondes en Mediterranée occidentale. *Rapports et Procès-verbaux CIESMM* **17**, 929–932.

SAINT-GUILY B. (1972) On the response of the ocean to impulse. *Tellus*, **24**, 344–349.

SANKEY T. (1973) The formation of deep water in the Northwestern Mediterranean. In: *Progress in Oceanography*, **6**, B. A. WARREN, editor, Pergamon Press, Oxford.

SAUNDERS P. M. (1973) The instability of a baroclinic vortex. *Journal of Physical Oceanography*, **3**, 61–65.

SEMTNER A. J. (1973) An oceanic general circulation model with bottom topography. Unpublished Ms., Princeton University, N.J.

STOMMEL H. (1962) On the smallness of sinking regions in the ocean. *Proceedings of the National Academy of Science, Washington*, **48**, 766–772.

STOMMEL H. (1972) Deep winter-time convection in the western Mediterranean Sea. In: *Studies in Physical Oceanography, a tribute to Georg Wüst on his 80th Birthday*, **2**, A. L. GORDON, editor, Gordon and Breach, pp. 207–218.

SWALLOW J. C. and G. F. CASTON (1973) The preconditioning phase of MEDOC 1969—I. Observations. *Deep-Sea Research*, **20**, 429–448.

VOORHIS A. and D. C. WEBB (1970) Large vertical currents observed in a western sinking region of the Northwestern Mediterranean. *Cahiers Océanographiques*, **22**, 571–580.

WELANDER P. (1973) Lateral friction in the oceans as an effect of potential vorticity mixing. *Geophysical Fluid Dynamics*, **5**, 173–190.

WOODS J. D. (1973) Space-time characteristics of turbulence in the seasonal thermocline. *Proceedings of the Hydrodynamics Colloquium Liège, Belgium.*

WORTHINGTON L. V. (1969) An attempt to measure the volume transport of Norwegian Sea overflow water through the Denmark Strait. *Deep-Sea Research*, **16** (suppl.), 421–432.

WORTHINGTON L. V. (1970) The Norwegian Sea as a mediterranean basin. *Deep-Sea Research*, **17**, 77–84.

*Prog. Oceanog.* Vol. 7, pp. 91–133, 1977. Pergamon Press. Printed in Great Britain

# Seawater: an explanation of differential isothermal compressibility measurements in terms of hydration and ion–water interactions*

IVER W. DUEDALL

Marine Sciences Research Center, State University of New York, Stony Brook, New York 11794

*(Received 12 February 1975; in revised form 2 February 1976; accepted 10 May 1976)*

**Abstract**—Hydration, ion–water interactions, and water structure effects in seawater were studied by determining differences ($\Delta\beta$) between the compressibilities of test salt solutions and the compressibilities of reference solutions. The reference solutions were distilled water and seawater (35‰), and the test salt solutions were either 0.13 m or 0.26 m with respect to one of the following test salts: LiCl, NaCl, KCl, CsCl, NaF, NaI, MgCl$_2$, CaCl$_2$, BaCl$_2$, Na$_2$SO$_4$, K$_2$SO$_4$, and MgSO$_4$. The compressibility measurements (to 900 bars) were carried out at 2°C and also at 15°C using a differential method in which a pressure increase or a temperature increase causes $\Delta\beta$ to become less negative. At 1 bar and 15°C, the $\Delta\beta$ (0.26 m, distilled water reference) values ranged from $-1.14 \times 10^{-6}$ bar$^{-1}$ for NaI to $-3.84 \times 10^{-6}$ bar$^{-1}$ for Na$_2$SO$_4$, and the $\Delta\beta$ (0.26 m, seawater reference) values ranged from $-1.30 \times 10^{-6}$ bar$^{-1}$ for NaCl to $-3.04 \times 10^{-6}$ bar$^{-1}$ for Na$_2$SO$_4$. The $\Delta\beta$ values were used to calculate hydration numbers. Entropy of transfer, excess hydrogen bond breaking (determined by NMR), and effective radii of ions are properties which can be used to describe the influence of ions on water structure. The extent to which these properties correlate with $\Delta\beta$ values depends upon whether the ion is an anion or a cation, and this correlation forms the thesis that anions alter water structure in a different way than do cations.

## I. INTRODUCTION

THIS paper describes a study of a number of the interactions of the water in seawater with the major sea salt components.

In the past, chemical oceanographers have been mainly concerned with the analytical composition of seawater and have often neglected its more fundamental aspects. Before 1960 virtually no attention was given to the role played by the water in seawater despite the obvious fact that seawater is composed mainly of water, and only in the past few years have chemical oceanographers begun the task of describing seawater in terms of ion–ion interactions and ion–water interactions. Indeed, I find it very significant that the first chemical oceanography textbook dealing with the molecular structure of seawater was published as recently as in 1969 (HORNE, 1969).

The problem of elucidating the structure of seawater is complicated for the following reasons: (1) there is no universally accepted theory for the structure of 'pure' liquid water; (2) the effect of

ion–water interaction on the water structure of simple two-component systems is not fully understood; and (3) the combined effects of ion–ion interactions and the multitude of ion–water interactions on the structure of water in a multicomponent system such as seawater are unknown and are perhaps beyond our present level of comprehension. In spite of these conditions, it is possible to learn something about the structure of seawater by discovering how the additions of certain salts affects its physical properties. In this approach the added salt acts as the agent which modifies the structure of the bulk water; the extent to which the water in seawater is then modified by the salt is related to a measurable physical property.

In this paper, a compressibility difference derived from compressibility measurements is used as a water structure parameter. This para-

---

*Contribution 174 of the Marine Sciences Research Center (MSRC) of the State University of New York at Stony Brook.

meter is essentially a reinterpretation of the hydration number originally described by PASSYNSKI (1938). He envisioned the hydration sheath surrounding an ion in water as being composed of layers of highly compressed water. The water molecules were considered compressed because of the ion's intense electrical field which orients the water dipoles around the ion. Passynski calculated that a water molecule within 1 Å of a monovalent ion is under an internal pressure of 79 000 atmospheres, and at this pressure he estimated the compressibility of the compressed water to be only 6% of the value for water at 1 atmosphere. For the water molecules situated at 5 Å (about two molecular layers, according to Passynski) from an ion, the internal pressure was calculated by Passynski to be only 126 atmospheres, and at this pressure the compressibility of water is nearly the same as at 1 atmosphere. Hence, the compressibility of water situated between the ion and a point 5 Å away changes very rapidly due to the change in the intensity of the electrical field. Because the compressibility of the water in the water sheath is small compared to the compressibility of the bulk water, the number of water molecules in the sheath can be estimated by assuming that the primary water sheath is essentially incompressible compared to the bulk water.

The following derivation, adapted in part from CONWAY and BOCKRIS (1954), is a mathematical approach to Passynski's hydration concept. In applying the derivation to seawater or other aqueous electrolyte solutions, the *reference* solution is defined either as pure water or as unperturbed seawater, and the solute is defined as the salt that is added either to pure water or to seawater. A *test* solution will contain $n_w$ moles of water (or when seawater is the reference solution, $n_w$ moles of water in seawater) and $n_s$ moles of solute whose effect on the $n_w$ moles of water (or $n_w$ moles of water in seawater) is to be determined.

These symbols are used in the derivation:

$\beta_{\text{ref.}}$ = compressibility of the unperturbed reference water

$\beta_{\text{test}}$ = compressibility of the perturbed refer-

ence water now containing $n_s$ moles of solute

$\Delta\beta = \beta_{\text{test}} - \beta_{\text{ref.}}$

$V$ = volume of the solution

$v^h$ = volume of the free water removed from the bulk solution to form the hydration sheath

$v^i$ = volume of the bare ions and ion-pairs

$\alpha = v^h/V$; volume fraction of the water, in the solution, used to form the hydration sheath.

The isothermal compressibility of the test solution containing $n_s$ moles of added solute is:

$$\beta_{\text{test}} = -\frac{1}{V}\left(\frac{\partial V}{\partial P}\right)_T. \qquad (1)$$

The isothermal compressibility of the reference solvent in the test solution is:

$$\beta_{\text{ref.}} = -\frac{1}{(V-v^h-v^i)}\left(\frac{\partial(V-v^h-v^i)}{\partial P}\right)_T \qquad (2)$$

where $(V-v^h-v^i)$ is the total compressible volume of the reference water in the test solution. The volume increment $-(v^h+v^i)$ is defined as not part of the compressible volume, and therefore eqn. (2) reduces to:

$$\beta_{\text{ref.}} = -\frac{1}{(V-v^h-v^i)}\left(\frac{\partial V}{\partial P}\right)_T. \qquad (3)$$

Dividing eqn. (1) by eqn. (3) gives:

$$\frac{\beta_{\text{test}}}{\beta_{\text{ref.}}} = 1 - \frac{v^h}{V} - \frac{v^i}{V}. \qquad (4)$$

The ratio, $v^h/V$, in eqn. (4) is the volume fraction of the water used to form the hydration sheath which surrounds the added solute; it is a parameter which can be used to describe the extent to which the major sea salts interact with the water in seawater. The ratio $v^i/V$ is small compared to $v^h/V$, because $v^h$ represents the volume, at the density of the free water, of that portion of water which was removed from the solution to form the hydration sheath. Therefore, neglecting $v^i/V$ and substituting the symbol $\alpha$ for $v^h/V$ into eqn. (4) and rearranging, and further substituting $\Delta\beta$ for $\beta_{\text{test}}$ minus $\beta_{\text{ref.}}$ gives the following approximation for $\alpha$ in terms of the difference between the compressibilities of a reference solution and a test solution whose compositions differ by $n_s$:

$$\alpha \cong -\frac{\Delta\beta}{\beta_{\text{ref.}}}. \tag{5}$$

The values $\Delta\beta$ and $\beta_{\text{ref.}}$ are measurable quantities, therefore $\alpha$ can be experimentally determined. $\Delta\beta$ is a negative quantity, thus $\alpha$ is positive. In the work by Passynski, $\alpha$ is interpreted as being proportional to the number of water molecules in the hydration sheath; thus the hydration number, $N_w$, for a particular solute can be calculated from the following equation:

$$N_w \cong \alpha \frac{n_w}{n_s}, \tag{6}$$

where $n_w$ is the number of moles of water and $n_s$ is the number of moles of added solute. Equation (6) is similar to that derived by ROBINSON and STOKES (1968, p. 61) although their equation is also an approximation because they assume that $V = n_w \bar{V}_w^\circ$ (where $\bar{V}_w^\circ$ equals the partial molal volume of pure water) which is an approximation for dilute solutions.

In this research I initially adopted this two-zone model for an electrolyte solution as a working model. However, on the basis of more recent models proposed by FRANK and WEN (1957) and SAMOILOV (1957, 1965, 1972) for ions in solution, the Passynski interpretation of an incompressible hydration sheath may only approximate the true state of water near an ion. In the Frank and Wen model, an ion is pictured as being surrounded by two concentric regions of altered water, instead of one 'solvation sheath' as Passynski visualized. Region A is the innermost layer of water, and region B is the layer of water which is situated between region A and the bulk or normally structured water. The Frank and Wen model is therefore a three-zone model. The water molecules in region A are partially immobilized and oriented because of the charge on the ion, and it is in region A that the water molecules are compressed, perhaps in the same way as Passynski visualized them to be. NEMETHY (1965) suggests that the water in this region is 'structured', although not in the same way as pure water. In region B the structure of the water is altered, but the ionic charge is shielded by region A; consequently, the water molecules in region B will not be oriented or structured to the same extent as those in region A. If an ion is small and doubly charged (e.g. $Mg^{2+}$), it will induce some additional structure to the water in region B; this ion is called a structure maker. But if an ion is large and weakly charged, the bulk water itself will compete with the ion for the water near an ion, with the net result that the water structure in region B is loosely organized; this type of ion is called a structure breaker.

Clearly, Frank and Wen's postulation of a region B adds a new dimension to the Passynski concept of hydration. Frank and Wen show that certain physical properties, as, for instance, the viscosity of aqueous solutions, can be interpreted in terms of a three-zone model of the water near an ion. For a small singly or doubly charged ion, the hydration number derived from compressibility data is probably a fairly accurate estimate of primary hydration because, if the Frank and Wen model is correct, the extent of region A would greatly predominate over region B. However, for a large singly charged ion, the number of water molecules in region B becomes important and may even predominate over the number found in region A. In this latter case, one has to be concerned with the relative compressibilities of the water in region B and the bulk water, if compressibility data are used to calculate a hydration number. The compressibility of region B may be similar to the compressibility of the bulk water, and therefore the so-called hydration number as calculated from eqn. (6) could be underestimated if region B is not considered an integral part of the hydration sphere. Recently, BOCKRIS and SALUJA (1972) calculated that the 'nonsolvational coordinated water' near ions is not totally incompressible, and therefore, to preserve the concept of an incompressible sphere, they did not consider this water as part of the hydration sphere. Furthermore, because of the electroneutrality law, one is restricted to determining changes in compressibility due only to the addition of salts and not to individual ions. Hence, the calculated hydration number may represent either the sum of two ionic values or some average value because of compensation effects.

SAMOILOV (1957, 1965, 1972) has developed a model for aqueous electrolyte solutions in which the ion–water interaction is described in terms of positive hydration and negative hydration. In his model the equilibrium exchange time, $\tau_i$, for water near an ion is compared with the equilibrium exchange time, $\tau$, for water in pure water. $\tau_i$ and $\tau$ were calculated from measurements of the self-diffusion of water. An ion firmly binds the nearby water molecules if the ratio $\tau_i/\tau$ for a particular ionic solution is high. This type of hydration is called positive hydration. But for a solution in which $\tau_i/\tau$ is less than one, the water molecules near the ions exchange more frequently with the bulk water than with the ions because the ions have in some way caused a weakening of the hydration bond. This type of hydration is called negative hydration. SAMOILOV (1957) calculated $\tau_i/\tau$ for the alkali and alkaline earth ions, and his calculations indicate that small or multiply charged ions are positively hydrated while larger singly charged ions are negatively hydrated. Samoilov's negative hydration concept implies a disruption of the water structure near large ions. This is not inconsistent with the Frank and Wen proposal of a significant B region for larger ions in which the water structure is believed to be broken and in a state of disorder. That the Frank and Wen model and the Samoilov model should lead to a common outcome is not too surprising because both models were derived, in part, from transport data (i.e. viscosity and diffusion).

The preceding discussion suggests that the interpretation of compressibility data in terms of the simple Passynski hydration number model may have to be modified to take account of the three-zone models described. These models for water near an ion do not detract from the practical usefulness of the compressibility approach for obtaining some ideas on how ions influence water structure. It is experimentally observed that a soluble salt added to an aqueous solution reduces the compressibility of the solution, and it can be shown by calculation that the decrease in

compressibility is much greater than one would expect when taking into account that a small volume of the water has been replaced by an equal volume of a salt with negligible compressibility at pressures of oceanic interest. It is possible that either (1) the salt acts in some way to make a portion of the water in an aqueous solution structurally more rigid than the bulk water and consequently this portion of water structure is not further broken by hydrostatic pressure, as is the bulk water in the solution, or (2) the salt acts to alter the kinetic behavior of the water near an ion.

In the discussions to follow I have considered the question of water structure in seawater by determining experimentally both (1) how dissolved single salts affect the compressibility of water and (2) how the major sea salts of seawater affect the compressibility of water in seawater. The compressibility approach to the problem is of course analogous to that taken by Passynski for determining hydration numbers, with the distinction that the results will be interpreted (1) in terms of a method for determining hydration numbers and (2) in terms of ion–water interactions. By comparing values of $\Delta\beta$ determined for salts of a common ion, it should be possible to compare the extent to which ion types influence water structure. The overall effect which sea salt ions have on the structure of water in seawater can be determined by comparing the values of $\Delta\beta$ for seawater with those for single salt systems. Most of the major ions in seawater also form ion-pairs to some extent, and the question of the compressibility of ion-pairs and the effect of pressure on the dissociation of ion-pairs (e.g. $MgSO_4^0$ and $NaSO_4^-$) will have to be considered in the interpretation of the data.

## II. METHODOLOGY AND EXPERIMENTAL PROCEDURE

### A. Apparatus

A specially designed high pressure differential densimeter was constructed to measure $\Delta\beta$. The theory, design and fabrication, and operation of the densimeter, which I call a bellows-type differential compressimeter (BDC), has been described by DUEDALL and PAULOWICH (1973).

The BDC is schematically illustrated in Fig. 1 and consists of three main parts: (1) a pair of closely matched stainless steel bellows units which are mounted independently of each other but

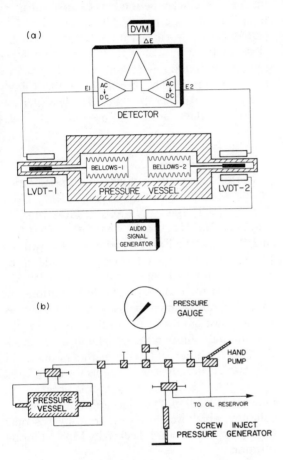

FIG. 1. (a) The bellows-type differential compressimeter; (b) the hydraulic system.

arranged back-to-back inside the same pressure vessel; (2) two closely matched electromechanical transducer assemblies; and (3) a specially constructed detector assembly. One bellows is filled with a reference solution and the other one is filled with a test solution; in our work the reference solution is seawater of salinity 35‰, and the test solution is the same as the reference solution except that it has been doped with an

additional amount of any one of the major sea salts. As hydrostatic pressure is applied on the bellows assembly, the resulting changes in compressed volumes of the solutions in the bellows are determined by sensing the differential axial compression of the bellows using linear variable differential transformers (LVDT). The core of each LVDT is mechanically linked to each bellows. The LVDT windings (primary and secondary), however, are situated outside the pressure vessel and are not subjected to hydrostatic pressure. The output voltages from the LVDT are fed into an operational amplifier which functions to rectify the voltage signals from the transducers and to subtract the two signals; this difference voltage is directly related to the difference between the compressibilities of the two solutions.

The advantage of this apparatus over the single bellows type is that the difference voltage of the BDC gives a direct measure of the effect of a salt on the compressibility of the reference solution. This effect is small, and variations in the compressibilities of solutions containing similar salt types are small. It is these latter changes that we are measuring. Use of the BDC eliminates the problem of exact replication of experimental conditions (i.e. pressure, temperature, electronics, etc.) which if not precisely duplicated when using a single bellows densimeter would mask the small compressibility changes that we measure.

## B. *Experimental*

The compressibility experiments were conducted in the following manner. An initial series of experiments was carried out to determine $\Delta\beta$ for systems in which the reference water was ordinary water (i.e. distilled water). The salts tested were the sea salts NaCl, $Na_2SO_4$, $MgSO_4$, and $MgCl_2$. The second series of experiments consisted of runs in which the reference solution was seawater and the salts tested were the same (with some exceptions as explained below) as those tested in the ordinary water runs. The $\Delta\beta$ measurements for the ordinary water provide the basis for a comparison of the water in seawater with ordinary water. The water in seawater is influenced by the background presence of sea salt ions; therefore, a comparison of the $\Delta\beta$ determined for a particular salt added to ordinary water with the $\Delta\beta$ determined for the same salt tested in seawater allows one to assess the extent to which the bulk water in seawater is modified by the presence of sea salt ions. The $\Delta\beta$ values in these two series of runs were determined at nominal concentrations of $0.13\,m$ ($m$ = molal) and $0.26\,m$ at temperatures of $2°C$ and $15°C$.

As mentioned in the introduction, hydration depends in some way, or combination of ways, on ion size, charge, or type. A study was therefore conducted to determine the effect of these parameters on $\Delta\beta$. In this series of experiments the reference solution was ordinary water. (The experiments could not be carried out for a seawater system because the addition of some of the salts to seawater would have caused the precipitation of insoluble salts.) These experiments were carried out at $15°C$, and the concentration of the salt added to water was nominally $0.26\,m$. The ion size effect for cations was estimated from determinations of $\Delta\beta$ for the group IA chlorides LiCl, NaCl, KCl, and CsCl. Similarly, the ion size effect for anions was estimated from determinations of $\Delta\beta$ for the series of group VIIA halides NaF, NaCl, and NaI. To determine the effect of ion charge on $\Delta\beta$, runs were made for the group IIA chlorides $MgCl_2$, $CaCl_2$, and $BaCl_2$. The $\Delta\beta$ values determined from these runs can be compared with the values obtained for selected group IA chlorides whose cation sizes are about the same as the group IIA cations.

## 1. The reference solutions

*Ordinary water runs.* The reference solution was once-distilled water which was recirculated through Millipore® ion exchange columns ('Super Q' system) until the conductivity of the delivered water was less than $18\,M\Omega^{-1}\,cm^{-1}$. This water was also used in preparation of the test solutions for this series of runs.

*Seawater runs.* The reference solution in these runs was Sargasso seawater collected at a depth of about 2000 meters, using Niskin water bottles, during a Bedford Institute of Oceanography cruise (Cruise Report No. 71-016 and 71-029). It was filtered through a $0.45\,\mu m$ Millipore® filter,

and a small amount of distilled water added to adjust salinity to nearly $35‰$ (final salinity = $35.007 \pm 0.005‰$). As a check on the relative composition of the seawater, a calcium and magnesium determination was carried out using the EDTA method reported by CULKIN and COX (1966). The Ca/Mg mass ratio was 0.3179, which is close to the value of 0.3177 determined for a sample of IAPSO (Copenhagen water) salinometer seawater. (The precision of the Ca/Mg analysis is $\pm 0.0002$ ratio units.) The seawater was stored in a glass carboy and no preservative was added.

## 2. The test solutions

The following salts were used in the preparation of the test solutions: LiCl, NaCl, KCl, CsCl, $MgCl_2$, $CaCl_2$, $BaCl_2$, $Na_2SO_4$, $K_2SO_4$, $MgSO_4$, NaF, and NaI. These salts were Fisher 'Certified' and 'ACS Certified' grade, and were used without purification. The test solutions were made up by weight at nominal concentrations of $0.13\,m$ and $0.26\,m$. For the ordinary water test solutions the molal ($m$) unit assumes its usual meaning, that is, moles of test salt per kg of distilled water. For the seawater test solution the molal unit refers specifically to the moles of test salt per kg of the water in seawater. Because the solutions were prepared by weight, the concentrations are accurate to better than $0.001\,m$. The concentrations of the test solutions which were prepared from hygroscopic $CaCl_2$ and $MgCl_2$ were checked by Mohr chloride titration.

The reference and test solutions were degassed prior to filling the bellows. The procedure as explained by DUEDALL and PAULOWICH (1973) involved placing a beaker, containing about 400 ml of solution, inside a vacuum desiccator and then pulling a vacuum on the desiccator for about 3 min while the solution was stirred with a magnetic stirrer.

## III. EXPERIMENTAL RESULTS

### A. Calculation of $\Delta\beta$ and $\beta_{ref.}$

The $\Delta\beta$ values were calculated from the $\Delta v$, $\Delta V_{ref.}$, $\Delta V_{test}$ and $V_b^1$ data using the following equation:

$$\Delta\beta = -\frac{1}{V_b^1 + \Delta V_{test}}\left(\frac{\partial \Delta v}{\partial P}\right)_T$$
$$+ \frac{\Delta v}{(V_b^1 + \Delta V_{test})(V_b^1 + \Delta V_{ref.})}\left(\frac{\partial \Delta V_{ref.}}{\partial P}\right)_T. \quad (7)$$

$\Delta\beta$ equals $\beta_{test}$ minus $\beta_{ref.}$, and the derivation of eqn. (7), along with definitions of the experimental quantities $\Delta v$, $\Delta V_{ref.}$, $\Delta V_{test}$, and $V_b^1$, are presented in a paper by DUEDALL and PAULOWICH (1973). The calculated values are negative quantities and represent a decrease in the compressibility of the test solution (relative to the reference solution) due to the effects caused by the presence of a test salt. In the derivation of eqn. (7) (see eqn. (2) through eqn. (3) in DUEDALL and PAULOWICH (1973)), the volume of the reference bellows was made equal to the volume of the test bellows (i.e. $V_{ref.}^1 = V_{test}^1$). This is true when the compressibility runs are carried out at the same temperature at which the bellows are filled. However, this was never the case in this work because the measurements were conducted at either 15°C or 2°C while the filling was done at room temperature which ranged from 23 to 26°C. In the most extreme case, where the composition of the test and reference solutions differed by 0.26 m and the compressibility run was conducted at 2°C, the differential volume contraction effect would cause only a $0.001 \times 10^{-6}$ bar$^{-1}$ change in $\Delta\beta$, which is about thirty times smaller than the precision of the $\Delta\beta$ measurement and can therefore be neglected as a source of uncertainty.

A tabulation of the $\Delta\beta$ values at pressures to 900 bars is presented in Appendix A.

$\beta_{ref.}$ is required to calculate $N_w$, the hydration number (see eqns. (5) and (6)). $\beta_{ref.}$ was calculated from the compressibility data for seawater and for ordinary water using the equation

$$\beta_{ref.} = \frac{-1}{V_b^1 + \Delta V_{ref.}}\left(\frac{\partial \Delta V_{ref.}}{\partial P}\right)_T. \quad (8)$$

A tabulation of the $\beta_{ref.}$ values is given in Appendix B.

B. *The variation of $\Delta\beta$ with pressure, temperature, and test salt concentration*

Figures 2 through 7 show the variation of $\Delta\beta$ with pressure, temperature, and test salt concen-

tration. The figures show the following: $\Delta\beta$ becomes less negative with increasing hydrostatic pressure (Figs. 2 and 3); $\Delta\beta$ becomes more negative with decreasing temperature (Figs. 4 and 5); and, $\Delta\beta$ becomes more negative with increasing test salt concentration (Figs. 6 and 7). Figures 2 through 7 will be referred to in later sections.

C. *Calculation of the apparent molal compressibility*

The $\Delta\beta$ values in this work may be used to determine the apparent molal compressibility ($\phi_K$) of salts. $\phi_K$ at atmospheric pressure (1 bar $\approx$ 1 atm) is calculated using the following equation (HARNED and OWEN, 1958):

$$\phi_K = \frac{1000}{md_{ref.}}\Delta\beta + \beta_{test}\,\phi_V. \quad (9)$$

In eqn. (9), $m$ is the molality of the test solution, $d_{ref.}$ is the density of the reference solution at atmospheric pressure, $\beta_{test}$ is the compressibility of the test solution at atmospheric pressure, and $\phi_V$ is the apparent molal volume of the test salt at atmospheric pressure. From an oceanographic standpoint, $\phi_K$ is an important thermodynamic quantity because it is the pressure derivative of the apparent molal volume. Apparent and partial molal volume changes have been used to calculate the effect of hydrostatic pressure on ionic equilibria that occur in the ocean (OWEN and BRINKLEY, 1941). However, the volume changes used in such calculations are usually approximations because $\phi_K$ is neglected for lack of data (MILLERO and BERNER, 1971; DUEDALL, 1972).

The concentration dependence of $\phi_K$ follows the equation (HARNED and OWEN, 1958; CONWAY and VERRALL, 1966; MILLERO, 1973):

$$\phi_K = \phi_K^0 + S_K\sqrt{C} \quad (10)$$

where $\phi_K^0$ is the apparent molal compressibility at infinite dilution. $S_K$ is an empirical constant that relates the concentration dependence of $\phi_K$ to the square root of the molar concentration, and $C$ is the molar concentration. MASSON (1929) was the first to use an empirical equation relating $\phi_V$ (partial molal volume) to $\sqrt{C}$. Later, GUCKER (1933a) used the same type of equation to relate

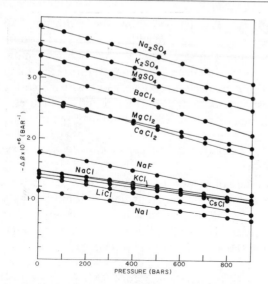

FIG. 2.   $\Delta\beta$, at 15°C, as a function of hydrostatic pressure. The reference solution is ordinary water, and the test salt concentration is 0.26 m.

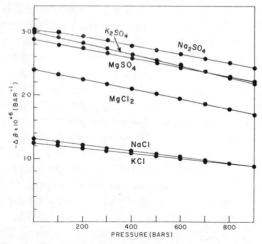

FIG. 3.   $\Delta\beta$, at 15°C, as a function of hydrostatic pressure. The reference solution is seawater (35‰), and the test salt concentration is 0.26 m.

$\phi_K$ to concentration; in a subsequent paper, GUCKER (1933b) derived a theoretical expression, based on the Debye–Hückel Theory, which predicted $\phi_K$ would vary with $\sqrt{C}$.

Values of $\phi_K^0$ have been published (HARNED and OWEN, 1958; OWEN and SIMONS, 1957; and OWEN and KRONICK, 1961), but they cannot be directly compared with apparent molal com-

pressibilities calculated from the $\Delta\beta$ values of this work because (1) the $\phi_K$ values calculated from $\Delta\beta$ cannot be accurately extrapolated to infinite dilution ($\Delta\beta$ measurements were made at two concentration points) and (2) most of the published $\phi_K^0$ values were calculated from data collected at 25°C ($\Delta\beta$ was measured at 15° and 2°C). MILLERO (1973), however, has reported single ion values of $\phi_{K(i)}^0$ and $S_{K(i)}$ which cover the temperature range 0–40°C. Therefore, $\phi_K^0$ and $S_K$ for salts common to this work were calculated from Millero's work, and from these values, $\phi_K$ (for ordinary water as the reference solution) was calculated, using eqn. (10), for NaCl, KCl, NaF, $MgCl_2$, $CaCl_2$, $Na_2SO_4$, $K_2SO_4$, and $MgSO_4$ at the nominal concentration 0.26 m. (Millero reports $S_K$ in units of volume ionic strength ($I_V$); therefore, in eqn. (10), $C$ was replaced by $I_V$.) Table 1 shows a comparison of the $\phi_K$ calculated from Millero's data and those derived (eqn. 9) from some of the measurements reported in this paper. In the calculation of $\phi_K$ from eqn. (9), the data for $\phi_V$ were taken from MILLERO's (1972) recent compilation of $\phi_V^0$ values. The $\phi_K$ values calculated from the $\Delta\beta$ measurements are systematically higher than those calculated from Millero's work. Also, there is significant disagreement in the $\phi_K$ values for NaF and $CaCl_2$. Millero, however, points out that his $\phi_{K(Ca^{2+})}^0$ and $\phi_{K(F^-)}^0$ values at 15°C are not based on direct experimental evidence at 15°C but on the temperature dependence of $\phi_{K(i)}^0$ and $S_{K(i)}$ for $Mg^{2+}$, $Na^+$, and $F^-$. Hence this lack of direct data may account for some of the difference shown in the NaF and $CaCl_2$ results. OWEN and KRONICK (1961) measured $\phi_K^0$ for NaCl and KCl using a sound velocity method. Therefore, as an additional check on the accuracy of $\phi_K$ determined from $\Delta\beta$ measurements, values of $\phi_K$ were calculated from eqn. (10) using the $\phi_K^0$ results of Owen and Kronick. Owen and Kronick did not report a $S_K$; therefore Millero's $S_K$ values were used. The $\phi_K$ derived from Owen and Kronick's data is shown in brackets (Table 1), below Millero's results. From this comparison it appears that the results of Owen and Kronick are also systematically high when compared with Millero's data. Owen and

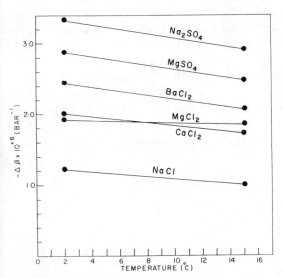

FIG. 4. $\Delta\beta$, at 900 bars, as a function of temperature. The reference solution is ordinary water, and the test salt concentration is 0.26 $m$.

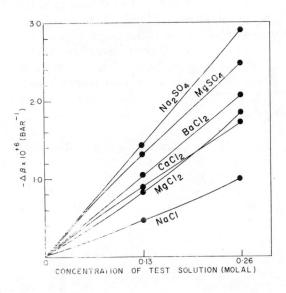

FIG. 6. $\Delta\beta$, at 900 bars and 15°C, as a function of test salt concentration. The reference solution is ordinary water.

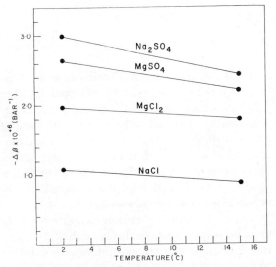

FIG. 5. $\Delta\beta$, at 900 bars, as a function of temperature. The reference solution is seawater (35‰), and the test salt concentration is 0.26 $m$.

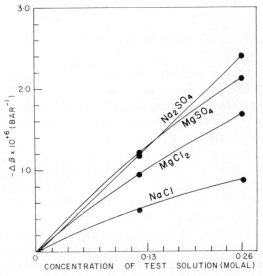

FIG. 7. $\Delta\beta$, at 900 bars and 15°C, as a function of test salt concentration. The reference solution is seawater (35‰).

Kronick mention that they used approximate $\phi_V$ values for NaCl and KCl to calculate $\phi_K$ from eqn. (9). However, it is unlikely that the use of uncertain $\phi_V$ values is the major source of the difference between Millero's results and those of Owen and Kronick because the $\phi_V$ values that Owen and Kronick used would have had to be too high by 75 and 30% for NaCl and KCl, respectively.

## IV. DISCUSSION OF $\Delta\beta$ RESULTS

In this section my aim is to relate the compressibility effects in seawater to the structural alterations brought about by sea salt ions. It has already been shown by calculation (PADOVA, 1963; DESNOYERS, VERRALL and CONWAY, 1965; WHALLEY, 1963) that the effective compressibility of the water in the primary hydration zone of small cations is small compared to the compressi-

Table 1.   Apparent molal compressibilities ($\phi_K$) of salts in ordinary water at 15°C

| Salt | Concentration $mol\,l^{-1}$ | (1) This work[a] | $\phi_K \times 10^{+4}$ $ml\,mol^{-1}\,bar^{-1}$ (2) Millero[b] | (2)−(1) |
|------|------|------|------|------|
| NaCl | 0.266 | −49.0 | −53.2 ($-48.7$)[c] | −4.2 |
| KCl | 0.266 | −43.1 | −46.3 ($-41.3$)[c] | −3.2 |
| NaF | 0.267 | −69.1 | −79.1 | −10.0 |
| MgCl$_2$ | 0.265 | −93.7 | −97.1 | −3.4 |
| CaCl$_2$ | 0.262 | −94.0 | −81.5 | −12.5 |
| Na$_2$SO$_4$ | 0.266 | −140 | −141.1 | −1.1 |
| K$_2$SO$_4$ | 0.265 | −123 | −126.5 | −3.5 |
| MgSO$_4$ | 0.268 | −133 | −133.1 | −0.1 |

[a] Calculated from eqn. (9).

[b] Calculated from eqn. (10) using MILLERO's (1973) $\phi_K^0$ and $S_K$ values. Millero (personal communication) says that $\phi_K$ values computed from eqn. (10) are reliable, at low concentrations, to $\pm 5 \times 10^{-4}\,ml\,mol^{-1}\,bar^{-1}$.

[c] Based on $\phi_K^0$ values reported by OWEN and KRONICK (1961).

In summary, an explanation for the systematic difference which exists between Millero's results and mine has not been found. The precision of a single $\Delta\beta$ determination is about $\pm 0.05 \times 10^{-6}$ bar$^{-1}$, which is equivalent to determining $\phi_K$ to $\pm 2.0 \times 10^{-4}\,ml\,mol^{-1}\,bar^{-1}$. The actual $\Delta\beta$ values in this work may be high by an average amount ranging from $-0.10 \times 10^{-6}$ to $-0.13 \times 10^{-6}\,bar^{-1}$ (DUEDALL and PAULOWICH, 1973). However, the very recent results for $\beta_{sw} - \beta_w$ reported by EMMETT and MILLERO (1974) would suggest that the $\Delta\beta$ values reported here are probably accurate to $\pm 0.07 \times 10^{-6}\,bar^{-1}$ which is in agreement with the recent error analysis performed by CHEN and MILLERO (1976). This error should be kept in mind if $\Delta\beta$ values from this work are used to calculate accurate thermodynamic quantities.

bility of the bulk water. Therefore, as a crude first approximation, it can be assumed that the water molecules very near ions have zero compressibility while all others have unchanged compressibility; hence, $\Delta\beta$ is proportional to that amount of water which is no longer involved in the compressibility of a solution. (As will be shown in section D, a portion of the so-called incompressible water is compressible.)

Anions probably do not interact with water in the same way that cations do (as we shall see later). It is fundamentally not possible to separate the compressibility effect caused by an anion from that caused by a cation, and it is probably fruitless to try to gain any absolute information from a $\Delta\beta$ value alone. However, some useful (but relative) information can be gained by comparing compressibility effects within a homologous series of salts which have common anions and cations. In the following paragraphs, under the heading 'Cation Effect', the effects of different alkali-metal chlorides and alkaline-earth chlorides on compressibility are compared and discussed. Similarly, this will be followed by a discussion on 'Anion Effect' in which the effects of some sodium halides and alkali-metal sulfates are compared. At the close of this section, I will compare the so-called hydration number determined by the compressibility method with that determined by other methods. It will be seen that the hydration number of a particular salt, or ion, is a reflection of the method used to determine hydration.

## A. *Cation Effect*

Figures 2 and 3 show that similar salt types produce similar compressibility changes. This behavior is especially true for the alkali-metal chlorides, where it can be seen that LiCl produces about the same decrease in compressibility as does CsCl in spite of the fact that the crystallographic radii of the two cations differ by almost a factor of three. The $\Delta\beta$ for $MgCl_2$ and $CaCl_2$ are about twice the values for the alkali-metal chlorides. If these data are considered on an equivalent concentration basis instead of a mole basis (to normalize the chloride concentration effect), then $0.5 \times \Delta\beta$ for $MgCl_2$, $CaCl_2$, or $BaCl_2$ is about the same as the $\Delta\beta$ value for each of the alkali-metal chlorides. This concentration effect is shown in the results presented in Fig. 6 which is a plot of $\Delta\beta$, at a pressure of 900 bars, versus concentration. Here it can be seen that the $\Delta\beta$ values for $0.13\,m$ $MgCl_2$, $0.13\,m$ $CaCl_2$, and $0.13\,m$ $BaCl_2$ are all about the same as the $\Delta\beta$ for $0.26\,m$ NaCl. Thus one mole of $Na^+$ added to water produces about the same decrease in compressibility as one-half mole of $Ca^{2+}$. In this particular case, $Ca^{2+}$ and $Na^+$ have about the same crystallographic radii; however, even when the $\Delta\beta$ results of two cations whose radii are different are compared (i.e. $Cs^+$ versus $Mg^{2+}$) roughly the same effect on compressibility is observed. The $BaCl_2$ results appear slightly anomalous because they do not follow quite the same trend as does $MgCl_2$ and $CaCl_2$. (It may be

that $Ba^{2+}$ has an optimum ionic radius, and that a combination of its size and double charge enables $Ba^{2+}$ to accommodate additional water in its hydration structure, thus producing a slightly greater compressibility effect.) As might be expected, charge is an important parameter in the hydration of small cations; this has been shown in the work reported by QUIST and MARSHALL (1968). They found that the net change in waters of hydration for the ionization of alkali metal chlorides (excepting LiCl) in a water–dioxane system was about one-half that found for divalent metal sulfates in the same solvent. Before making these compressibility measurements, I would have expected $\Delta\beta$ to be roughly inversely proportional to ionic radius because large cations should have a lower charge density (and hence a smaller electric field at the closest approach distance) than small cations, and therefore the rearranging and restructuring of nearby water by large cations would be less than that by small cations. Such is not the case, at least for the alkali-metal and alkaline-earth cations. In the Frank and Wen model it is suggested that cations smaller or more highly charged than $K^+$ are net structure makers, that is, region A of their model containing restructured water predominates over region B. In their scheme, $K^+$ through $Cs^+$ are classed as structure breakers; for these cations, region B becomes important although region A still exists.

Assuming that the $\Delta\beta$ I measure for the homologous series of alkali-metal halides is interpretable in terms of the degree to which ions influence some portion of the nearby bulk water, and considering that for $Li^+$ and $Na^+$ this portion must be the nearest water (i.e. region A water), it becomes clear, as suggested by BOCKRIS (1949), that the compressibility method for studying hydration of cations must give a measure of primary hydration for such small single charged cations as $Li^+$ and $Na^+$, and also for such large singly and doubly charged cations as $K^+$ and $Cs^+$, $Mg^{2+}$, $Ca^{2+}$, and $Ba^{2+}$. I can now infer, as ALLAM and LEE (1966) have done, that the compressibility of water in region B around larger cations such as $Cs^+$ and $K^+$ is probably not too different from the compressibility of the bulk water. (Later this will be shown in the calculation of the compressibility of the hydration sphere.) Otherwise, $\Delta\beta$ for these ions would be significantly smaller than the value for a strong structure maker such as $Li^+$, where region A is supposed to predominate over region B. This finding eliminates the problem of having to assume a low compressibility value for region B. (According to FRANK and WEN (1957), region B has a disordered and broken water structure and therefore one might expect region B to have negligible compressibility because of the absence of the void spaces which occur in normally structured (tetrahedral) water.) It now appears, however, that either the compressibility of region B water is not too different from that of the bulk water in an aqueous solution, or that region B for larger cations is not a significant feature. Therefore the compressibility method is probably only accurate for the study of such cation-water interactions as primary hydration.

## B. Anion Effect

Figures 2 and 3 also show that the sulfate salts affect compressibility more than the chlorides and other halide salts. The interpretation of the sulfate results is, however, not straightforward for several reasons. First, $SO_4^{2-}$ by virtue of its tetra-oxy structure and its doubly negative charge would not be expected to interact with water in the same way as $Cl^-$ does. Second, $SO_4^{2-}$ forms ion-pairs with cations and it is usually assumed that $Cl^-$ does not form ion-pairs in solutions whose compositions are similar to that studied in this work. The structure of the water near or within sulfate metal ion-pairs may be very different from that near free $SO_4^{2-}$. That sulfate salts produced the greatest effect on compressibility must be due in some way to a relatively large interaction of the sulfate ion with neighboring water molecules. Nearby water may hydrogen bond with the oxygen atoms of the sulfate ion; it is known that hydrogen bonds are present in the crystalline hydrates of sulfate salts (FALK and KNOP, 1973). SAMOILOV (1972), however, implies that the sulfate ion does not bond appreciably with water. He bases his reasoning on the reported

shift in the IR absorption spectra of the hydrated salts of sodium perchlorate and sodium sulfate. Apparently when these hydrates are dissolved in water, the spectra indicate hydrogen bonding undergoes a shift in the direction of decreased hydrogen bonding. SUBRAMANIAN and FISHER (1972) also concluded (partly on the basis of the work reported by Samoilov) that perchlorate does not hydrogen bond with water in solution. However, such an interpretation may not be entirely correct. The small shifts of the sulfate (and perchlorate) bands may be due to causes other than the breaking of $OH \cdots$ sulfate bonds. KARYAKIN and MURADOVA (1968) have shown that shifts in the IR spectra could be due to the breaking of $H_2O \cdots H_2O$ bonds, which are postulated (by Karyakin and Muradova) to occur in a hydrate, rather than the breaking of $OH \cdots$ sulfate bonds when a salt is dissolved in water. The calculated (Karyakin and Muradova) bond energy of the $H_2O \cdots H_2O$ bond in a hydrate is less than the bond energy of the $OH \cdots$ anion bond.

The effect of different halides on the compressibility of water is shown (Figs. 2 and 3) in the comparison of the $\Delta\beta$ results for NaF, NaCl, and NaI. The crystallographic radii of the anions of these salts varies from 1.36 Å for $F^-$ to 2.16 Å for $I^-$. There seems to be a trend in the data showing that the smaller the halide the greater the effect on compressibility, indicating that smaller halides such as $F^-$ or $Cl^-$ interact a bit more strongly with the water than does a larger halide ion such as $I^-$.

As mentioned in the preceding discussion of cations, the charge on a cation appeared to play the important role in determining $\Delta\beta$ for a series of salts containing a similar anion. But in the case of anions, that is, in a series of salts containing a similar cation, size as well as charge has to be considered in the interpretation of $\Delta\beta$.

There is plenty of experimental evidence (summarized by KAVANAU, 1964) to show that anions do not interact with water in the same way as do cations. It is believed that anions break down water structure, while cations reinforce water structure. (Actually, cations also break down water structure, but the subsequent reorientation of the water by the cation is thought of as structure making.) These effects are commonly referred to as structure breaking and structure making. The terms structure breaking and structure making originated in the work of FRANK and EVANS (1945) and of FRANK and WEN (1957). According to the work of Frank *et al.*, certain solutions (e.g. 0.1 *m* CsCl) have entropies which are in excess of those calculated from the assumptions of an 'immobilized' primary hydration sphere and polarization of distant water. Such solutions contain ions which are called structure breakers. For other solutions (e.g. 0.1 *m* LiF), however, the observed entropies are not too different from those calculated; these solutions contain ions which are called structure makers. The problem with defining structure breaking and structure making in terms of thermodynamic properties is that such definitions provide very little insight into the mechanisms of structure breaking and structure making. The mechanisms are important and will be discussed in section V. GREYSON's (1967) work, however, is worth mentioning now since it very nicely contrasts the difference, with respect to structure breaking, between anions and cations. Greyson used an electro-chemical technique to determine the entropy of transfer of a salt from $D_2O$ to $H_2O$. He found that the values for a series of halide ions ($F^-$, $Cl^-$, $Br^-$, $I^-$) varied considerably more than the $\Delta S$ for a series of alkali-metal ions ($Li^+$, $Na^+$, $K^+$). Greyson's inference was that anions, which in general are larger than cations, are more easily polarized by water than are cations and that the polarized anion leads to a water-anion bond which is less compatible with the surrounding water structure.

## C. *The Effect of Sulfate Speciation on* $\Delta\beta$

Figure 3 showed a plot of $\Delta\beta$ versus hydrostatic pressure for seawater runs ($\Delta\beta_{sw}$). The trends (i.e. grouping of salt types) in these results are similar to those in Fig. 2 for the ordinary water runs ($\Delta\beta_w$). The main difference between $\Delta\beta_{sw}$ and $\Delta\beta_w$ is that the $\Delta\beta_{sw}$ data are *less* negative than the $\Delta\beta_w$ data, thus indicating that the compressibility of seawater is less affected by

the addition of a test salt than is ordinary water. This difference is especially noticeable for the sulfate data; this effect is shown in Fig. 8 where a plot of $[\Delta\beta_{sw} - \Delta\beta_w]$ versus pressure is presented. The displacement in the two sets of data could be interpreted either in terms of the added salt in seawater competing for the less available free (not hydrating) water in the seawater matrix, therefore making $\Delta\beta$ less negative, or in terms of some ion-pair effect. In the seawater runs, this ion-pair effect may be important in the sulfate salt test solutions and may act to make $\Delta\beta$ lower (less negative) than if ion-pairing were not important in seawater because either some hydrated water is released when an ion-pair is formed, or more important, because of the possible effect on water structure due to the partial or total charge cancellation when an ion-pair is formed. Ion-pairing is an important consideration in the behavior of seawater, especially with respect to $NaSO_4^-$ and $MgSO_4^0$ ion-pairs whose concentrations in seawater have been discussed and argued in recent literature (KESTER and PYTKOWICZ, 1970; MILLERO, 1971; FISHER, 1972; and PYTKOWICZ, 1972). The existence of $MgSO_4^0$ and $NaSO_4^-$ ion-pairs in aqueous solutions has been substantiated by the work of EIGEN and TAMM (1962). MILLERO and MASTERTON (1974) have recently reported the values for the apparent molal volume of $MgSO_4^0$. Therefore, it is not unreasonable to assume that some of the effect shown in Fig. 8 is due to metal–sulfate ion-pairs. Because of the nearly complete dissociations of NaCl, KCl, and $MgCl_2$ in aqueous solutions (DAVIES, 1927; STOKES, 1945), the effects of metal–chloride ion-pairs would not be expected to be as great as those caused by metal–sulfate ion-pairs.

The concentrations of the free and ion-paired sulfate species in 0.26 *m*(w) $Na_2SO_4$, 0.26 *m*(sw) $Na_2SO_4$, 0.26 *m*(w) $MgSO_4$, and 0.26 *m*(sw) $MgSO_4$ were kindly calculated by Dana Kester (private communication) to demonstrate the $SO_4^{2-}$ speciation in the different matrix. Kester used the association constants for $NaSO_4^-$ and $MgSO_4^0$ which were reported by KESTER (1970) KESTER and PYTKOWICZ (1970); the association constants were adjusted to take into account

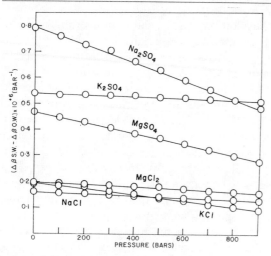

FIG. 8. $[\Delta\beta(\text{seawater}) - \Delta\beta(\text{water})]$ as a function of hydrostatic pressure. The differences in $\Delta\beta$ refer to 15°C and the test salt concentration is 0.26 m.

$[\text{SO}_4^{2-}]_w^{Na}$ = free sulfate in 0.26 m(w) $\text{Na}_2\text{SO}_4$

$[\text{SO}_4^{2-}]_w^{Mg}$ = free sulfate in 0.26 m(w) $\text{MgSO}_4$

$[\text{SO}_4^{2-}]_{sw}^{Na}$ = free sulfate in 0.26 m(sw) $\text{Na}_2\text{SO}_4$

$[\text{SO}_4^{2-}]_{sw}^{Mg}$ = free sulfate in 0.26 m(sw) $\text{MgSO}_4$

$[\text{NaSO}_4^-]_w^{Na}$ = sodium sulfate ion-pair in 0.26 m(w) $\text{Na}_2\text{SO}_4$

$[\text{MgSO}_4^0]_w^{Mg}$ = magnesium sulfate ion-pair in 0.26 m(w) $\text{MgSO}_4$

$[\text{NaSO}_4^-]_{sw}^{Na}$ = sodium sulfate ion-pair in 0.26 m(sw) $\text{Na}_2\text{SO}_4$

$[\text{MgSO}_4^0]_{sw}^{Na}$ = magnesium sulfate ion-pair in 0.26 m(sw) $\text{Na}_2\text{SO}_4$

$[\text{MgSO}_4^0]_{sw}^{Mg}$ = magnesium sulfate ion-pair in 0.26 m(sw) $\text{MgSO}_4$

$[\text{NaSO}_4^-]_{sw}^{Mg}$ = sodium sulfate ion-pair in 0.26 m(sw) $\text{MgSO}_4$

the increased ionic strength of the altered sea water composition (0.26 m in either $\text{Na}_2\text{SO}_4$ or $\text{MgSO}_4$). Table 2 summarizes the concentration of the various sulfate species. These results (Table 2) are expressed in two ways: (1) the concentration of free sulfate; and (2) the concentrations of the metal-sulfate ion-pairs. (The concentration unit is molal.) The following list gives the definitions, as they pertain to Table 2, of the sulfate species.

An ion-pair model for sulfate speciation shows that there are significant differences among the distributions of sulfate species in the seawater solutions and the distributions of sulfate species in the ordinary water solutions. It is, however, difficult to resolve the question of whether it is the free sulfate effects or the metal–sulfate ion-pair effects which influence the compressibility values. An interpretation based on an ion-pair effect is further complicated when one considers the unexplained convergence of the $\text{Na}_2\text{SO}_4$ and $\text{K}_2\text{SO}_4$ results shown in Fig. 8. The problem is to

Table 2. Concentrations (molal) of free sulfate ($[\text{SO}_4^{2-}]$) and $\text{NaSO}_4^-$ and $\text{MgSO}_4^0$ ion-pairs, at 15°C, in 0.26 m salt solutions

| P (Bars) | (1) Water $[\text{SO}_4^{2-}]$ | (2) Seawater $[\text{SO}_4^{2-}]$ | (3) (1)–(2) | (4) Water $[\text{NaSO}_4^-]$ $[\text{MgSO}_4^0]$ | (5) Seawater $[\text{NaSO}_4^-]$ | (6) $[\text{MgSO}_4^0]$ | (7) Unaltered $[\text{SO}_4^{2-}]$ | (8) $[\text{Na}^+]$ | (9) Seawater $[\text{Mg}^{2+}]$ | (10) $[\text{NaSO}_4^-]$ | (11) $[\text{MgSO}_4^0]$ |
|---|---|---|---|---|---|---|---|---|---|---|---|
| **Set A.** | $\text{Na}_2\text{SO}_4$ solutions: | | | | | | | | | | |
| 1 | 0.1312 | 0.0747 | 0.0565 | 0.1288 | 0.1888 | 0.0208 | 0.0107 | 0.4735 | 0.0480 | 0.0121 | 0.0058 |
| 201 | 0.1387 | 0.0810 | 0.0577 | 0.1213 | 0.1825 | 0.0208 | 0.0108 | 0.4742 | 0.0480 | 0.0114 | 0.0058 |
| 401 | 0.1462 | 0.0875 | 0.0587 | 0.1138 | 0.1759 | 0.0207 | 0.0116 | 0.4750 | 0.0481 | 0.0107 | 0.0057 |
| 601 | 0.1535 | 0.0943 | 0.0592 | 0.1065 | 0.1692 | 0.0206 | 0.0123 | 0.4757 | 0.0482 | 0.0100 | 0.0057 |
| 801 | 0.1607 | 0.1013 | 0.0594 | 0.0993 | 0.1624 | 0.0205 | 0.0131 | 0.4764 | 0.0483 | 0.0093 | 0.0056 |
| **Set B.** | $\text{MgSO}_4$ solutions: | | | | | | | | | | |
| 1 | 0.0931 | 0.0738 | 0.0193 | 0.1669 | 0.0947 | 0.1160 | | | | | |
| 201 | 0.1003 | 0.0789 | 0.0214 | 0.1597 | 0.0906 | 0.1149 | | | | | |
| 401 | 0.1078 | 0.0842 | 0.0236 | 0.1522 | 0.0865 | 0.1137 | | | | | |
| 601 | 0.1154 | 0.0897 | 0.0257 | 0.1446 | 0.0826 | 0.1122 | | | | | |
| 801 | 0.1232 | 0.0953 | 0.0279 | 0.1368 | 0.0785 | 0.1106 | | | | | |

learn what effect the individual ion-pairs ($MgSO_4^0$, $NaSO_4^-$) themselves have on $\Delta\beta$. As we shall see later, ion-pairs may be considered dipolar molecules with large dipole moments and therefore it is not unreasonable to assume that they will have some disruptive effect on the structure of water.

### D. *The Effect of Pressure on $\Delta\beta$*

Figures 2 and 3 show the effect of pressure on $\Delta\beta$. In all cases the result is that $\Delta\beta$ becomes less negative as pressure increases. $\beta_{test}$ and $\beta_{ref.}$ decrease with increasing pressure, but, as can be shown from the data in Appendix B, $\beta_{ref.}$ decreases more rapidly with increasing pressure than does $\beta_{test}$. One would expect this behavior because a small volume of the total test solution volume is occupied by a hydrated solute whose compressibility is small (but probably not negligible) compared to the water it replaced. The test solution is essentially precompressed due to the structure breaking effect of the test salt, and therefore it would undergo compression at a slower rate than the reference solution. This would have the effect of causing $\Delta\beta$ to increase (become less negative) with increasing pressure (as shown in Figs. 2 and 3), because $\beta_{test}$ would include a contribution due to the compressibility of the hydration sphere. According to BOCKRIS and SALUJA (1972), there is a portion of water near an ion which is not totally oriented and is left behind when an ion moves. They call this water the 'nonsolvational coordinated water (NSCW)'. Bockris and Saluja calculated that the compressibilities of the NSCW near $Na^+$ and $Cl^-$ are $7 \times 10^{-6}\,bar^{-1}$ and $30 \times 10^{-6}\,bar^{-1}$, respectively. As a check on Bockris and Saluja's results, the compressibility of the 'incompressible' hydration sphere was calculated from the $\Delta\beta$ values by assuming the hydration sphere has some finite compressibility ($\beta_{hyd.}$). The following equation (MOELWYN-HUGHES, 1954) may be used to relate $\beta_{hyd.}$ to $\beta_{test}$

$$\beta_{test} = \alpha_0\beta_{ref.} + \alpha\beta_{hyd.}, \qquad (17)$$

where $\alpha_0$ is the volume fraction of the free water in a solution, and $\alpha$ is the volume fraction of the hydrated water. $\Delta\beta$, as mentioned in section I, is

the difference between $\beta_{test}$ and $\beta_{ref.}$, i.e.

$$\Delta\beta = \beta_{test} - \beta_{ref.}. \qquad (18)$$

Substituting eqn. (17) into eqn. (18) gives

$$\Delta\beta = -\beta_{ref.} + (\alpha_0\beta_{ref.} + \alpha\beta_{hyd.})$$
$$= (\alpha_0 - 1)\beta_{ref.} + \alpha\beta_{hyd.} \qquad (19)$$

$\alpha = 1 - \alpha_0$, therefore eqn. (19) reduces to the following

$$\Delta\beta = -\alpha\beta_{ref.} + \alpha\beta_{hyd.}. \qquad (20)$$

A plot of $\Delta\beta$ versus $\beta_{ref.}$ will have a slope of $-\alpha$ and an intercept of $\alpha\beta_{hyd.}$; hence, $\beta_{hyd.}$ can be calculated. Figure 9 shows a plot of $\Delta\beta$ versus $\beta_{ref.}$ for the ordinary water results (0.26 m) at 15°C.

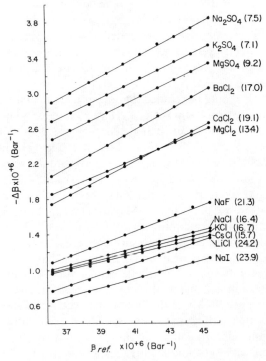

FIG. 9. $\Delta\beta$, at 15°C, as a function of $\beta_{ref.}$. The number in brackets is $\beta_{hyd.}$ ($\times 10^{-6}\,bar^{-1}$).

The $\beta_{hyd.}$ values ranged from $7.1 \times 10^{-6}\,bar^{-1}$ for $K_2SO_4$ to $24.5 \times 10^{-6}\,bar^{-1}$ for LiCl. It is probably not practical to attach much physical significance to differences in the individual $\beta_{hyd.}$ values, although the fact that the $\beta_{hyd.}$ values for the sulfate salts are systematically lower than the $\beta_{hyd.}$ values for the other salts may be indicative of a metal-sulfate ion-pair effect. However, the significant result of the calculation is that the hydration sphere is partially compressible. (In the hydration sphere I also include the coordinated water referred to by Bockris and Saluja.) The $\beta_{hyd.}$ for NaCl is $16.4 \times 10^{-6}$ bar$^{-1}$ which is in good agreement with the theoretical value of $18 \times 10^{-6}$ bar$^{-1}$ (average of Na$^+$ and Cl$^-$ values) calculated by Bockris and Saluja. Bockris and Saluja mention that the compressibility of the NSCW amounts to no more than 0.5 hydration units in the hydration number of ions. They point out that the NSCW cannot be counted as part of the hydration sphere because the NSCW maintains some translational freedom. Such a scheme appears rather arbitrary, especially when one considers it in relation to the Samoilov model in which even the water closest to an ion can exchange positions with the water outside the immediate hydration zone.

It should be recalled that the above calculation and interpretation of $\beta_{hyd.}$ is based on the assumptions that (1) an increase in pressure does not cause the dehydration of a hydrated ion in solution and (2) an increase in pressure does not result in ion-pair dissociation (which would subsequently lead to a net increase in hydration). However, according to MARSHALL'S (1972) work, the latter assumption is probably false. If it were demonstrated that pressure can cause dehydration, then the calculation of $\beta_{hyd.}$ would have to be based both on the volume change for the dehydration step and on the volume change due to the compressibility of that fraction of the ion–water complex which may not have undergone dehydration. Unfortunately, at the present time there appears to be no definitive answer as to whether or not increased pressure will dehydrate ions. (Conceptually, the dehydration due to the application of pressure may be viewed in terms of Samoilov's hydration concept: The residence time for a water molecule near an ion is less at high pressure than it is at low pressure.) Infrared and Raman spectroscopic methods which have been

adapted for examining electrolyte solutions under pressure would probably be very useful in studying the effects of pressure on ion hydration or ion–water interactions.

### E. *The Effect of Temperature on $\Delta\beta$*

Figures 4 and 5 show that the effect of increasing temperature is to make $\Delta\beta$ less negative. This cannot be interpreted in terms of a hydration change because, as shown from the data in Appendix B, $\beta_{ref.}$ decreases more rapidly with increasing temperature than does $\beta_{test}$. SAMOILOV's (1965) work, however, does indicate that an increase in temperature will increase the hydration of ions. The Walden product (HARNED and OWEN, 1958) varies little with temperature and thus it would seem that temperature has little affect on that part of hydration that is carried with an ion when it moves.

QUIST and MARSHALL (1968) have shown that the net *change* in hydration for an ion-pair dissociation reaction increases when temperature is increased (to 400°C), while, as just mentioned, the total hydration (as inferred from Samoilov's work) for dissolved species probably increases when temperature is increased. Therefore, based on the relationship in which the net hydration change (as determined by Quist and Marshall) equals the difference between the hydrations of the products and the reactants, it can be postulated that an increase in temperature has a greater effect in increasing the hydration of an ionic species than it does in increasing the hydration of an ion-pair.

It is often pointed out that decreasing temperature and increasing solute concentration produce similar effects on water structure. This concept comes from the classic work of BERNAL and FOWLER (1933) who coined the term 'structural temperature' to indicate some sort of fundamental feature of water structure. The structural temperature has been defined by NEMETHY (1970) in the following way: 'A solution at a given temperature, $T_s$, ought to be structurally equivalent to water at a different temperature $T_w$, and $\Delta T = T_s - T_w$ should be a measure of the extent of structural changes.' Such a definition is misleading. The fact that $\Delta\beta$ approaches zero as either temperature or solute addition increases does not mean (although it is sometimes implied) that temperature increase and solute addition alter water structure in the same way. Salts (i.e. ions) cause structural changes which are complicated in a different way (ion–dipole interactions) than the structural changes caused by temperature changes. The addition of a salt to water will cause some of the water structure to break up into unbounded water molecules (hydrogen bond breaking), but some of these newly formed monomers (single water molecules) may be partially or temporarily oriented towards the ions. The orientation process is, itself, dependent upon whether the ion is large or small, or positively or negatively charged. If the solution is already moderately concentrated, such as sea-water, the addition of further salt may cause ion-pairing which would probably have some effect on water structure because of the dipolar nature of ion-pairs.

## F. *The Hydration Number*

In this section the hydration number calculated from compressibility results will be compared with hydration numbers determined by other methods. By bringing hydration results from other experiments together with results from compressibility experiments, it may be possible to learn more about that portion of hydration actually measured by the compressibility method. A comparative approach, such as this, should be useful in understanding how ions alter water structure.

The compressibility hydration number, $N_w$, was calculated from eqn. (6); the $\Delta\beta$ and $\beta_{ref.}$ values used in the calculation are those given in Appendix A and B, respectively. Table 3 shows the $N_w$ values (1 bar), at $2°C$ and $15°C$, for salts in ordinary water and salts in seawater. Also shown in Table 3 are some recent literature values which were determined either by ultrasonic compressibility and/or isothermal compressibility methods (BOCKRIS and SALUJA, 1972; and MILLERO, WARD, LEPPLE and HOFF, 1974).

Most of the compressibility $N_w$ values, for a particular salt, are the average of $N_w$ (0.13 m) and $N_w$ (0.26 m). This averaging method was used because there were insufficient data ($\Delta\beta$ deter-

mined at two concentration points) to extrapolate the $\Delta\beta$ data to infinite dilution. Actually the scatter in the $\Delta\beta$ data for ordinary water was such that $N_w$ (0.13 m) differed from $N_w$ (0.26 m) by $\pm 0.9$ hydration units. Bockris and Saluja showed that $N_w$ (0.05 m) was, on the average, 0.6 hydration units greater than $N_w$ (0.10 m); hence, no serious

Table 3.  Hydration numbers (1 bar) computed from compressibility data

| Salt | $N_w{}^a$ | | $N_w{}^b$ |
|---|---|---|---|
| | 2°C | 15°C | 25°C |
| Reference solution: Ordinary water | | | |
| LiCl | | 6.4 | 5.3 |
| NaCl | 7.9 | 6.7,  6.6[c] | 6.5 |
| KCl | | 6.8 | 6.2 |
| CsCl | | 6.6 | 3.6 |
| NaF | | 8.3 | 8.5 |
| NaI | | 5.4 | 5.5 |
| MgCl₂ | 14.3 | 12.0,  12.3[c] | 12.3 |
| CaCl₂ | 13.8 | 12.3 | 11.1 |
| BaCl₂ | 16.0 | 15.4 | 13.5 |
| Na₂SO₄ | 21.9 | 18.6,  18.2[c] | |
| K₂SO₄ | | 16.7 | |
| MgSO₄ | 18.0 | 16.2,  17.3[c] | |
| Reference solution: Seawater (35⁰/₀₀) | | | |
| NaCl | 7.9 | 6.9 | |
| KCl | | 6.3 | |
| MgCl₂ | 14.5 | 13.1 | |
| Na₂SO₄ | 19.2 | 15.9 | |
| K₂SO₄ | | 15.3 | |
| MgSO₄ | 17.6 | 15.4 | |

[a] Computed from compressibility data of this work.
[b] Computed from compressibility data obtained from ultrasonic velocity determinations (BOCKRIS and SALUJA, 1972).
[c] Recently reported by MILLERO, WARD, LEPPLE and HOFF (1974); $N_w$ are infinite dilution values based on isothermal compressibility measurements.

error is made by averaging the $N_w$ (0.13 m) and $N_w$ (0.26 m) results. For the seawater data, $N_w$ (0.13 m) was on the average 0.8 hydration units greater than $N_w$ (0.26 m). The $N_w$ values of Bockris and Saluja, which are values in 0.1 m solutions, appear to agree very well with the results from this work. A direct comparison of the two sets of data, however, is not strictly valid because Bockris and Saluja's results are referenced to 25°C while the results of this work are 15°C values. Except for MgSO$_4$, the $N_w$ values of MILLERO, WARD, LEPPLE and HOFF (1974) are in excellent agreement with the work reported here.

Not much can be learned from the $N_w$ results as they stand in Table 3. This is, of course, because anions and cations alter water structure in different ways. The main trend of the results (Table 3) is that divalent salts are two to three times more 'hydrated' than monovalent salts. Also, it is interesting that the $N_w$ values for sea salts in ordinary water are about the same as the $N_w$ values for sea salts in seawater. This is to be expected because seawater is relatively dilute and thus ions are not competing for primary waters of hydration.

## 1. Comparison with previous data

A summary of hydration data from other methods is given in Fig. 10. The methods used to determine hydration are referenced on the vertical axis, and the hydration numbers are plotted on the horizontal axis. No attempt was made to normalize all data; the hydration numbers as they were reported are plotted in Fig. 10. The hydration observed depends on the method used. By bringing together several types of hydration data, it should be possible to sort out systematically the main features of hydration.

Hydration numbers determined from an equilibrium or bulk property of a solution give an indication of the number of water molecules whose physical properties have been modified by an ion. It provides no information as to the lifetime of the ion–water complex. Some bulk properties used to determine hydration are compressibility, partial molal volume, and vapor pressure.

Hydration numbers may also be determined from experiments which measure a dynamic or transport property of water or the ion in solution. These hydration numbers give an indication of the hydrated water which is bound firmly enough to an ion so that the water remains associated with the ion for a longer period of time than it remains associated with the bulk water. (This description

is essentially the one given by SAMOILOV (1965) for positive hydration.)

A classification of hydration based on the kind of measurement technique employed has already been given by AZZAM (1962). However, Azzam's scheme includes neither the recent NMR work on electrolyte solutions, nor Samoilov's positive-negative hydration concept. These two recent approaches, along with Raman studies of electrolyte solutions, are particularly important in

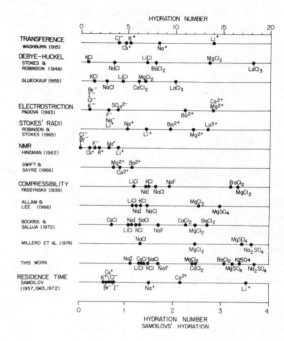

FIG. 10.    Hydration determined by different methods.

demonstrating that anions alter water structure differently than do cations. The following paragraphs are summaries of the hydration methods given in Fig. 10. From these summaries, it is hoped that a consensus can be reached as to the utility of the compressibility method as a way of determining what effects ions have on the structure of water.

(a) *Hittorf transference method: Washburn* (1915). The principle of this method is that ions in a transference cell will migrate to electrodes under the force of an external applied voltage. As the ions move, they will carry with them some portion of their hydrated water; this hydrated water will change the concentration of an inert reference compound (raffinose in Washburn's experiment). The main assumption in the method is that the inert compound does not migrate. However, BOCKRIS (1949) cites work which shows that the inert material used in such transference experiments is transported. LONGSWORTH (1947) also demonstrated, using a moving boundary technique, that some of the inert material moves in a Washburn experiment. However, as Longsworth points out, Washburn's results are not greatly affected by a moving reference compound. In principle, the Washburn method gives a measure of the number of water molecules bound firmly enough to an ion to overcome drag effects as the ion moves toward an electrode.

(b) *Activity coefficient calculations: Stokes and Robinson* (1948) *and Glueckauf* (1955). This method allows the calculation of the hydration number from an extended form of the Debye–Hückel equation. Stokes and Robinson (also ROBINSON and STOKES, 1965, p. 238–246) state that the total free energy of a solution must be the same regardless of whether it is assumed the components are hydrated ions plus water or unhydrated ions plus water. From this they derive a two-parameter extended form of the Debye–Hückel equation which relates the activity coefficient of an ion to the activity of the water, the ion size parameter, and the hydration number $N_w$. Stokes and Robinson then determine $N_w$ by back calculating from their Debye–Hückel equation to get a best agreement between the observed and calculated activity coefficient. Stokes and Robinson mention that $N_w$ does not represent the number of water molecules in the primary or first hydration layer, but that $N_w$ is an average effect of all ion-solvent effects, which may include water from outside the first hydration layer. They are '...inclined to accept the idea that it is the cations rather than the anions which are hydrated,...' but they also admit that such a statement contradicts their results which showed an increase in $N_w$ with increasing anion size, for a given cation. (Table 2 shows that $N_w$ decreases with increasing

anion (halide) size.) GLUECKAUF (1955) essentially used Stokes and Robinson's basic idea except that Glueckauf expressed his revised Debye–Hückel equation in terms of volumes of the hydrated ions instead of mole fraction, as used by Robinson and Stokes. Glueckauf was critical of the Stokes and Robinson treatment because it neglected co-volume interactions which are important at higher concentrations. Co-volume interactions are forces of attraction or repulsion, superimposed on the normal coulombic forces (those used in the derivation of the Debye–Hückel equation), due to overlapping of ion spheres in concentrated solutions (GURNEY, 1953). For the monovalent salts (except for LiCl), Glueckauf's hydration numbers are in fair agreement with those of Stokes and Robinson. For the divalent and trivalent salts, however, the agreement is not good. The reason may be, as Glueckauf points out, that Stokes and Robinson did not consider co-volume interactions. Such interactions could be important for multivalent cations.

In their calculations, Stokes and Robinson, and Glueckauf used activity coefficient data which were derived from isopiestic vapor pressure measurements. Glueckauf's results (Fig. 10) show that divalent salts are roughly twice as hydrated as monovalent salts.

The use of the extended form of the Debye–Hückel equation to account for hydration effects has been very successful in predicting activity coefficients in moderately concentrated solutions (ROBINSON and STOKES, 1965, Fig. 9.3). This would seem to imply that hydration numbers change very little with concentration. It would be interesting to see if the extended Debye–Hückel equation is as successful in predicting the activity coefficients of ions that are known to ion-pair extensively.

(c) *Electrostriction: Padova* (1963). In this method, the theoretical electrostriction per mole of water and the partial molal volume are used to calculate the hydration number. Padova assumed that the ion–water complex is a hard incompressible sphere. An assumed value for the partial molal volume of $I^-$ permits the calculation of hydration numbers for ions. The results showed that all halides (except $F^-$) have the same hydration number (hydration number of 1); the hydration numbers for the alkali-metal ions are roughly comparable to those reported by GLUECKAUF (1955) who used activity data to calculate hydration numbers.

Padova also calculated the effective radii ($r_e$) of ions in solution and then compared $r_e$ with the theoretical minimum radius ($r_o$) of an ion situated in a small sphere of dielectrically saturated water. If, according to Padova, $r_e < r_o$, then the ion is considered hydrated in the traditional sense. If, however, $r_e > r_o$, then the ion is a structure breaker. According to this scheme, $Cl^-$, $Br^-$, $I^-$, $Rb^+$, and $Cs^+$ are structure breakers. This approach to the problem of structure breakers is interesting, and I will return to it in section V.

(d) *Stokes' Law Radii: Robinson and Stokes* (1965). Robinson and Stokes calculated hydration numbers using a 'corrected' Stokes' law radii for ions. Stokes' law says that the velocity of a small sphere in an ideal hydrodynamic medium is proportional to the external force on the particle and inversely proportional to the radius of the sphere and to the viscosity of the medium. By assuming Stokes' law is applicable to ions in solution, the radii of ions in solution can be calculated from ionic mobility and viscosity data. Robinson and Stokes showed that the Stokes' law is obeyed for large ions, but is not for small ions. Small ions interact strongly with water and therefore one might expect some uncertainty in the calculated Stokes radii for small ions. (Also, water cannot really be considered an ideal liquid.) Robinson and Stokes obtained a correction factor for the Stokes' law radii from the calculated radii (from partial molal volume data) of tetra-alkyl ammonium ions. They assumed this correction could be applied to Stokes' radii of small ions. The hydration number was calculated as the ratio of the hydrated ion volume (calculated from the corrected Stokes' law radius) to the volume of ordinary nonelectrostricted water.

(e) *Compressibility: Passynski* (1938), *Allam and Lee* (1966), *Bockris and Saluja* (1972), *Millero, Ward, Lepple and Hoff* (1974), this work (1977). The compressibility method for determining hydration has already been described and discussed (see sections I and II). The main assumption in the method is that the compressibility of the water in the primary hydration zone is small compared to the compressibility of the bulk water. (This assumption is considered valid, according to the calculations of PADOVA (1963) and DESNOYERS, VERRALL and CONWAY (1965); although, as we have seen, a portion of the hydration sphere is compressible.) Therefore, the amount of water in the primary hydration zone can be calculated from compressibility measurements. Admittedly, however, one cannot simply assume that a compressibility discontinuity exists near an ion. In reality, the compressibility probably varies with distance in some continuous but nonlinear manner. Hence, the compressibility of the hydrated water in the outer periphery of the hydration zone will not be too different from the compressibility of the bulk solution. Therefore peripheral hydration will not be determined *in toto* by the compressibility method. This has already been shown to be the case from the calculations of the compressibility of the hydration sphere. (See part C of this section.)

Of the five sets of compressibility hydration numbers shown in Fig. 10, Passynski's results were calculated directly from adiabatic compressibility measurements rather than from isothermal compressibilities. (Passynski used a sound velocity method to determine the adiabatic compressibility.) Allam and Lee also determined the adiabatic compressibility (using a sound velocity method), but they converted their adiabatic data to isothermal compressibilities and then calculated the hydration numbers. The compressibility data for my work were determined with a PVT cell (a bellows densimeter) and therefore the data are isothermal compressibilities. Allam and Lee's hydration numbers agree fairly well with mine. However, a direct comparison of the two sets of data is not strictly valid because (1) Allam and Lee's hydration numbers are true infinite dilution values, while my values are only approximately so, and (2) Allam and Lee's values are referred to 25°C, while my values are referred to 15°C. The infinite dilution effect is probably not too important; as I stated previously $N_w$ (0.13 $m$) differed from $N_w$ (0.26 $m$) by $\pm 0.9$ hydration units. The temperature effect may be crudely estimated from $N_w$ (15°C) and $N_w$ (2°C), given in Table 3: for the monovalent salts (based on $N_w$ for NaCl), $N_w$ decreases an average of 0.9 hydration units per 10°C increase; for the divalent salts, $N_w$ decreases an average of 1.3 hydration units per 10°C increase. This temperature effect is in agreement with the work of MILLERO, WARD, LEPPLE and HOFF (1974). Therefore the temperature effect does not appreciably change the agreement between the different sets of hydration numbers. The compressibility hydration numbers of BOCKRIS and SALUJA (1972), shown directly in Table 3, appear to be the most extensive so far published. Except for a few cases, the results of Bockris and Saluja agree very well with the results of this work and the other published values.

The compressibility method for determining hydration gives an estimate of primary hydration because compressibility is a bulk property of a solution. The hydration numbers derived from the compressibility method are probably an underestimate of the true number of water molecules associated with the ion–water complex because peripherally hydrated water is not completely determined from the compressibility measurement. A serious limitation to the method is that it is not possible to separate compressibility effects caused by anions from compressibility effects caused by cations.

(f) *NMR Methods: Hindman* (1962), *Swift and Sayre* (1966). Hindman was the first person to interpret the chemical shift in terms of a hydration number. He considered the following four contributions as the main cause of the chemical shift: (1) a hydrogen bond breaking contribution which causes a high field shift; (2) a structural contribution which relates the ability of ions to restructure (or destructure) that portion of water beyond the primary hydration zone; (3) a polarization contribution which is due to the resonance, caused by the electrostatic field of the ion, of the water molecules in the primary hydration zone; and (4) a nonelectrostatic contribution which is considered negligible for alkali-metal ions. Hind-

man derived an equation for each contribution and then equated the sum of the equations to his observed chemical shift. Each contribution to the chemical shift (with the exception of the nonelectrostatic contribution, which is neglected anyway) is a function of several variables including the hydration number. The hydration number is determined by inserting the various parameters into the summation equation until a best agreement is obtained between the observed and calculated chemical shift. Hindman's hydration numbers appear a bit low when compared to the other values shown in Fig. 10. It is difficult to judge the reliability of his values because the final equation used contained a large number of parameters.

In the derivation of the structural contribution term, Hindman introduced an interesting new parameter, $n_i$, which gives the number of additional hydrogen bonds formed in excess of those involved in the reorientation process to form the hydrated ion. It was very interesting that $n_i$ was found to be negative for the following anions ($n_i$ values in brackets): $Cl^-$ ($-0.6$), $Br^-$ ($-1.2$), $I^-$ ($-1.5$), $NO_3^-$ ($-1.0$), $ClO_4^-$ ($-2.0$). The implication of the minus sign is that these anions are structure breakers. I will return to these data in section V when GREYSON'S (1967) results on entropy of transfer are discussed.

SWIFT and SAYRE (1966) developed an NMR technique to determine the number of water molecules associated with an ion for a time which is long compared to the diffusion (molecular) time of ordinary water. Clearly, this approach to hydration is a break from the traditional hard sphere model. The method is based on a comparison of the line widths of the proton resonance signals from a reference solution and a test solution. The reference solution contains a cation whose hydration is to be determined. (The anion of the test salt and the reference salt is the same.) Both solutions are doped with equal amounts of a 'probe' reagent which will react with the bulk water and with the ion–water complex. The reactions of the bound and free waters with the probe reagent will proceed according to first-order kinetics. Swift and Sayre reported only results for divalent cations. Those cations of interest to the present work are (hydration numbers in brackets): $Mg^{2+}$ (3.8), $Ca^{2+}$ (4.3), and $Ba^{2+}$ (5.7). The hydration numbers are low in comparison with values determined by other methods. The results, however, are not directly comparable with the other data because they are hydration numbers in a 1.0 molar solution of the salt. I see no reason why the method couldn't be used in dilute solutions; however, it may be that in order to get a reasonable signal for all the shifts, a solution must have a high proportion of ion-associated water to bulk water. But in spite of this possible drawback, the method appears to be a desirable approach to hydration because it distinguishes the water associated with the ion–water complex from the bulk water.

(g) *Residence Times: Samoilov* (1957, 1965, 1972). Samoilov has developed a general approach to hydration which does not depend on the preconception that a given number of water molecules are firmly bound to an ion. Samoilov makes the point that the problem of deciding if and how many water molecules are bound to an ion should only follow as a possible consequence of the generalized model. As a basis for his model, Samoilov considers that water molecules near an ion can exchange with the bulk water because all water (bound or otherwise) has some translational freedom. If the exchange of water near an ion is frequent, the ion cannot be hydrated to any great extent. If, in fact, the water near an ion is exchanging more frequently with nearby water, then negative hydration is said to occur. Positive hydration occurs when the hydrated water does not exchange frequently with the bulk water. The parameter that Samoilov uses to quantify his model is $\tau$, the residence time of water molecule. The value $\tau$ is the mean time that a water molecule remains close to another water molecule. If a water molecule is close to an ion, $\tau$ (the residence time) will not be the same as that for pure water; as Samoilov states 'an ion is, energetically, not equivalent to a molecule of water'. In this case, the mean time that a water molecule is in its equilibrium position is $\tau_i$. The energy necessary to activate a water molecule in pure water is $E$. The energy necessary to activate a water molecule near an ion is $E + \Delta E$. If $\Delta E \gg 0$ then the water molecules are bound, in the traditional sense, to an ion. If, however, $\Delta E < 0$ then obviously one cannot speak of bound water. $E$ and $\tau$ are determined from self-diffusion and Raman frequency data, respectively (NEMETHY, 1970).

The ratio $\tau_i/\tau$ can be used as an indicator of hydration: an ion is positively hydrated if $\tau_i/\tau > 1$; an ion is negatively hydrated if $\tau_i/\tau < 1$. Figure 10 shows SAMOILOV's (1972) recent $\tau_i/\tau$ values for some of the alkali-metal ions and some of the halide ions. The results are interesting because (1) the halides all produce about the same hydration (negative) in spite of differences in crystallographic radii, (2) $K^+$ and $Cs^+$ are negatively hydrated (other hydration data show that these ions are hydrated), and (3) $Na^+$ is only marginally hydrated. $Li^+$ is positively hydrated, which is what one would have predicted. It is difficult to compare Samoilov's results with the traditional hydration numbers because Samoilov's hydration concept is fundamentally different from the hydrated hard sphere models. One point, however, seems fairly evident: Samoilov's hydration values for the halide ions are consistent with the low (and sometimes zero) hydration numbers determined by other methods. This gives added support to the idea that anions probably interact with water differently than do cations. The real value of the Samoilov scheme is that he has replaced the traditional hydration concept by a residence time concept.

## 2. Summary of hydration number concept

The purpose of bringing together various hydration data was (1) to see if the hydration number can be related to the techniques used, (2) to see if the compressibility approach to hydration gave results which were consistent with other methods, and (3) to see if hydration numbers are, in fact, useful in describing how ions alter water structure.

With the exception of Samoilov's work and possibly Swift and Sayre's work, it appears that the various hydration methods reviewed in this paper succeeded in giving a relative estimate of primary hydration. Because of the differences in methods used and the assumptions made in the models, it is not practical to attach much significance to the actual differences in the hydration numbers among the various methods. Hydration numbers determined by methods which depend on a kinetic property or measurement are by and large similar to those determined by measuring a bulk or thermodynamic property. (For instance, a multiply charged ion appears to be more hydrated than a singly charged ion.)

In section V the problem of ions and water structure will be examined by relating the compressibility results to properties which have already been used to describe aspects of water structure. The question of structure breaking by ions will be of particular interest.

## V. STRUCTURE BREAKING BY IONS

The hydration number may be looked upon as a qualitative indicator of the number of water molecules in the 'primary hydration zone'; at best, the hydration number is proportional to the average time that a hydration shell may be said to exist around most ions (DESNOYERS and JOLICOEUR, 1969). In the present section, I will attempt to relate the $\Delta\beta$ values themselves to other physical properties which have been used to describe how ions might influence water structure.

Before proceeding with the discussion, however, I should mention that this approach will lead mainly to a 'rationalization' of the $\Delta\beta$ results, rather than to a new 'predictive' model for the structure of water near an ion. In this regard, HOLTZER and EMERSON (1969) demonstrated, in a very dialectical style, that 'rationalization' of experimental results can lead to contradictory conclusions. As Holtzer and Emerson put it, 'indeed, in many cases even a single given argument is sufficiently indeterminate as to be capable of producing several contradictory conclusions'. I shall now discuss, keeping in mind the wisdom expressed by Holtzer and Emerson, how ions affect water structure. The theme of the discussion will deal with the idea that anions interact with water differently than do cations. This is particu-

larly relevant to seawater chemistry because seawater contains several different anions and cations (i.e. $Cl^-$, $SO_4^{2-}$, $Na^+$, $Mg^{2+}$, $Ca^{2+}$, $K^+$) each of which probably affects the structure of water in seawater in a different way.

### A. Entropy of Transfer and Compressibility

GREYSON (1967) determined entropies of transfer ($\Delta S$), from $D_2O$ to $H_2O$, for most of the alkali-metal halides. His results showed that the halides have a greater structure altering effect on water than do the alkali-metal ions. Greyson's method (GREYSON, 1962) was to measure the EMF of an electrochemical cell in which salt solutions

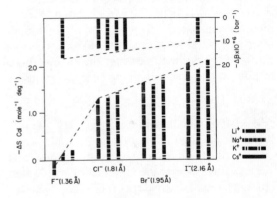

FIG. 11.   Comparison of GREYSON's (1967) entropies of transfer with $\Delta\beta$ values.

of normal and of heavy water were separated by a cation exchange membrane. The EMF of the cell was shown by Greyson to be related to the free energy change when 1 equivalent of salt in $D_2O$ was transferred to $H_2O$. Greyson reasoned that the entropy of transfer for the salt is due entirely to the difference between the structure of $D_2O$ and $H_2O$. A large negative entropy of transfer may be interpreted as due to structure breaking of $D_2O$ by the salt. Greyson's entropy results are summarized in Fig. 11; for comparison (later), the $\Delta\beta$ for the appropriate salts are also shown in Fig. 11. Greyson's results are particularly interesting because they show that the change in $\Delta S$ along the series of halide salts having the same cation is greater than the corresponding change in $\Delta S$

along the series of alkali-metal salts having the same anion. Greyson's interpretation of these trends is that the halides (except for LiF), because of their size, disrupt water structure to a greater degree than do the alkali-metal ions.

I now turn my attention to Fig. 11: the comparison of Greyson's $\Delta S$ values with $\Delta \beta$ values from the present research. In section IV (parts A and B), it was concluded that $\Delta \beta$ for alkali-metal ions did not appear to be related to crystallographic ionic radii, but that the $\Delta \beta$ values for the halide ions did show a small but measurable variation which could be correlated with the radii of the halide ions. The effect of ion size on $\Delta S$ and on $\Delta \beta$ is shown (Fig. 11) by the dashed lines. The fact that both lines are sloping in the same direction could indicate that the variation of $\Delta \beta$ among different salts is due to the variation of the amount of broken water structure. It is obvious from the slopes of the dashed lines that the two methods respond very differently to the effect of anions.

## B. Hydrogen Bond Breaking by Anions

HINDMAN (1962), who used an NMR method to study the hydration of ions in water, also discovered a difference between cations and anions with respect to their interaction with water. Hindman found that, except for $F^-$, the calculated hydration numbers for anions were negative rather than positive. Negative hydration was thought to be due to the breaking of hydrogen bonds over and above those broken when an ion coordinates water (the primary bond breaking process) in its first coordination shell. Negative hydration is analogous to negative contributions to the structural entropy (FRANK and EVANS, 1945) in that it indicates the failure of an ion, hydrated or otherwise, to fit into the normal water structure. Hindman attributed the structure broken region of water as due not to the disorganization of water in the outer coordination shell, but rather to thermally disoriented water in the immediate vicinity of the anion. This interpretation is very similar to the view held by Samoilov, in which the action of ions on the thermal and translational motion of water molecules should form the general basis for hydration.

Hindman used a structural parameter, which he called $n_i$, in his chemical shift formula to account for the number of additional hydrogen bonds either formed or broken due to the action of an ion. $n_i$ may be considered an adjustable parameter which permits the assignment of zero values for the hydration number of anions. Although the introduction of $n_i$ provided Hindman with a convenient way to overcome the problem of dealing with a negative hydration number, the initial

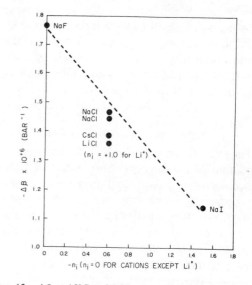

FIG. 12. $\Delta \beta$, at 15°C and 0.26 $m$ test salt concentration, and HINDMAN'S (1962) $n_i$ values for halides. $-n_i$ equals the number of excess hydrogen bonds broken.

outcome of a negative hydration number demonstrates rather clearly that (1) anions are not hydrated (at least not in the classical sense where hydration is viewed as the number of water molecules which are geometrically arranged around a charged ion), and (2) anions are structure breakers to a much greater extent than are cations. Figure 12 shows a plot of $\Delta \beta$ versus $n_i$. As previously mentioned, $n_i$ is the number of hydrogen bonds formed in excess of those involved in the reorientation of water near an ion; $n_i$ is zero for cations ($Li^+$ is an exception: $n_i(Li^+) = +1.0$), but $n_i$ is negative for anions ($F^-$ is an exception: $n_i(F^-) = 0$) because 'excess' hydrogen bonds are broken by anions. $\Delta\beta$ for a series of halides having a common cation should correlate with the $n_i$ for anions because the compressibility of a solution containing a strong structure breaker, such as $I^-$, will not be as great as for a solution containing a weak structure breaker (e.g. $Cl^-$ or $F^-$). A strong structure breaker will, in effect, produce a minimal amount (if any) of reoriented water, but will cause additional collapse of local water structure because of excess hydrogen bond breaking. However, a weak structure breaker will produce a greater degree of reoriented local water but a minimal amount of hydrogen bond breaking over and above that which occurred in the re-orientation process. Therefore, it would seem from Fig. 12 that $\Delta\beta$ is a more sensitive measure of the reorientation process than is the excess hydrogen bond breaking process because $\Delta\beta$ for weak structure breakers (i.e. $n_i \geqslant 0$) is more negative than the $\Delta\beta$ for strong structure breakers (i.e. $n_i < 0$).

The structure breaking ability of anions is related to the size of the ion. This is evident in Hindman's work where it can be seen (Fig. 12) that there is a greater excess hydrogen bond breaking by $I^-$ than by $F^-$; the crystallographic radii of $I^-$ and $F^-$ are 2.16 Å and 1.36 Å, respectively. GREYSON (1967) also found that the structure breaking ability of anions was strongly correlated with ionic size. As we shall see later, the size effect may be due to the ease with which large anions become polarized by the water molecules.

### C. Padova's Effective Radii of Ions in Solution

That size is an important factor in the propensity of ions to cause structure breaking is also shown in the work by PADOVA (1963) who calculated the intrinsic 'effective' radii ($r_e$) of ions in aqueous solutions. Padova's approach was to calculate the minimum radius, $r_o^o$, necessary to cause complete dielectric saturation of the local water. For a monovalent ion, $r_o^o$ is calculated to be 1.80 Å. If, according to Padova, $r_e < r_o^o$ then the ion is a structure maker. However, if $r_e > r_o^o$ then the ion is a structure breaker. (The terms structure maker and structure breaker, as used by

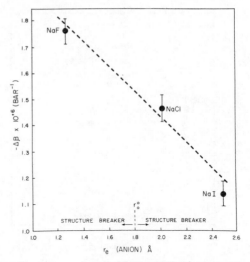

FIG. 13. The correlation of $\Delta\beta$ (at 15°C, 1 bar pressure, and 0.26 m test salt concentration) and PADOVA's (1963) $r_e$ values for halide ions.

Padova, are in reference to the Frank and Wen model. As previously mentioned, any ion added to water will, to a certain degree, break down water structure. The term structure making essentially refers to the reorientation of the broken water structure.) Figure 13 shows $\Delta\beta$ plotted as a function of $r_e$ for the series NaF, NaCl, and NaI. The $\Delta\beta$ correlate with $r_e$. However, this was not unexpected because it has already been shown that the $\Delta\beta$ values for the sodium halides correlate with the degree to which sodium halides break hydrogen bonds (Fig. 12). According to Padova's scheme, $Cl^-$ and $I^-$ are structure breakers ($r_e > r_o^o$) while $F^-$ is a structure maker ($r_e < r_o^o$). Figure 14 shows a similar plot ($\Delta\beta$ versus $r_e$) for the cation series LiCl, NaCl, KCl, and CsCl. Here there is no obvious correlation between the $\Delta\beta$ values for the solutions and the $r_e$ for the cations. Figure 15 shows another plot of $\Delta\beta$ versus $r_e$, but for the series of alkaline earth chlorides $MgCl_2$, $CaCl_2$, and $BaCl_2$. Here there exists a good correlation between $\Delta\beta$ and $r_e$. It is particularly interesting that for the alkaline earth salts the slope of $\Delta\beta$ versus $r_e$ is positive, but for the halide salts the slope of $\Delta\beta$ versus $r_e$ is negative. Both $F^-$ and $Mg^{2+}$ are usually considered strong structure makers; yet they appear to affect $\Delta\beta$ in relatively

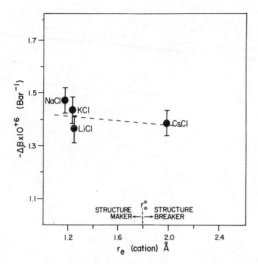

FIG. 14. The correlation of $\Delta\beta$ (at 15°C, 1 bar pressure, and 0.26 m test salt concentration) and PADOVA's (1963) $r_e$ values for alkali-metal ions.

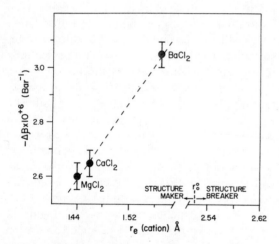

FIG. 15. The correlation of $\Delta\beta$ (at 15°C, 1 bar pressure, and 0.26 m test salt concentration) and PADOVA's (1963) $r_e$ values for alkaline earth ions.

opposite ways. The same analogy can, of course, be applied to $I^-$ and $Ba^{2+}$, which are generally classed as structure breakers. (However, $Ba^{2+}$, according to Padova, is a structure maker.) Such inverse effects, as shown in Figs. 13 and 14, give added support to the belief that anions alter water structure in a very different way than do cations.

### D. *The Polarization of Ions*

The previous discussion gave evidence that anions (that is, the halides, with the possible exception of $F^-$) disrupt water structure to a greater extent than do cations. The size of an anion appears to be the factor which is related, in some way, to structure breaking. A possible explanation for this observation is that anions are more easily polarized by water molecules than are cations. Polarization of an ion is the deformation of the outer electron cloud of an ion and is caused by either ion-ion or ion-dipole (water) interactions (KORTUM and BOCKRIS, 1951). Negative ions are more polarizable than positive ions because the positive charge on cations acts to stabilize the electron clouds surrounding an ion. When an anion is polarized by a water molecule, the OH bond involved in $O-H\cdots$ anion interaction is less polar than the OH bond involved in water$\cdots$water bonding. According to GREYSON (1967), the decreased polarity of the OH bond caused by the $O-H\cdots$ anion interaction leads to a disruption of water structure. However, the OH bond involved in an $O-H\cdots$ cation interaction has a polarity similar to the OH bond in water$\cdots$water bonding; therefore, cation-$H_2O$ species are more compatible with nearby water structure and are less disruptive to water structure than anion-$H_2O$ species (GREYSON, 1967).

Evidence for ion polarization in aqueous solutions can be found in the work of WALRAFEN (1962), who determined the Raman spectra for pure water and for various alkali metal halide solutions. Walrafen's approach was to determine the intensity of the Raman bands at different temperatures and for different salts. Shifts in the position and intensity of the Raman bands were interpreted in terms of the polarity of the anion-$H_2O$ bond and the cation-$H_2O$ bonds. According to Walrafen, the intensity of the Raman scattering should increase with decreasing bond polarity. Presumably this is because polarizability of the electrons in the bond varies inversely with bond polarity. Thus the greater Raman intensity of the stretching of the $O-H\cdots$ anion bonds is due to their lower polarity compared with the OH$\cdots$ water bonds. Walrafen essentially found that the intensity of the librational bands increased with increasing anion size, and this effect was attributed to the polarization effect. The alkali metal ions, however, did not produce large changes in Raman intensities, and hence these ions were assumed not to be as polarizable as anions.

## VI. APPLICATIONS AND DISCUSSIONS

In this section the observations and interpretations presented in the previous sections will be related to seawater. The previous discussions dealt with the breaking of water structure by ions. For the most part, the extent of the disruption of water by an ion depends on whether the ion is a cation or an anion and also on specific properties (e.g. size and charge) of the ion. It appears established, based on what was said in section V, that structure breaking by an anion (especially the larger halides) represents some sort of steady state disruption process in which the broken water structures are not reoriented by the anion. ($F^-$ may be an exception, because of its small size.) This was implicit in most of the work I reviewed, although Samoilov expressed the hydration concept for anions (and also for cations) in terms of the exchangeability of the water molecules in a solution. That water molecules can exchange positions with each other is a direct consequence of the kinetic (thermal) properties of the molecules (SAMOILOV, 1972). Cations also disrupt water structure, but the disruption process is accompanied by a more or less instantaneous subsequent reorganization of the broken water structure. But, even for cations, there is still an exchange of the water molecules near an ion with those outside the immediate vicinity of the ion (SAMOILOV, 1957, 1965, 1972). By the same token, the bulk water, usually assumed to be outside the influence of an ion (FRANK and WEN, 1957), will be affected by the presence of an ion. 'Immobilization' of water molecules near cations (i.e. the water molecules within the hydration zone) can only be accepted on an average time basis.

I now focus my attention on seawater, that salty liquid that fills the oceans. For this discussion, seawater is assumed to be a single phase multi-component system. (The effects of gas bubbles

and suspended particulate matter no doubt influence water structure (HORNE, DAY, YOUNG and YU, 1968; HORNE, 1969), but these effects will not be considered in this discussion.) The concentrations of the ten most abundant ions found in seawater are given in Table 4. To a first approximation, seawater is a solution containing essentially $0.5\,m$ NaCl, $0.03\,m$ MgSO$_4$, and $0.02\,m$ MgCl$_2$. The main questions, with regard to this work, are: what is the extent to which sea salt ions

Table 4. Composition$^a$ of 1 kg of seawater (based on salinity = 35‰).

| Ion | g | mole |
|---|---|---|
| Anions | | |
| Cl$^-$ | 19.35 | 0.537 |
| SO$_4^{2-}$ | 2.71 | 0.0282 |
| HCO$_3^-$ | 0.142 | 0.0023 |
| Br$^-$ | 0.066 | 0.00083 |
| F$^-$ | 0.001 | 0.00005 |
| | | |
| Cations | | |
| Na$^+$ | 10.77 | 0.468 |
| Mg$^{2+}$ | 1.30 | 0.0535 |
| Ca$^{2+}$ | 0.414 | 0.00373 |
| K$^+$ | 0.307 | 0.0099 |
| Sr$^{2+}$ | 0.008 | 0.00009 |

$^a$ Composition of seawater based on the formula of KESTER et al. (1967).

alter the structure of water in seawater, and why is it important to know something about the water in seawater? An answer to the first question is apparent from what has already been said in this paper. The second question will result in a more or less subjective answer, depending on the scale of events being examined and also on one's own philosophical approach to chemical oceanography.

## A. The Effect of Sea Salt Ions on the Structure of Seawater

As just mentioned, seawater is a solution containing nearly equal amounts ($0.5\,m$) of Na$^+$ and Cl$^-$ and small amounts of Mg$^{2+}$ ($0.05\,m$) and SO$_4^{2-}$ ($0.03\,m$). The fact that the concentrations of Na$^+$ and Cl$^-$ in seawater are nearly balanced does not mean that the structure breaking effect of Cl$^-$ is counterbalanced by some sort of structure 'making' effect of $Na^+$. One might get this impression from a comparison of the entropy changes ($\Delta S$) for water near $Na^+$ and for water near $Cl^-$, because, according to work cited by SAMOILOV (1972), the $\Delta S$ of water near $Na^+$ is about $+2$ entropy units whereas the $\Delta S$ of water near $Cl^-$ is about $-2$ entropy units. (Here the standard state is pure water.) However, entropy is a thermodynamic property, and therefore it provides essentially no information about the specific structural effects which result from either structure breaking or structure making effects. It is important, therefore, to establish the effect of sea salt ions on water structure either by means other than a direct comparison of different thermodynamic properties, or by correlations of thermodynamic properties and specific structural parameters which have been used successfully to describe some aspect of water structure. The latter approach was used in section V where the $\Delta \beta$ values for the series of sodium halides were correlated with the extent of excess hydrogen bond breaking by the halide ions. Those qualitative results are directly applicable to seawater. The $Cl^-$ in seawater must be responsible for disrupting more of the normal water structure than is disrupted during the primary hydrogen bond breaking process. Further extra-thermodynamic support for $Cl^-$ as a structure breaker comes from the X-ray work of BRADY and KRAUSE (1957), who found that the number of nearest water neighbors increases from 4.6 for normal pure water (MORGAN and WARREN, 1938) to around 7 for KCl solutions. This was interpreted in terms of an apparent increased packing of water molecules in which the $Cl^-$ destroys some of the normal tetrahedral water structure.

$Na^+$ interacts differently with water than does $Cl^-$. The addition of $Na^+$ to water also results in a primary hydrogen bond breaking process, but with $Na^+$, unlike $Cl^-$, there is a reordering or some type of steady state reorganization of the broken water structure. This is evident from entropy considerations, but what is usually not appreciated is that the local order near $Na^+$ is probably very different from that of any water molecules in its normal tetrahedron. However,

from entropy considerations alone, it is not possible to determine internal organization. According to HINDMAN (1962), the most probable orientation of a $Na^+$—$H_2O$ complex is a co-planar orientation (negative end of the water dipole directed toward the $Na^+$) while that of a $Cl^-$—$H_2O$ complex is an angular orientation (one hydrogen of the water dipole pointing toward $Cl^-$). The essential difference between these two modes is that the rotational freedom of water molecules at the surface of a negative ion is supposed to be much greater than it is at the surface of a positive ion. (This interpretation is based on dielectric constant data in which cations were found to depress the dielectric constant of water while anions were found to have only a negligible effect on the dielectric constant (HAGGIS, HASTED and BUCHANAN, 1952).) From this consideration alone, one could infer that the forces holding together the $Cl^-$—$H_2O$ complex are more easily overcome by the thermal energy of a solution than those forces holding together the $Na^+$—$H_2O$ complex. This is entirely consistent with the Samoilov hydration model in which the activation energy ($\Delta E$) necessary for water to escape from the immediate vicinity of $Na^+$ is greater than that necessary for water to escape from the immediate vicinity of $Cl^-$. In connection with the effect of cations on water structure, it is sometimes emphasized that there exists a region or layer of highly electrostricted immobile water around such ions as $Na^+$ (HORNE, 1966; HORNE and BIRKETT, 1967). Presumably, this is because calculations show that there exists a strong coulombic force between an ion and a water dipole. If such calculations are based on the total amount of energy necessary to remove a dipolar molecule an infinite distance from the ion then one might visualize a rigid electrostricted zone. However, as Samoilov points out, a water molecule does not instantaneously separate from an ion over an infinite distance and therefore only a fraction of the total interaction energy is necessary to remove water from the so-called electrostricted zone to outside the hydration zone. In this sense, any given water molecule in the vicinity of $Na^+$ can only be considered temporarily bound to $Na^+$.

The residence time of a water molecule co-ordinated with $Na^+$ has been estimated to be $10^{-11}$ s in solution (Friedman, 1971).

So far, I have considered only the effects of ion-associated water on the structure of water in seawater. The important question arises as to what effect ion-pairs have on the structure of water in seawater. The question of ion-pairing and its effect on water structure is important because the sulfate ion is known to ion-pair with $Mg^{2+}$ and also $Na^+$ (EIGEN and TAMM, 1962). The fact that a neutral ion-pair has no net charge is not, *a priori*, a reason for discounting ion-pairs as structure breakers. An ion-pair, while not having a net charge, does have two centers of opposite charge and is, therefore, a dipolar molecule. The dipole moment of a dipolar molecule can be very large. For example, glycine has a dipole moment of about 15 Debye units (KIRKWOOD, 1939), whereas the dipole moment of water is 1.84 Debye units (EISENBERG and KAUZMAN, 1939). Therefore, one might expect dipolar molecules to have a strong structure breaking effect on water as a result of dipole alignment. The actual extent of structure breaking by an ion-pair would be difficult to calculate because the ion-pairs themselves are probably hydrated; this would give rise to quad-rupole and other multipole moments, all of which would cause inhomogeneous electric fields in and around the ion-pair. Another difficulty would be the size and asymmetry of the ion-pair itself. A long or narrow ion-pair (perhaps the solvent-separated $MgSO_4{}^0$) may contain an effective non-polar region near which a local reinforcement of the hydrogen bonding of water may exist in a similar way proposed for the nonpolar regions of tetraalkylammonium ions (FRANK, 1965; DESNOYERS and JOLICOEUR, 1969). To my know-ledge, no theory or quantitative model for the effects of an ion-pair on the structure of water has been published, although models for dipolar molecules have been proposed by KIRKWOOD (1934, 1939), ROSS and LEVINE (1968), and ROSS (1968). These models simulate the ion-pair by some type of cavity, such as a spheroid or ellipsoid, whose boundary conditions can be defined mathematically. Ross's model differs from that of Kirkwood in that Ross treats the ion-pair as the interaction of two hydrated ions whose hydration spheres are overlapped and truncated, while Kirkwood considers a dipolar molecule as simply two spheres which are rigidly connected. It is almost useless to apply such models to the problem of ion-pairs and water structure because the models were developed with the view that the ion-pair, or dipolar molecule, is a simple geome-trical cavity which is embedded in a sea of continuum solvent. The water in seawater can hardly be called a continuum solvent.

## B. *The Statistical Approach*

The purpose of the work described herein was to establish the thesis that the water in seawater is broken by the action of sea salt ions. This was accomplished, but the fundamental problems of water structure in aqueous solutions still remain unanswered. It is apparent that the most important problem deals with hydrogen bond breaking by ions. It seems clear from the NMR work that both anions and cations cause hydrogen bonds to break, but as yet there is no satisfactory model or theory which can be used to predict the extent or number of hydrogen bonds broken by ions during the primary bond breaking process. The problem becomes further complicated when one considers that water molecules near ions have a finite rather than an infinite residence time (SAMOILOV, 1972).

The terms 'structure breaking' and 'structure making' have been profusely used throughout this work, but an answer to the question, 'What is really meant by water structure?' is by no means straightforward. This is probably because a great deal of past chemistry has been based on pure Gibbsian thermodynamics in which the intersti-tial properties of a liquid are not considered in the interpretation of thermodynamic properties. (Most standard physical chemistry textbooks deal mainly with the structure or geometry of pure water and treat the structure of water in electrolyte solutions in a very rudimentary way.) Water is a highly *structured* liquid (EISENBERG and KAUZ-MANN, 1966), but one whose structure may in fact be represented by some sort of equilibrium between clusters of hydrogen bonded water and non-hydrogen bonded water (NEMETHY and SCHERAGA, 1972). The acceptance of this description complicates any proposed model or theory for the structure of water in an aqueous solution such as seawater. VASLOW (1972) has recently questioned the 'concept of structure' in electrolyte solutions. Basically his point is this: In crystals, each unit of structure is referenced to an unchanging coordinate system. In liquids, however, the relative position of any two molecules at macroscopic distances are random, and even at close range (e.g. the hydration zone) there is at best only a temporary spatial arrangement between the two molecules. Hence, there is an inherent difference between the concept of structure in a liquid versus that in a solid. Vaslow goes on to discuss the structure of a liquid in terms of the probability of finding a molecule at a certain distance from another molecule and in a given orientation. A probability model would be particularly appealing because it can be related to radial distribution functions which are derived from X-ray data. Although Vaslow doesn't develop a probability model *per se*, he does suggest several applications of such a model. The interaction of molecules (ions and water molecules) are determined by correlation functions. The correlation function is a measure of the effect of one type of molecule in repelling or attracting another molecule to its neighborhood (FRIEDMAN, 1971). In a solution such as seawater, several situations may occur which could be described by these functions. If, according to Vaslow, two electrical fields (e.g. one from a water dipole and one from an ion) are comparable in size but approximately cancel each other there will be little correlation between position and angle of the two molecules. In such a situation, the entropy would be high because there are no net orienting forces to coordinate the local or nearby water. It is not too difficult to imagine that in seawater there exists an entire hierarchy of weakly dipolar molecules (ion-pairs), some of which have net electrical fields which are balanced by the electrical fields of other molecules and would therefore lead to local regions of disordered or broken water

structure.

Such probability models, which use correlation functions, form the basis of the statistical-mechanics approach to electrolyte solutions. The value of this approach is in predicting thermodynamic properties from a calculation of the cumulative pair-wise interactions (ion-water) in a solution. The problem of computing these pair-wise potential forces for all possible interactions in an electrolyte solution such as seawater is a formidable task. So far, statistical mechanics has only been applied to simple 1–1 electrolyte solutions (i.e. the 'primitive model') using the Monte Carlo method of computation (CARD and VALLEAU, 1970). In this method, the mean potential energy of a system is determined by summing the calculated pair-wise potentials of the positive and negative species which interact with water.

In summary, statistical models which deal with the degrees of randomness and orientation may provide the most practical approach to ion–water interactions. This is because of the transient behavior of water either as a pure substance or in an ionic solution and also because of the absence of permanent order which prevents any other description of the system.

## C. Chemical Oceanography

As a result of this work it seems clear to me that water structure in seawater will have to be considered by chemical oceanographers if they want to comprehend better the role of water in the oceans. The water in seawater is *not* inert; it interacts with sea salt ions, and the type of interaction (i.e. ion–water and ion-pair, etc.) and pathways of interaction are very complicated. The most acceptable picture of the water in seawater, to come from the present work, is that of a mixture including several different types of structure: (1) broken and disrupted water around $Cl^-$; (2) temporarily and partially oriented water around $Na^+$. (The concentration of $Mg^{2+}$ is about ten times less than $Na^+$ and therefore its effect on the structure of water in seawater is less than $Na^+$, although not proportionally less since the residence time of water near $Mg^{2+}$ is twice that for

$Na^+$ (SAMOILOV, 1972).) (3) broken and disrupted water near $NaSO_4^-$ $MgSO_4^0$ ion-pairs; and (4) the normal tetrahedral water structures which themselves are continuously being broken and rebuilt. Each of the above structural elements exists only on an average time basis; but on the whole, all elements must cooperate with each other. Admittedly, such a picture of water in seawater is very qualitative. But until satisfactory models for hydrogen bond breaking by ions are developed, we are left with only qualitative ideas.

It is certainly safe to say that the water in seawater takes part in all processes that occur in the oceans. Oceanography has benefited from the recent determinations of activities and thermodynamic properties of seawater (summarized by MILLERO, 1974). Such research has been very productive (in terms of describing seawater as an entity in itself), but so far the role of the water in seawater has only been tacitly considered in the interpretation of the thermodynamic results. That water plays a role in chemical equilibria has been shown by MARSHALL and QUIST (1967) who considered the water in an aqueous solution to be an ingredient of a chemical reaction. Recently, NORTH (1974) and MACDONALD and NORTH (1974) have shown that the solubility behavior of certain weakly soluble salts in aqueous solutions can be interpreted in terms of hydrate phases formed at surfaces of the crystal. If all interfaces are considered there must be a very large number of chemical reactions occurring in the oceans. WANGERSKY (1972) has listed twenty-three reactions which alone are linked with the carbon dioxide system in seawater.

## 1. Salinity, temperature, and depth effects

In the preceding sections, I put forth the thesis that the water structure in seawater is essentially broken as a result of the action of sea salt ions. This was inferred from the compressibility measurements in which the $\Delta\beta_{sw}$ (seawater as the solvent) values were found to be less than the $\Delta\beta_w$ (water as the solvent) values. (The extent to which a particular salt acts to disrupt water structure (either in seawater or ordinary water) is related to the magnitude of the measured $\Delta\beta$ value.) For this

idea to be relevant to oceanography it has to be shown that certain aspects of seawater can be described qualitatively, if not quantitatively, from the interpretations of this research. I will now discuss the possible effects of salinity, temperature, and depth on the structure of the water in seawater. The aim of the discussions will be to relate the conclusions of this work to some physicochemical properties of seawater.

(a) *Salinity effect.* The salinity of the world ocean is slightly less than 35‰; the range is from 33 to 37‰. (This range omits regions in coastal zones where salinities are significantly less than 33‰ and also isolated regions in the Red Sea where salinities are reported to be very high.) As seen previously (Fig. 8), the effect on $\Delta\beta$ in going from a distilled water reference to a seawater reference was to make $\Delta\beta$ less negative by an amount which ranged from $0.2 \times 10^{-6}$ bar$^{-1}$ to $0.8 \times 10^{-6}$ bar$^{-1}$. For example, at 15°C the $\Delta\beta$ for 0.26 m(w) NaCl is $-1.47 \times 10^{-6}$ bar$^{-1}$ whereas the $\Delta\beta$ for 0.26 m(sw) NaCl is $-1.30 \times 10^{-6}$ bar$^{-1}$. Therefore, in effect, a 35‰ salinity change will produce a $0.17 \times 10^{-6}$ bar$^{-1}$ change in $\Delta\beta$. (Actually, the difference between $\Delta\beta_{sw}$ and the corresponding $\Delta\beta_w$ is not all entirely due to ion–water effects. In the calculation of $\Delta\beta_{sw}$, using eqn. (7), $V_b^1$ contains a volume contribution due to the presence of sea salt. This will have the effect of making $\Delta\beta_{sw}$ smaller than if the volumes of the water in the seawater reference solution and the water in ordinary water reference solution were the same. However, this effect (calculated from partial molal volume data) would lower $\Delta\beta_{sw}$ by only 1%, and therefore it is small compared to the $[\Delta\beta_{sw} - \Delta\beta_w]$ values which, shown in Fig. 8, ranged from 15 to 27% of $\Delta\beta_{sw}$.) Hence, a 4‰ salinity change (equivalent to the salinity range of the world ocean) would cause $\Delta\beta_{sw}$ to change by only $0.002 \times 10^{-6}$ bar$^{-1}$. This calculation is based on the results for NaCl as the test salt; however, the salinity effect would be similar for other salts. Based on these calculations, one might conclude that the normal salinity range of the oceans would have only a negligible influence on ion–water interactions such as 'hydration'. Salinity changes can, however, cause changes in the distribution of ion-pairs in seawater, although the smaller the salinity change the smaller the ion-pair redistribution. The cause of the redistribution may be due to a change in the dielectric constant of seawater which decreases with increasing salinity. (The effect of salinity on the dielectric constant of water has not been measured directly, but, based on data for single salt solutions, I would expect an inverse salinity dependence.) The Bjerrum equation and other equations (Nancollas, 1966; Gilkerson, 1970) for ion-pair formation predict that ion-pair formation is inversely related to the dielectric constant of the water in solution. The decrement in the dielectric constant is due to structure breaking and subsequent dielectric polarization by ions. So far, oceanographers have not interpreted ion-pair formation in terms of structure breaking by ions, although it is sometimes considered self-evident that the existence of ion-pairs in seawater plays a fundamental role in chemical oceanography.

(b) *Temperature effect.* Temperature can also influence ion–water interactions which occur in seawater. The temperature of the world ocean varies from a high of about 30°C to a low of about $-1.5$°C. The effect of temperature may be demonstrated by comparing the thermal energy necessary to maintain water molecules in random motion (Brownian movement) with the potential energy of an ion–water interaction. At 30°C, the thermal energy, $kT$ ($k$ is Boltzman's constant and $T$ is in degrees Kelvin) is about 0.59 kCal, whereas at $-1.5$°C, $kT$ equals 0.54 kCal. The electrostatic energy of $Na^+$ and a water molecule is about $-24$ kCal (Nancollas, 1966). Therefore, by decreasing the temperature, the Brownian movement of a water molecule will be more easily overcome by the presence of a positive ion. If ion–water interactions vary inversely with temperature, then $\Delta\beta$ may also vary inversely with temperature. This is shown in Figs. 4 and 5 where $\Delta\beta$ (15°C) are greater (less negative) than values of $\Delta\beta$ (2°C) by amounts ranging from $0.3 \times 10^{-6}$ to $0.9 \times 10^{-6}$ bar$^{-1}$. Not all the temperature effect on $\Delta\beta$ is due to changes in ion–water interactions. The compressibilities ($\beta_w$ and $\beta_{sw}$) of pure water and seawater increase with decreasing tempera-

ture. This is probably due to the increased amount of opened (voids) water structure as temperature is decreased.

An interesting feature of the temperature dependency of $\Delta\beta$ is that the $[\Delta\beta_{sw}(2°C) - \Delta\beta_{sw}(15°C)]$ value for any particular salt system is the same (within experimental error) as the corresponding $\Delta\beta_w(2°C) - \Delta\beta_w(15°C)$ value for the same salt system. (The data for this comparison can be found in Table A1 of Appendix A.) That the $[\Delta\beta(2°C) - \Delta\beta(15°C)]$ values are independent of salinity indicates further that the kinetic behavior of the water molecule plays a very important role in ion–water interactions. (This is in agreement with Samoilov's generalized theory in which the extent of ion–water interactions is related to the thermal mobility of water molecules.)

Water structure effects due to temperature changes may also influence the formation of ion-pairs. Kessler, Povarova and Garbanev (1962) calculated the short-range energy of interaction of an ion-pair. They suggested that cations which are strongly positively hydrated (i.e. Samoilov type hydration) will repel anions much more strongly at 0°C than at 25°C. However, cations which are strongly negatively hydrated show a greater tendency toward ion-pair formation at lower temperatures than at higher temperatures. Kessler et al. offer the following explanation as support for their claim: As the temperature of a solution is lowered, the negatively hydrated ions produce a relatively greater disruption (or perturbation) of the surrounding water structure than positively hydrated ions. (Negatively hydrated ions may be thought of as Frank and Wen 'structure breakers'.) If the Kessler et al. interpretation is correct then one would expect cold ocean waters to favor the formation of ion-pairs whose parent cations are negatively hydrated. They suggest that $NaCl^0$ and $KCl^0$ are more important at low temperature (0°C) whereas $MgCl^+$ and $MgCl_2^0$ are important at higher temperatures (i.e. around 25°C). It is generally believed that NaCl and KCl are strong electrolytes and therefore almost completely dissociated in seawater. However, from the present discussion, it appears that water structure effects may play some role in the dissociation process. Experiments to determine the effect of temperature on ion-pairing will have to be performed to test the theory of Kessler *et al.*

(c) *Depth* (*pressure effect*. The hydrostatic pressure found in the depths of the ocean may play an important role in chemical processes which occur in the deep sea. The average depth of the ocean is close to 4000 m; ocean trenches such as the Marianas, however, will exceed 10,000 m. These depths are equivalent to 400 and 1000 bars, respectively, of hydrostatic pressure. From the P–V–T properties of seawater, we know that seawater is compressible; the specific volume of seawater decreases at a rate of about 4% per 1000 bars of pressure (WILSON and BRADLEY, 1966). There are conflicting views on the effect of pressure on ion–water interactions such as 'hydration'. Horne (summarized by HORNE, 1969) has generally taken the view that hydrated ions are dehydrated as pressure is increased. Marshall's work, however, indicates that the hydration change due to the dissociation of ion-pairs is independent of pressure. Because $\Delta\beta$ becomes less negative with increasing pressure (Figs. 2 and 3), one might conclude that hydration does indeed decrease with increasing pressure (in agreement with Horne). Such a conclusion, however, would have to be based on the assumption that the 'solvation' sphere surrounding an ion is incompressible. But it has already been shown (in part C of section IV) that this sphere (if such a sphere does exist) is probably compressible and therefore the variation of $\beta_{test}$ (both for the ordinary water runs and the seawater runs) with pressure contains a component due to the compressibility of the altered water. (It will be recalled that $\beta_{hyd.}$ for 0.26 $m$(w) NaCl is about $16 \times 10^{-6}$ bar$^{-1}$ and that $\beta_{test}$ for the same solution is about $43 \times 10^{-6}$ bar$^{-1}$.)

Viscosity is a parameter which can be used qualitatively to assess the effect of pressure on the structure of water in seawater. According to HARNED and OWEN (1958), viscosity is 'the stress transferred per unit velocity gradient per unit area from each layer of liquid to the layer beneath [or above] it.' Viscous effects are oceanographically

significant because of their role in the transmission of sound in seawater and in high pressure electrical conductivity of seawater. The viscosity of seawater is dependent upon the structure of the water in seawater, and therefore the compressibility results of this work should be useful in interpreting the pressure dependency of viscosity. That composition (salinity) affects water structure, which plays a role in viscosity, is evident from the high pressure viscosity measurements made by STANLEY and BATTEN (1966). They found that the viscosities of seawater and ordinary water decreased as pressure was increased. At higher pressures, however, the viscosities passed through separate minima and then increased with pressure. It is significant that the viscosity minimum for seawater, which is at about 700 bars, occurs before the viscosity minimum for pure water which is present at about 900 bars. According to Stanley and Batten, the presence of a viscosity minimum is due to (1) the breakdown of some of the normally structured water which therefore causes a decrease in viscosity, and (2) the compaction of the water structure which causes an increase in viscosity. The simultaneous presence of a decreasing component of viscosity and an increasing component leads to the viscosity minimum. The fact that the viscosity minimum for seawater occurs before the minimum for pure water can be interpreted in terms of sea salt acting to break down a portion of the water which would otherwise have been broken by the action of increased pressure.

The viscosity effects described above could have been qualitatively predicted from the compressibility results since the $\Delta\beta_{sw}$ values are greater (less negative) than the corresponding $\Delta\beta_{w}$ values. The fact that $\Delta\beta_{sw} > \Delta\beta_{w}$ indicates that sea salt initially present in seawater has disrupted a portion of the open or normally structured water which would have contributed to further compaction due to the application of pressure.

In summary, the main effect of pressure (i.e. pressure to 1000 bars) on seawater appears to be the breakdown of some of the existing tetrahedral water structures which had survived the initial structure breaking effect of sea salt ions.

Because $\Delta\beta$ contains a component due to the compressibility of the 'solvation' sphere, it is not possible from this work alone to determine how pressure might influence hydration effects. As we saw earlier (SAMOILOV, 1957, 1965, 1972), the traditional view of 'hydration' is itself the subject of different interpretations and therefore it does not seem worthwhile at this time to speculate further on the pressure effect of a mechanism which may not have firm theoretical grounding.

## 2. Concluding remarks and future research

In this paper I have taken the view that the structure of water in seawater is altered due to the breakdown of water structure by sea salt ions. So far, unfortunately, theoretical descriptions of the state of water in ion–water interactions have been either empirical or qualitative (DESNOYERS and JOLICOEUR, 1969; MILLERO, 1974). This, however, does not detract from the practical importance of water structure effects in understanding seawater chemistry. One might ask: How does oceanography benefit from a knowledge of the water structure aspects of seawater? To answer this question, one may consider those phenomena in the oceans which depend primarily upon the solvent (water) properties of seawater. The solubility of gases in seawater is probably a function of the structure of water in seawater. A knowledge of the distribution and solubility of gases in the ocean is important for understanding geochemical and biochemical processes (BENSON, 1965). The solubilities of oxygen, carbon monoxide, nitrogen, and argon decrease with increasing salinity (GREEN and CARRITT, 1967; DOUGLAS, 1965, 1967). Because these gases have no ionic charge, their salinity dependence of solubility is most likely a result of water structure effects rather than specific ion effects (i.e. ion–gas interactions). A fundamental problem in the dissolution of a gaseous solute is for the gas molecules to find, or create, suitable positions ('cavities') within the structure of the existing water structure. The existence of broken water structure would be expected to influence the solubility of gases. BEN-NAIM and EGEL-THAL (1965) have shown that the effect of ionic species

in aqueous solutions is to break down the structure of water and consequently to reduce the number of 'cavity' sites for the solute gas. Recently, there have been several excellent determinations of the solubilities of gases in seawater, although virtually none of the solubility results has been interpreted in terms of water structure effects.

Another application of water structure effects deals with the solubility of nonpolar organic compounds in seawater. This is closely related to the gas solubility problem in the sense that the solute is without charge and therefore neither polarizes nearby water nor reacts with electrolytes. In general, the presence of electrolytes in an aqueous solution decreases the solubility of a nonpolar organic compound; this is referred to as the salting-out effect, and historically it was attributed to the loss of the water (presumably the 'hydration' water) from its solvent role. However, BOCKRIS and REDDY (1971) calculated that the loss of hydrated water has only a negligible effect on the solubility of a non-electrolyte. Bockris and Reddy derived an equation which relates the solubility change to the difference between the polarizabilities of a nonpolar organic molecule and a water molecule. Implicit in their theory is the fact that a cavity must be found for the organic molecule to fit into. McDEVIT and LONG (1952), who were the first to derive a theoretical equation to predict the salting-out effect, said that 'when salt is added to an aqueous solution of a non-electrolyte, the increase in internal pressure resulting from ion-solvent interaction *squeezes out* the non-electrolyte molecule.' One could also explain the decrease in solubility as due to the exclusion of the organic molecule from that portion of water structure which has already been broken, or disrupted, by the effects of ions.

In a series of laboratory experiments, GORDON and THORNE (1967) demonstrated that sea salt was very effective in reducing the solubility of naphthalene in water. Gordon and Thorne determined systematically the Setschenow constant for each sea salt. (The Setschenow constant is a proportionality constant which relates the decrease in solubility of a non-electrolyte to the concentration of the salting-out electrolyte.) The Setschenow constant for $Na_2SO_4$ was reported to be twice as great as that found for the other sea salts. (Unfortunately, the Setschenow constant for $MgSO_4$ was not determined.) The strong salting-out effect caused by $Na_2SO_4$ correlates well with the fact that $Na_2SO_4$ was the sea salt most effective in decreasing the compressibility of solutions.

The concentration of natural nonpolar organic compounds in seawater has been guessed to range from parts per million to parts per billion (Blumer, private communication). The concentration of hydrocarbons from oil spills could be much greater than this. It is not known if the saturation limit of nonpolar organics (either in surface water or in deep water) has been reached. Water structure effects, which are fundamentally responsible for the salting-out of such compounds, serves as an effective mechanism against the build up of an appreciable dissolved fraction. This would be particularly important in estuaries where large salinity changes exist and where active deposition of nonpolar organics might be expected to take place.

This work has been concerned with the water structure effects in seawater. In reality, seawater must *not* be viewed as an idealized single phase system: It is at least a three phase system composed of (1) an aqueous phase, which has been the subject of this work, (2) a gas phase (the bubbles in seawater), and (3) a number of phases in the form of particulate matter which eventually becomes sediment. With respect to the solid phases, I am particularly intrigued by the possibility that water structure effects near a solid surface may be very different from those in the liquid phase. Very recent work by DAYAL (1975) has shown that increasing hydrostatic pressure

applied to different clay–sea water systems inhibits the sorption of cations by clay minerals. According to Dayal, decreasing cation sorption by clay with increasing pressure is attributed to the effect of pressure on the ion-exchange properties of the structured water in the vicinity of the clay particles.

It is important to know as much as possible about structure effects near surfaces since it is at interfaces (i.e. liquid–solid and gas–liquid) that 'action', such as charge transfer and mass transfer, takes place. The rate of a particular event, such as the dissolution of a particle, may depend upon how the water is structured on the surface of the material. The type of structure will itself depend on the kind of material present. Oceanographers will undoubtedly have to examine surface–chemical effects in order to solve particular existing and future problems.

**Acknowledgments:** This paper was taken from a thesis in partial fulfillment of the requirements of the Ph.D. degree, Dalhousie University, Halifax, Nova Scotia. I thank Dr. Lloyd Dickie, past Director of the Marine Ecology Laboratory (MEL) of the Fisheries Research Board of Canada at the Bedford Institute of Oceanography (BIO), Dartmouth, Nova Scotia, for supporting this work. I also thank Mr. Stephen Paulowich of MEL for giving a lot of time to the design and development of the instrumentation used in this research. The experimental work was conducted at BIO, and I thank Dr. Alan Walton (Director of the Division of Chemical Oceanography at BIO) and his staff for their assistance.

Dr. Peter J. Wangersky, Dalhousie University, was my thesis advisor. His interest in water structure effects in oceanography led me to undertake this research. I am grateful to him for suggestions on the interpretation of the experimental results and for the written criticisms of the drafts. I am also grateful to the following people who gave criticisms of the drafts: Drs. Robert Cooke and Jan Kwak, Dalhousie University, Dr. Michael Falk, National Research Council of Canada (Halifax), and Dr. William L. Marshall, Oak Ridge National Laboratory, Oak Ridge, Tennessee. I thank Jacqueline Restivo for her assistance in the preparation of the manuscript and Dr. Dana Kester for performing the $SO_4^{2-}$ speciation calculations.

APPENDIX A

Tables A1 through A10. Tabulation of $\Delta\beta$ values as a function temperature, pressure, and test salt concentration.

Table A1. Summary of $\Delta\beta$ values as a function of temperature and test salt concentration; $\Delta\beta(\text{Bar}^{-1})$ = table values $\times 10^{-6}$. $P = 1$ Bar.

| Reference Solution | Ordinary Water | | | | Seawater | | | |
|---|---|---|---|---|---|---|---|---|
| Test Salt Concentration (m) | 0.13 | | 0.26 | | 0.13 | | 0.26 | |
| Temperature (°C) | 2 | 15 | 2 | 15 | 2 | 15 | 2 | 15 |
| LiCl | | | | −1.36 | | | | |
| NaCl | −0.88 | −0.66 | −1.83 | −1.47 | −0.82 | −0.70 | −1.63 | −1.30 |
| KCl | | | | −1.43 | | | | −1.24 |
| CsCl | | | | −1.39 | | | | |
| NaF | | | | −1.76 | | | | |
| NaI | | | | −1.14 | | | | |
| MgCl₂ | −1.52 | −1.25 | −3.20 | −2.60 | −1.56 | −1.36 | −2.90 | −2.40 |
| CaCl₂ | −1.58 | −1.30 | −3.12 | −2.65 | | | | |
| BaCl₂ | −1.73 | −1.72 | −3.80 | −3.05 | | | | |
| Na₂SO₄ | −2.43 | −2.01 | −4.72 | −3.84 | −2.01 | −1.58 | −3.96 | −3.04 |
| K₂SO₄ | | | | −3.53 | | | | −2.99 |
| MgSO₄ | −2.09 | −1.74 | −3.99 | −3.33 | −1.88 | −1.58 | −3.57 | −2.86 |

Table A2. Summary of $\Delta\beta$ values as a function of temperature and test salt concentration; $\Delta\beta(\text{Bar}^{-1})$ = table values $\times 10^{-6}$. $P = 101.2$ Bars.

| Reference Solution | Ordinary Water | | | | Seawater | | | |
|---|---|---|---|---|---|---|---|---|
| Test Salt Concentration (m) | 0.13 | | 0.26 | | 0.13 | | 0.26 | |
| Temperature (°C) | 2 | 15 | 2 | 15 | 2 | 15 | 2 | 15 |
| LiCl | | | | −1.30 | | | | |
| NaCl | −0.84 | −0.64 | −1.77 | −1.42 | −0.82 | −0.68 | −1.57 | −1.26 |
| KCl | | | | −1.38 | | | | −1.20 |
| CsCl | | | | −1.34 | | | | |
| NaF | | | | −1.70 | | | | |
| NaI | | | | −1.08 | | | | |
| MgCl₂ | −1.46 | −1.20 | −3.08 | −2.52 | −1.52 | −1.32 | −2.80 | −2.33 |
| CaCl₂ | −1.53 | −1.25 | −3.01 | −2.56 | | | | |
| BaCl₂ | −1.67 | −1.66 | −3.66 | −2.95 | | | | |
| Na₂SO₄ | −2.36 | −1.95 | −4.56 | −3.74 | −1.95 | −1.53 | −3.86 | −2.98 |
| K₂SO₄ | | | | −3.44 | | | | −2.91 |
| MgSO₄ | −2.03 | −1.70 | −3.88 | −3.24 | −1.84 | −1.55 | −3.47 | −2.79 |

Table A3. Summary of $\Delta\beta$ values as a function of temperature and test salt concentration; $\Delta\beta(\text{Bar}^{-1})$ = table values $\times 10^{-6}$. $P = 201.2$ Bars.

| Reference Solution | Ordinary Water | | | | Seawater | | | |
|---|---|---|---|---|---|---|---|---|
| Test Salt Concentration (m) | 0.13 | | 0.26 | | 0.13 | | 0.26 | |
| Temperature (°C) | 2 | 15 | 2 | 15 | 2 | 15 | 2 | 15 |
| LiCl | | | | −1.23 | | | | |
| NaCl | −0.82 | −0.62 | −1.70 | −1.37 | −0.81 | −0.66 | −1.51 | −1.22 |
| KCl | | | | −1.34 | | | | |
| CsCl | | | | −1.30 | | | | |
| NaF | | | | −1.62 | | | | |
| NaI | | | | −1.03 | | | | |
| MgCl₂ | −1.40 | −1.16 | −2.95 | −2.44 | −1.49 | −1.28 | −2.70 | −2.26 |
| CaCl₂ | −1.48 | −1.20 | −2.89 | −2.46 | | | | |
| BaCl₂ | −1.60 | −1.59 | −3.52 | −2.85 | | | | |
| Na₂SO₄ | −2.30 | −1.89 | −4.44 | −3.64 | −1.89 | −1.49 | −3.76 | −2.92 |
| K₂SO₄ | | | | −3.36 | | | | −2.82 |
| MgSO₄ | −1.97 | −1.65 | −3.76 | −3.15 | −1.79 | −1.51 | −3.37 | −2.72 |

Table A4. Summary of $\Delta\beta$ values as a function of temperature and test salt concentration; $\Delta\beta(\text{Bar}^{-1})$ = table values $\times 10^{-6}$. $P = 301.4$ Bars.

| Reference Solution | Ordinary Water | | | | Seawater | | | |
|---|---|---|---|---|---|---|---|---|
| Test Salt Concentration (m) | 0.13 | | 0.26 | | 0.13 | | 0.26 | |
| Temperature (°C) | 2 | 15 | 2 | 15 | 2 | 15 | 2 | 15 |
| LiCl | | | | −1.17 | | | | |
| NaCl | −0.78 | −0.60 | −1.64 | −1.32 | −0.80 | −0.64 | −1.46 | −1.15 |
| KCl | | | | −1.29 | | | | −1.12 |
| CsCl | | | | −1.25 | | | | |
| NaF | | | | −1.55 | | | | |
| NaI | | | | −0.98 | | | | |
| MgCl₂ | −1.34 | −1.12 | −2.82 | −2.36 | −1.45 | −1.24 | −2.60 | −2.18 |
| CaCl₂ | −1.42 | −1.17 | −2.77 | −2.36 | | | | |
| BaCl₂ | −1.53 | −1.53 | −3.37 | −2.74 | | | | |
| Na₂SO₄ | −2.23 | −1.83 | −4.29 | −3.56 | −1.83 | −1.45 | −3.66 | −2.85 |
| K₂SO₄ | | | | −3.26 | | | | −2.73 |
| MgSO₄ | −1.91 | −1.60 | −3.64 | −3.06 | −1.74 | −1.48 | −3.27 | −2.66 |

Table A5.   Summary of $\Delta\beta$ values as a function of temperature and test salt concentration; $\Delta\beta(\text{Bar}^{-1}) = $ table values $\times 10^{-6}$. $P = 401.6$ Bars.

| Reference Solution | Ordinary Water | | | | Seawater | | | |
|---|---|---|---|---|---|---|---|---|
| Test Salt Concentration (m) | 0.13 | | 0.26 | | 0.13 | | 0.26 | |
| Temperature (°C) | 2 | 15 | 2 | 15 | 2 | 15 | 2 | 15 |
| LiCl | | | | −1.11 | | | | |
| NaCl | −0.75 | −0.58 | −1.57 | −1.27 | −0.79 | −0.62 | −1.40 | −1.12 |
| KCl | | | | −1.24 | | | | −1.09 |
| CsCl | | | | −1.20 | | | | |
| NaF | | | | −1.48 | | | | |
| NaI | | | | −0.93 | | | | |
| MgCl$_2$ | −1.27 | −1.07 | −2.68 | −2.28 | −1.41 | −1.19 | −2.50 | −2.10 |
| CaCl$_2$ | −1.37 | −1.13 | −2.65 | −2.26 | | | | |
| BaCl$_2$ | −1.46 | −1.46 | −3.23 | −2.63 | | | | |
| Na$_2$SO$_4$ | −2.16 | −1.76 | −4.14 | −3.44 | −1.77 | −1.40 | −3.55 | −2.78 |
| K$_2$SO$_4$ | | | | −3.18 | | | | −2.64 |
| MgSO$_4$ | −1.84 | −1.56 | −3.52 | −2.97 | −1.69 | −1.44 | −3.17 | −2.58 |

Table A6.   Summary of $\Delta\beta$ values as a function of temperature and test salt concentration; $\Delta\beta(\text{Bar}^{-1}) = $ table values $\times 10^{-6}$. $P = 501.6$ Bars.

| Reference Solution | Ordinary Water | | | | Seawater | | | |
|---|---|---|---|---|---|---|---|---|
| Test Salt Concentration (m) | 0.13 | | 0.26 | | 0.13 | | 0.26 | |
| Temperature (°C) | 2 | 15 | 2 | 15 | 2 | 15 | 2 | 15 |
| LiCl | | | | −1.04 | | | | |
| NaCl | −0.72 | −0.56 | −1.50 | −1.21 | −0.78 | −0.60 | −1.33 | −1.07 |
| KCl | | | | −1.18 | | | | −1.05 |
| CsCl | | | | −1.15 | | | | |
| NaF | | | | −1.40 | | | | |
| NaI | | | | −0.87 | | | | |
| MgCl$_2$ | −1.21 | −1.02 | −2.54 | −2.20 | −1.36 | −1.15 | −2.40 | −2.02 |
| CaCl$_2$ | −1.31 | −1.08 | −2.52 | −2.16 | | | | |
| BaCl$_2$ | −1.39 | −1.39 | −3.08 | −2.52 | | | | |
| Na$_2$SO$_4$ | −2.08 | −1.70 | −3.98 | −3.34 | −1.71 | −1.36 | −3.44 | −2.71 |
| K$_2$SO$_4$ | | | | −3.08 | | | | −2.55 |
| MgSO$_4$ | −1.78 | −1.51 | −3.40 | −2.87 | −1.63 | −1.39 | −3.07 | −2.51 |

Table A7.   Summary of $\Delta\beta$ values as a function of temperature and test salt concentration; $\Delta\beta(\text{Bar}^{-1}) = $ table values $\times 10^{-6}$.
$P = 601.4$ Bars.

| Reference Solution | Ordinary Water | | | | Seawater | | | |
|---|---|---|---|---|---|---|---|---|
| Test Salt Concentration $(m)$ | 0.13 | | 0.26 | | 0.13 | | 0.26 | |
| Temperature (°C) | 2 | 15 | 2 | 15 | 2 | 15 | 2 | 15 |
| LiCl | | | | −0.98 | | | | |
| NaCl | −0.68 | −0.54 | −1.43 | −1.17 | −0.77 | −0.59 | −1.27 | −1.03 |
| KCl | | | | −1.13 | | | | −1.01 |
| CsCl | | | | −1.10 | | | | |
| NaF | | | | −1.32 | | | | |
| NaI | | | | −0.82 | | | | |
| $MgCl_2$ | −1.14 | −0.98 | −2.40 | −2.11 | −1.32 | −1.10 | −2.29 | −1.94 |
| $CaCl_2$ | −1.26 | −1.04 | −2.40 | −2.05 | | | | |
| $BaCl_2$ | −1.31 | −1.32 | −1.92 | −2.41 | | | | |
| $Na_2SO_4$ | −2.01 | −1.63 | −3.82 | −3.23 | −1.64 | −1.31 | −3.33 | −2.65 |
| $K_2SO_4$ | | | | −2.98 | | | | −2.46 |
| $MgSO_4$ | −1.72 | −1.46 | −3.27 | −2.78 | −1.59 | −1.36 | −2.96 | −2.43 |

Table A8.   Summary of $\Delta\beta$ values as a function of temperature and test salt concentration; $\Delta\beta(\text{Bar}^{-1}) = $ table values $\times 10^{-6}$.
$P = 701.2$ Bars.

| Reference Solution | Ordinary Water | | | | Seawater | | | |
|---|---|---|---|---|---|---|---|---|
| Test Salt Concentration $(m)$ | 0.13 | | 0.26 | | 0.13 | | 0.26 | |
| Temperature (°C) | 2 | 15 | 2 | 15 | 2 | 15 | 2 | 15 |
| LiCl | | | | −0.90 | | | | |
| NaCl | −0.65 | −0.52 | −1.36 | −1.11 | −0.76 | −0.56 | −1.21 | −0.98 |
| KCl | | | | −1.08 | | | | −0.96 |
| CsCl | | | | −1.04 | | | | |
| NaF | | | | −1.25 | | | | |
| NaI | | | | −0.76 | | | | |
| $MgCl_2$ | −1.07 | −0.93 | −2.26 | −2.02 | −1.28 | −1.06 | −2.18 | −1.86 |
| $CaCl_2$ | −1.20 | −0.99 | −2.26 | −1.94 | | | | |
| $BaCl_2$ | −1.24 | −1.24 | −2.77 | −2.30 | | | | |
| $Na_2SO_4$ | −1.93 | −1.56 | −3.66 | −3.13 | −1.58 | −1.27 | −3.22 | −2.57 |
| $K_2SO_4$ | | | | −2.88 | | | | −2.37 |
| $MgSO_4$ | −1.65 | −1.41 | −3.14 | −2.68 | −1.53 | −1.32 | −2.85 | −2.36 |

Table A9. Summary of $\Delta\beta$ values as a function of temperature and test salt concentration; $\Delta\beta(\text{Bar}^{-1}) = $ table values $\times 10^{-6}$. $P = 801.2$ Bars.

| Reference Solution | Ordinary Water | | | | Seawater | | | |
|---|---|---|---|---|---|---|---|---|
| Test Salt Concentration (m) | 0.13 | | 0.26 | | 0.13 | | 0.26 | |
| Temperature (°C) | 2 | 15 | 2 | 15 | 2 | 15 | 2 | 15 |
| LiCl | | | | −0.84 | | | | |
| NaCl | −0.61 | −0.50 | −1.28 | −1.06 | −0.75 | −0.54 | −1.14 | −0.93 |
| KCl | | | | −1.03 | | | | −0.92 |
| CsCl | | | | −1.00 | | | | |
| NaF | | | | −1.17 | | | | |
| NaI | | | | −0.71 | | | | |
| MgCl$_2$ | −1.00 | −0.88 | −2.11 | −1.94 | −1.24 | −1.01 | −2.07 | −1.77 |
| CaCl$_2$ | −1.15 | −0.94 | −2.13 | −1.84 | | | | |
| BaCl$_2$ | −1.16 | −1.17 | −2.61 | −2.18 | | | | |
| Na$_2$SO$_4$ | −1.86 | −1.49 | −3.50 | −3.02 | −1.51 | −1.22 | −3.10 | −2.50 |
| K$_2$SO$_4$ | | | | −2.79 | | | | −2.27 |
| MgSO$_4$ | −1.58 | −1.36 | −3.00 | −2.58 | −1.48 | −1.28 | −2.75 | −2.28 |

Table A10. Summary of $\Delta\beta$ values as a function of temperature and test salt concentration; $\Delta\beta(\text{Bar}^{-1}) = $ table values $\times 10^{-6}$. $P = 901.2$ Bars.

| Reference Solution | Ordinary Water | | | | Seawater | | | |
|---|---|---|---|---|---|---|---|---|
| Test Salt Concentration (m) | 0.13 | | 0.26 | | 0.13 | | 0.26 | |
| Temperature (°C) | 2 | 15 | 2 | 15 | 2 | 15 | 2 | 15 |
| LiCl | | | | −0.77 | | | | |
| NaCl | −0.58 | −0.48 | −1.21 | −1.00 | −0.74 | −0.52 | −1.08 | −0.87 |
| KCl | | | | −0.97 | | | | −0.88 |
| CsCl | | | | −0.98 | | | | |
| NaF | | | | −1.09 | | | | |
| NaI | | | | −0.65 | | | | |
| MgCl$_2$ | −0.94 | −0.83 | −1.96 | −1.85 | −1.19 | −0.96 | −1.96 | −1.69 |
| CaCl$_2$ | −1.08 | −0.89 | −2.00 | −1.72 | | | | |
| BaCl$_2$ | −1.09 | −1.05 | −2.44 | −2.06 | | | | |
| Na$_2$SO$_4$ | −1.78 | −1.42 | −3.31 | −2.90 | −1.45 | −1.17 | −2.98 | −2.42 |
| K$_2$SO$_4$ | | | | −2.68 | | | | −2.17 |
| MgSO$_4$ | −1.51 | −1.31 | −2.86 | −2.48 | −1.42 | −1.24 | −2.62 | −2.20 |

## APPENDIX B

Values of $\beta_{ref.}(bar^{-1})$ determined in this work; $\beta_{ref.}$ = table values $\times 10^{-6}$.

| $P$ Absolute Pressure (bar) | Ordinary Water | | Seawater (35‰) | |
|---|---|---|---|---|
| | 2°C | 15°C | 2°C | 15°C |
| 1 | 48.99 | 45.51 | 44.83 | 42.14 |
| 101.2 | 47.85 | 44.53 | 43.85 | 41.27 |
| 201.2 | 46.70 | 43.54 | 42.85 | 40.37 |
| 301.4 | 45.53 | 42.54 | 41.84 | 39.50 |
| 401.6 | 44.34 | 41.52 | 40.82 | 38.56 |
| 501.6 | 43.13 | 40.50 | 39.75 | 37.64 |
| 601.4 | 41.93 | 39.46 | 38.74 | 36.71 |
| 701.2 | 40.68 | 38.41 | 37.68 | 35.77 |
| 801.2 | 39.43 | 37.34 | 36.60 | 34.82 |
| 901.2 | 38.16 | 36.27 | 35.52 | 33.85 |

## REFERENCES

ALLAM D. S. and W. H. LEE (1966) Ultrasonic studies of electrolyte solutions. Part II. Compressibilities of electrolytes. *J. chem. Soc.* (Sec. A), 5–9.

AZZAM A. A. (1962) Theoretical Studies on Solvation. VI. The Evaluation of the Maximum Secondary Solvation Number for the Mono- and Divalent Metal and Halide Ions at 25°C. *Z. Phys. Chem.* (*Neue Folge*) **33**, 320–336.

BENSON B. B. (1965) Some thoughts on dissolved gases in the ocean. *In*: D. R. SCHINK and J. T. CORLESS, Eds., *Symposium on Marine Geochemistry.* Graduate School of Oceanography, University of Rhode Island, Occasional Publ. No. 3, 1965, pp. 91–107.

BERNAL J. D. and R. H. FOWLER (1933) A theory of water and ionic solution, with particular reference to hydrogen and hydroxyl ions. *J. chem. Phys.* **1**, 515–548.

BOCKRIS J. O'M. (1949) Ionic solvation. *Chem. Soc. Q. Rev.* **3**, 173–180.

BOCKRIS J. O'M. and A. K. N. REDDY (1970) *Modern Electrochemistry.* Plenum Press, New York.

BOCKRIS J. O'M. and P. P. SALUJA (1972) Ionic solvation numbers from compressibilities and ionic vibration potentials measurements. *J. phys. Chem.* **76**, 2140–2151.

BRADY G. W. and J. T. KRAUSE (1957) Structure in ionic solutions. *J. chem. Phys.* **27**, 304–308.

CARD D. N. and J. P. VAILEAU (1970) Monte Carlo study of the thermodynamics of electrolyte solutions. *J. chem. Phys.* **52**, 6232–6240.

CHEN C.-T. and F. J. MILLERO (1976) The specific volume of seawater at high pressures. *Deep-Sea Res.* **23**, 583–612.

CONWAY B. E. and J. O'M. BOCKRIS (1954) Ionic solvation. *In*: J. O'M. BOCKRIS, Ed., *Modern Aspects of Electrochemistry.* Vol. I., Chap. 2. Butterworth, London.

CONWAY B. E. and R. E. VERRALL (1966) Partial molar volumes and adiabatic compressibilities of tetraalkylammonium and aminium salts in water. I. Compressibility behavior. *J. phys. Chem.* **70**, 3952–3961.

CULKIN F. and R. A. COX (1966) Sodium, potassium, magnesium, calcium and strontium in sea water. *Deep-Sea Res.* **13**, 789–804.

DAVIES C. W. (1927) The extent of dissociation of salts in water. *Trans. Faraday Soc.* **23**, 351–356.

DESNOYERS J. E. and C. JOLICOEUR (1969) Hydration effects and thermodynamic properties of ions. *In*: J. O'M. BOCKRIS and B. E. CONWAY, Eds., *Modern Aspects of Electrochemistry*, Chap. 1. Plenum Press, New York.

DESNOYERS J. E., R. E. VERRALL and B. E. CONWAY (1965) Electrostriction in aqueous solutions of electrolytes. *J. chem. Phys.* **43**, 243–250.

DOUGLAS E. (1965) Solubilities of argon and nitrogen in sea water. *J. phys. Chem.* **69**, 2608–2610.

DOUGLAS E. (1967) Carbon monoxide solubilities in sea water. *J. phys. Chem.* **71**, 1931–1933.

DUEDALL I. W. (1972) The partial molal volume of calcium carbonate in seawater. *Geochim. cosmochim. Acta* **36**, 729–734.

DUEDALL I. W. and S. PAULOWICH (1973) A bellows-type differential compressimeter for determining the difference between the compressibilities of two seawater solutions to 900 bars. *Rev. Sci. Instr.* **44**, 120–127.

EIGEN M. and K. TAMM (1962) Schallabsorption in Elektrolytlosungen als Folge chemischer Relaxation. II. Messergebnisse und Relaxationsmechanismen für 2-2wertige Elektrolyte. *Z. Elektrochemie* **66**, 107–121.

EISENBERG D. and W. KAUZMANN (1969) *The Structure and Properties of Water.* Clarendon Press, Oxford.

EMMET R. T. and F. J. MILLERO (1974) Direct measurement of the specific volume of seawater from −0.2° to 40°C and from 0 to 1000 bars, Preliminary results. *J. geophys. Res.* **79**, 3463–3472.

FALK M. and O. KNOP (1973) Water in stoichiometric hydrates. *In*: F. FRANKS, Ed., *Water: A Comprehensive Treatise.* Vol. II. Plenum Press, New York.

FISHER F. H. (1972) Effect of pressure on sulfate ion association and ultrasonic absorption in sea water. *Geochim. cosmochim. Acta* **36**, 99–101.

FRANK H. S. (1965) Structural influences on activity coefficients in aqueous electrolytes. *Z. Phys. Chem.* (Leipzig), **228**, 364–372.

FRANK H. S. and M. W. EVANS (1945) Free volume and

entropy in condensed systems. III. Entropy in binary liquid mixtures; partial molal entropy in dilute solutions; structure and thermodynamics in aqueous electrolytes. *J. chem. Phys.* **13**, 507–532.

FRANK H. S. and W. Y. WEN (1957) Structural aspects of ion–solvent interaction in aqueous solutions: A suggested picture of water structure. *Disc. Faraday Soc.* (No. 24), 133–140.

FRIEDMAN H. L. (1971) Computed thermodynamic properties and distribution functions for simple models of ionic solutions. *In*: J. O'M. BOCKRIS and B. E. CONWAY, Eds., *Modern Aspects of Electrochemistry* (No. 4), Chap. 1. Plenum Press, New York.

GILKERSON W. R. (1970) The importance of the effect of the solvent dielectric constant on ion-pair formation in water at high temperatures and pressures. *J. phys. Chem.* **74**, 746–750.

GLUECKAUF E. (1955) The influence of ionic hydration on activity coefficients in concentrated solutions. *Trans. Faraday Soc.* **51**, 1235–1244.

GORDON J. F. and R. L. THORNE (1967) Salt effects on non-electrolyte activity coefficients in mixed aqueous electrolyte solutions. II. Artificial and natural sea waters. *Geochim. cosmochim. Acta* **31**, 2433–2443.

GREEN E. J. and D. E. CARRITT (1967) Oxygen solubility in sea water: thermodynamic influence of sea salt. *Science* **157**, 191–193.

GREYSON J. (1962) Transfer free energies for some univalent chlorides from $H_2O$ to $D_2O$ from measurements of ion exchange membrane potentials. *J. phys. Chem.* **66**, 2218–2221.

GREYSON J. (1967) The influence of the alkali halides on the structure of water. *J. phys. Chem.* **71**, 2210–2213.

GUCKER F. T. (1933a) The compressibility of solutions. The apparent molal compressibility of strong electrolytes. *J. Am. chem. Soc.* **55**, 2709–2718.

GUCKER F. T. (1933b) The apparent molal heat capacity, volume, and compressibility of electrolytes. *Chem. Rev.* **13**, 111–130.

GURNEY R. W. (1953) *Ionic Processes in Solution.* Dover, New York.

HAGGIS G. H., J. B. HASTED and T. J. BUCHANAN (1952) The dielectric properties of water in solutions. *J. chem. Phys.* **20**, 1452–1465.

HARNED H. S. and B. B. OWEN (1958) *The Physical Chemistry of Electrolyte Solutions.* 3rd Edn. Reinhold, New York.

HINDMAN J. C. (1962) Nuclear magnetic resonance effects in aqueous solutions of 1-1 electrolytes. *J. chem. Phys.* **36**, 1000–1015.

HOLTZER A. and M. F. EMERSON (1969) On the utility of the concept of water structure in the rationalization of the properties of aqueous solutions of proteins and small molecules. *J. phys. Chem.* **73**, 26–33.

HORNE R. A. (1966) Electrostriction and the dehydration of ions under pressure. Technical Report No. 26, Arthur D. Little, Cambridge, Mass.

HORNE R. A. (1969) *Marine Chemistry.* Wiley, New York.

HORNE R. A. and J. D. BIRKETT (1967) The total hydration atmospheres of the alkali–metal cations in aqueous solutions. *Electrochim. Acta* **12**, 1153–1160.

HORNE R. A., A. F. DAY, R. P. YOUNG and N. T. YU (1968) Interfacial water structure: the electrical conductivity under hydrostatic pressure of particulate solids permeated with

aqueous electrolyte solution. *Electrochim. Acta* **13**, 397–406.

KARYAKIN A. V. and G. A. MURADOVA (1968) State of water in crystal hydrates. *Zh. Fis. Khim.* **42**, 2735–2740 (in Russian).

KAVANAU J. L. (1964) *Water and Solute–Water Interactions.* Holden-Day, San Francisco.

KESSLER YU. M., YU. M. POVAROVA and A. I. GARBANEV (1962) Interactions between neighboring ions in solutions. *J. Structural Chem.* (Consultants Bureau) **3**, 82–83.

KESTER D. R. (1970) Ion association of sodium, magnesium, and calcium with sulfate in aqueous solution. Ph.D. Thesis. Corvallis, Oregon State University.

KESTER D. R., I. W. DUEDALL, D. N. CONNORS and R. M. PYTKOWICZ (1967) Preparation of artificial seawater. *Limnol. Oceanogr.* **12**, 176–179.

KESTER D. R. and R. M. PYTKOWICZ (1970) Effect of temperature and pressure on sulfate ion association in sea water. *Geochim. cosmochim. Acta* **34**, 1039–1051.

KIRKWOOD J. G. (1934) Theory of solutions of molecules containing widely separated charges with special application to zwitterions. *J. chem. Phys.* **34**, 351–361.

KIRKWOOD J. G. (1939) Theoretical studies upon dipolar ions. *Chem. Rev.* **24**, 233–251.

KORTUM G. and J. O'M. BOCKRIS (1951) *Textbook of Electrochemistry.* Elsevier, New York.

LONGSWORTH L. G. (1947) A moving boundary method for the measurement of non-electrolyte transport in mixed solvents. *J. Am. Chem. Soc.* **69**, 1288–1291.

MACDONALD R. W. and N. A. NORTH (1974) The effect of pressure on the solubility of $CaCO_3$, $CaF_2$, and $SrSO_4$ in water. *Can. J. Chem.* **52**, 3181–3186.

MARSHALL W. L. (1972) A further description of complete equilibrium constants. *J. phys. Chem.* **76**, 720–731.

MARSHALL W. L. and A. S. QUIST (1967) A representation of isothermal ion–ion-pair solvent equilibria independent of changes in dielectric constant. *Proc. natn Acad. Sci. U.S.A.* **58**, 901–906.

MASSON D. O. (1929) Solute molecular volumes in relation to solvation and ionization. *Phil. Mag.* **8**, 218–235.

MCDEVITT W. and F. LONG (1952) The activity coefficient of benzene in aqueous salt solutions. *J. Am. Chem. Soc.* **74**, 1773–1777.

MILLERO F. J. (1971) Effect of pressure on sulfate ion association in sea water. *Geochim. cosmochim. Acta* **35**, 1089–1098.

MILLERO F. J. (1972) The partial molal volume of electrolytes in aqueous solutions. *In*: R. A. HORNE, Ed., *Water and Aqueous Solutions*, Chap. 13. Wiley–Interscience, New York.

MILLERO F. J. (1973) Theoretical estimates of the isothermal compressibility of seawater. *Deep-Sea Res.* **20**, 101–105.

MILLERO F. J. (1974) Seawater as a multicomponent electrolyte solution. *In*: E. GOLDBERG, Ed., *The Sea.* Volume 5, Chap. 4, pp. 3–80.

MILLERO F. J. and R. A. BERNER (1972) Effect of pressure on carbonate equilibria in seawater. *Geochim. cosmochim. Acta* **36**, 92–98.

MILLERO F. J. and W. L. MASTERON (1974) Volume change for the formation of magnesium sulfate ion pairs at various temperatures. *J. phys. Chem.* **78**, 1287–1294.

MILLERO F. J., G. K. WARD, F. K. LEPPLE and E. V. HOFF (1974) Isothermal compressibility of aqueous sodium chloride, magnesium chloride, sodium sulfate, and magnesium sulfate solutions from 0 to 45° at 1 atm. *J. phys. Chem.*

**78**, 1636–1643.

MOELWYN-HUGHES E. A. (1964) *Physical Chemistry.* Macmillan, New York.

MORGAN J. and B. E. WARREN (1938) X-ray analysis of the structure of water. *J. chem. Phys.* **6**, 666–673.

NANCOLLAS G. H. (1966) *Interactions in Electrolyte Solutions.* Elsevier, New York.

NEMETHY G. (1965) Comparison of models for water and aqueous solutions. *Fedn Proc. Fedn Am. Soc. exp. Biol.* **24**, 38–41.

NEMETHY G. (1970) The structure of water and the thermodynamic properties of aqueous solutions. *Annali D'Ins. Superiore di Sanita* **6** (Special Issue), 487–592.

NEMETHY G. and H. A. SCHERAGA (1962) Structure of water and hydrophobic bonding in proteins. I. A model for the thermodynamic properties of liquid water. *J. chem. Phys.* **36**, 3382–3400.

NORTH N. A. (1974) Pressure dependence of $SrSO_4$ solubility. *Geochim. cosmochim. Acta* **38**, 1075–1081.

OWEN B. B. and S. R. BRINKLEY, JR. (1941) Calculation of the effect of pressure upon ionic equilibria in pure water and salt solutions. *Chem. Rev.* **29**, 461–474.

OWEN B. B. and P. L. KRONICK (1961) Standard partial molal compressibilities by ultrasonics. II. Sodium and potassium chlorides and bromides from 0 to 30°C. *J. phys. Chem.* **65**, 84–87.

OWEN B. B. and H. L. SIMONS (1957) Standard partial molal compressibilities by ultrasonics. I. Sodium chloride and potassium chloride at 25°. *J. phys. Chem.* **61**, 479–482.

QUIST A. S. and W. L. MARSHALL (1968) The independence of isothermal equilibria in electrolyte solutions on changes in dielectric constant. *J. phys. Chem.* **72**, 1536–1544.

PADOVA J. (1963) Ion–solvent interaction. II. Partial molar volume and electrostriction: A thermodynamic approach. *J. chem. Phys.* **39**, 1552–1557.

PASSYNSKI A. (1938) Compressibility and solvation of electrolytes. *Acta Physicochim. URSS* **8**, 385–418.

PYTKOWICZ R. M. (1972) The status of our knowledge of sulfate association in sea water. *Geochim. cosmochim. Acta* **36**, 631–633.

ROBINSON R. A. and R. H. STOKES (1965) *Electrolyte Solutions.* 2nd ed., revised. Butterworth, London.

ROSS D. K. (1968) The interaction energy of an ion-pair with overlapping hydration shells. *Aust. J. Phys.* **21**, 587–595.

ROSS D. K. and S. LEVINE (1968) The interaction energy of an ion-pair with overlapping hydrated shells. *Aust. J. Phys.* **21**, 571–585.

SAMOILOV O. YA. (1957) A new approach to the study of hydration ions in aqueous solutions. *Disc. Faraday Soc.* (No. 24), 141–146.

SAMOILOV O. YA (1965) *Structure of Aqueous Electrolyte Solutions and the Hydration of Ions.* Consultants Bureau, New York.

SAMOILOV O. YA. (1972) Residence times of ionic hydration. *In*: R. A. HORNE, Ed., *Water and Aqueous Solutions*, Chap. 14. Wiley–Interscience, New York.

STANLEY E. M. and R. C. BATTEN (1969) Viscosity of sea water at moderate temperatures. *J. Geophys. Res.* **74**, 3415–3420.

STOKES R. H. (1945) A thermodynamic study of bivalent metal halides in aqueous solutions. Part XIV. Concentrated solutions of magnesium chloride at 25°C. *Trans. Faraday Soc.* **41**, 642–645.

STOKES R. H. and R. A. ROBINSON (1948) Ionic hydration and activity in electrolyte solutions. *J. Am. chem. Soc.* **70**, 1870–1878.

SUBRAMANIAN S. and H. F. FISHER (1972) Near-infrared spectral studies on the effects of perchlorate and tetrafluoroborate ions on water structure. *J. phys. Chem.* **76**, 84–89.

SWIFT T. J. and W. G. SAYRE (1966) Determination of hydration numbers of cations in aqueous solutions by means of proton NMR. *J. chem. Phys.* **44**, 3567–3574.

WALRAFEN G. E. (1962) Raman spectral studies of the effects of electrolytes on water. *J. chem. Phys.* **36**, 1035–1042.

WANGERSKY P. J. (1972) The control of seawater pH by ion-pairing. *Limnol. Oceanogr.* **17**, 1–6.

WASHBURN E. W. (1915) The hydration of ions determined by the transference experiment in the presence of a non-electrolyte. *J. Am. chem. Soc.* **37**, 694–699.

WHALLEY E. (1963) Some comments on electrostatic volumes and entropies of solvation. *J. chem. Phys.* **38**, 1400–1405.

WILSON W. and D. BRADLEY (1966) Specific volume, thermal expansion, and isothermal compressibility of sea water. Technical Report NOLTR 66-103, United States Naval Ordnance Laboratory, White Oak, Md.

VASLOW F. (1972) Thermodynamics of solutions of electrolytes. *In*: R. A. HORNE, Ed., *Water and Aqueous Solutions*, Chap. 12. Wiley–Interscience, New York.

Prog. Oceanog., 1977, Vol. 7, pp. 135–162   Pergamon Press.   Printed in Great Britain

# Observations of Rossby waves near site D[1]

RORY O. R. Y. THOMPSON[2]

*Woods Hole Oceanographic Institution*

(*Received* 24 *March* 1976; *in revised form* 6 *September* 1976; *accepted* 12 *November* 1976)

**Abstract**—A theory of Rossby waves makes a number of predictions about motions below the thermocline at Site D (39°10'N, 70°W). An experiment was made to test these predictions. Kinetic energy spectra show most of the energy is associated with periods of a week to a few months. The Reynolds' stress is negative for these periods. There is high coherence and no phase shift between 1000 m and 2000 m depths. Most of the wave-number estimates point to the south-west quadrant, consistent with westward propagation and with momentum flux into the Gulf Stream and energy flux out of it. There is enough consistency between statistics based on successive 8-month records to conclude that statistics based on a single 8-month record near Site D are meaningful. In an 8-month array of twenty-six current meters, for periods of a week to a month, the divergence is small compared to the vorticity, and the motion is transverse. The energy increases toward the bottom. The observed wave-numbers and frequencies fit the theoretical dispersion relationship satisfactorily. The kinetic energy of the fluctuations is much larger near the Gulf Stream than farther upslope. The vorticity is in quadrature with the upslope velocity. I conclude there is strong evidence that topographic Rossby wave mechanics are dominant below the thermocline on the continental rise north of the Gulf Stream.

## 1. INTRODUCTION

THIS PAPER may be regarded as a follow-up and extension to WEBSTER (1969) and to THOMPSON (1971a) (hereafter called 'T'). The Site D area is north of the Gulf Stream and south of the continental shelf, on topography which can be broadly described as having east–west isobaths and a gentle slope, getting shallower to the north. According to Gordon Volkmann (personal communication), the mean density surfaces along 70°W are essentially flat between 37°N and 40°N, so baroclinicity outside the Gulf Stream may be ignored. While there are deviations from this general picture, the situation is perhaps as close to the idealized models of topographic Rossby waves given by RHINES (1970) as can be expected in the real ocean. Site D is also convenient to Woods Hole, so an array of current meters was put out to test T's hypothesis that the low frequency motions below the thermocline can be described as topographic Rossby waves. T further suggested that these waves should carry energy away from the Gulf Stream, and consequently, should carry momentum into the Gulf Stream, so the wave-number vectors should point into the southwest quadrant. It should be emphasized that here are reported the results of an experiment to test a previously stated hypothesis, rather than rationalization of data gotten before the hypothesis. This is rather unusual in oceanography. The reader will realize that the aim here is not originality, but to provide solid support to some concepts commonly just assumed to be true.

---

[1] Contribution number 3678 of the Woods Hole Oceanographic Institution.
[2] Present address: CSIRO, Aspendale, Victoria, Australia.

## 2. REVIEW OF MODEL

RHINES (1970) gave a linear theory of potential vorticity conserving motions of a stratified rotating fluid on a slope, that is, a theory of topographic Rossby waves. The portion of the theory for small slope and low frequencies will be applied here to Site D. Perhaps the weakest part of the comparison is in approximating the topography around Site D by a constant slope. SUAREZ (1971) discusses some of the effects of uneven topography. Nonetheless, the topographic charts of UCHUPI (1965) and a local chart (Fig. 1) prepared from Buoy Project data [with the help of Marlene Noble] suggest that,

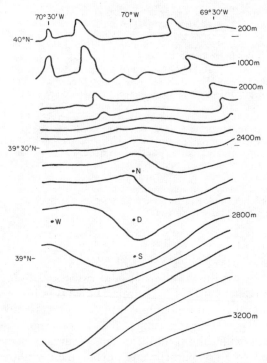

FIG. 1.   Chart of local bathymetry in Site D region. Note there is an abrupt slope between 2000 m and 200 m depth; otherwise the contour interval is 100 m.

averaged over distances comparable with a Rossby deformation radius ($\approx 35$ km), the bottom may be approximated around Site D by a plane with east–west isobaths, to within oceanographic accuracy. The north–south distance between the 2400 m and 2800 m isobaths averaged between 69°45′W to 70°30′W is about 55 km, giving a slope $\alpha = 0.0073$. RHINES (1970) in essence suggested that slow curvature of the isobaths will not invalidate the model for low frequencies, in that slow currents will tend to cling to isobaths. SUAREZ (1971) showed that the short waves, of relatively high frequency, will remain in the vicinity of generation, so will also not be influenced by slow curvature of isobaths.

The model adopted from RHINES (1970) has a rigid horizontal lid at $z = 0$ and a sloping bottom at $z = -H + \alpha y$, where we use $H = 2.7$ km to correspond to Site D. The Coriolis parameter $f$ is taken constant at $10^{-4}$ sec$^{-1}$. Define $N^2 = -g(\partial \bar{\rho}/\partial z)/\rho_0$. Figure 2 is a graph of average values of the Brunt–Väisälä period, $2\pi/N$, at Site D (from Gordon

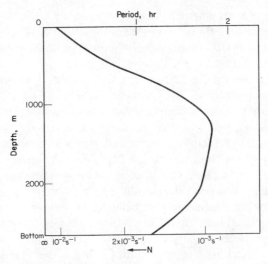

FIG. 2. Average values of the Brunt–Väisälä period, from numerous stations at Site D (70°00′W, 39°10′N), in hours. The inverse scale at the bottom gives $N$ in sec$^{-1}$.

Volkmann, personal communication). With suitable scaling, and assuming a wave $\exp[i(kx+ly-\omega t)]$, and $\kappa^2 = k^2 + l^2$, RHINES (1970) derived the equations:

$$\frac{d}{dz}\left(\frac{f^2}{N^2}\frac{dp}{dz}\right) = \kappa^2 p, \tag{2.1}$$

with
$$\frac{dp}{dz} = 0 \quad \text{at} \quad z = 0, \tag{2.2}$$

and
$$\frac{dp}{dz} = \frac{\alpha k}{f\omega}N^2 p \quad \text{at} \quad z = -H. \tag{2.3}$$

These equations are real, so there should be no phase shift with depth for a given wave: this is tested in the observations below. If $N$ is constant, RHINES (1970) gave the solution:

$$\omega = -\frac{k\alpha N}{\kappa^2}\coth\left(\frac{NH\kappa}{f}\right), \tag{2.4}$$

$$p(z) = p_0 \cosh\left(\frac{N\kappa z}{f}\right). \tag{2.5}$$

Numerical solutions of these equations using the actual $N(z)$ from Site D show that the effect of the thermocline is mainly to prevent disturbances from penetrating to the surface. One can see this in the WKB approximate solution already given by WEBSTER (1969), $p(z) \approx p_0 \cosh\left[f^{-1}\kappa\int_z^0 N(\xi)d\xi\right]$: there is effectively a lot more distance $\int N(\xi)d\xi$ to decay where $N$ is large. Oddly, numeric trial showed the WKB approximation to be inferior to (2.4, 5) with adjusted $\bar{N}$, especially in giving $\omega$ from (2.3).

One of the features of (2.5) is that the horizontal motion is bottom intensified. This is significant, for WEBSTER (1969) and T found that the observed kinetic energy at Site D decreases with depth. By use of a Sturm–Liouville zero-crossing argument, it is not hard to show that all solutions of (2.1, 2) are bottom intensified, requiring only $\kappa^2 > 0$ and $N^2 \geqslant 0$ (stable stratification). We will look for this rather surprising feature in the observations. Using (2.3), the argument further shows that $\omega/k < 0$, so for any stable strati-fication, the phase velocity is always to the left of the upslope. This 'westward propagation' also follows from RHINE's (1970) equation (2.6), but does not seem to have been remarked there. This prediction will be observed in the data.

Momentum fluxes into the Jet Stream due to eddies are important for the dynamics of the atmosphere, as pointed out by STARR, PEIXOTO, and GAUT (1970). WEBSTER (1965) and T found Reynolds' stresses indicating a similar momentum flux into the Gulf Stream. THOMPSON (1971b) suggested a simple explanation, based on barotropic topographic Rossby wave dynamics. This can be generalized, again by a Sturm–Liouville argument, to show that for any stable stratification in (2.1, 2, 3), if the energy flux of a wave has an upslope component, then the momentum flux is downslope (negative). This can be extended to include $\beta$, if upslope is poleward. Including $\beta$, but not topography, HELD (1975) has shown that westward momentum will be transported out of a passive region and into a source of disturbances. Since Site D is north of the Gulf Stream, we will look for a negative correlation between $u$ and $v$.

Thompson tried to establish that a random forcing at the deep ('Gulf Stream') end of a wedge in a numeric model leads to negative correlation between $u$ and $v$. RHINES (1971) questioned the numerics, in particular suggesting that truncating the wedge (to simulate the continental escarpment north of Site D) would cause reflections that would cancel the correlation, since energy would be propagating downslope as fast as up. Bill Schmitz and I re-ran T's model with a wall and found that a strong negative correlation still occurred. There is a simple explanation: the field started from rest, so energy from the forcing had to propagate in, causing a negative correlation. It is just that equilibrium took longer to establish than seemed reasonable. When we ran the model yet longer, the energy did start reaching equilibrium, and the correlation decayed, both with and without a wall. Therefore T's model result was fortuitous. However, putting in friction causes a permanent correlation. It might be noted that the correlation between $u$ and $v$ of $-0.6$ observed by T at Site D only requires the amplitude of the reflected wave to be about a third smaller than the upslope wave.

### 3. DATA

Standard moorings (HEINMILLER and WALDEN, 1973) were put out by the Woods Hole Oceanographic Institution's Moored Array Program at locations 'D', 'W', 'S', and 'N', indicated in Fig. 1. The spacings were chosen to be on the order of a Rossby radius of deformation. Early moorings used surface flotation; in view of GOULD and SAMBUCO's (1975) misgivings, these were not used in the following analyses, except for some phase estimates in the DWS triangle. The rest of the moorings had subsurface flotation. Current meters were typically placed at 1000 m and 2000 m below the surface.

The first array was put out in August 1971 (HEINMILLER and MOLLER, 1974). The data recovered is indicated in Fig. 3. By merging replacement moorings, continuous records up to $2\frac{1}{2}$ years long are available. The broad line from 11 April to 6 December 1974 represents twenty more current meters from an array extending south from Site D to the Gulf Stream, which are discussed in section 6.

Inertial oscillations, with period 19 hours, can be strong near Site D (WEBSTER, 1968), so seemed worth special filtering to avoid aliasing. Originally, a narrow Gaussian notch-filter centered at (rotary) frequency $-f$ was used, but a simple 19-hour running average was found to be quite adequate and to lose less record at the ends. Then the records

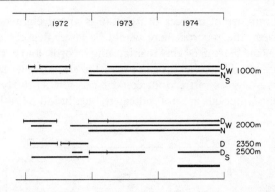

FIG. 3. Current-meter records successfully obtained from the Site D array, represented by continuous lines. Long records were gotten by merging abutting shorter records from successive moorings in the same area. 'D' means 39°10'N, 70°00'W ($\pm$ about 3 km); 'W' is west of D at 39°10'N, 70°30'W; 'N' is north at 39°23.5'N, 70°00'W; 'S' is south at 39°00'N, 70°00'W. The current meters were mainly at 1000 m and 2000 m depths. The lines in parentheses are records from moorings with surface flotation, and these records are little used in these analyses. The very heavy line at lower right represents twenty more current meter records of the large array extending south of Site D.

were low-passed by a gaussian filter with 95% pass at 5 days, and the low-passed record sub-sampled twice a day (00Z and 12Z).

Twelve records of 512 day lengths were made by using the first 512 days and the last 512 days of each of the six long records. The harmonograms for each of $u$ and $v$ for the 12 records were squared, and the average of these 24 spectral estimates is plotted in Fig. 4, multiplied by half to get kinetic energy. The first and last records overlap, so

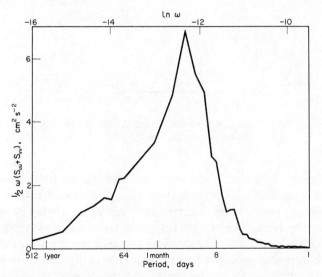

FIG. 4. Spectrum of the kinetic energy on area-preserving log plot. Vertical coordinate is $\frac{1}{2}\omega[S_{uu}(\omega)+S_{vv}(\omega)]$ with $S_{uu}$ the spectral (variance) estimate for the eastward component of velocity, and $S_{vv}$ for the northward. Units of the vertical coordinate are cm²/sec²; $\omega$ is in sec⁻¹, with $2\pi/\omega$ the period, from 512 days down to 1 day. The horizontal coordinate is $ln\omega$, running from near $-16$ to near $-10$. The total kinetic energy is 11.7 cm²/sec². The estimates were made by averaging over the 1000 m and 2000 m depth records from sub-surface moorings at D, W, and N. For periods longer than 64 days, there are about 17 degrees of freedom; for periods of 64 days or less, there are about 138 degrees of freedom.

an estimate of the number of degrees of freedom is about eighteen; under a hypothesis of correlation between the records there would be fewer degrees of freedom. The right half of the figure was made from 64-day overlapping records, and each point has nominally 140 degrees of freedom. The coherence between the six current meter records is rather high, so the effective degrees of freedom could be as few as thirty or forty. Due to this ambiguity, confidence intervals are not indicated, but should be rather narrow.

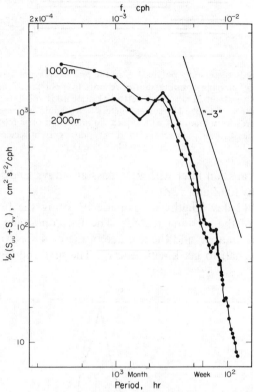

FIG. 5.   Kinetic energy spectra for depths of 1000 m and 2000 m in the Site D area, on log–log plot, in $cm^2/sec^2/cph$ vs cycles/hour. The straight line has a slope of $-3$.

The kinetic energy spectrum is definitely not red. Most of the kinetic energy of the perturbations is associated with periods short enough for the dynamics reasonably to be described by linear theory. There is a remarkable peak at 16-day period. The graph is area-preserving; the average kinetic energy per mass of the six records is $11.7\,cm^2/sec^2$. The effect gauss low-pass filter is to decrease the estimate by 10% at 120 hours. Since the spectrum is dropping rapidly, recoloring would have little effect on Fig. 4.

RHINES (1971) compared T's kinetic energy spectra for several depths, and found them nearly coincident for periods of several days to a month, suggesting quasi-barotropic waves. Figure 5 here is similar, for depths of 1000 m and 2000 m. The data here is independent of T, and processed rather differently; for instance, these records are from subsurface moorings and have no gaps, like T's records did. Yet the spectral estimates here are not significantly different than those in T (a large part of which were supplied by Ferris Webster). In particular, the rather striking '−3' slope in the log–log plot again appears. [Though T confused this frequency spectrum slope with the '−3' wave number spectral slope appearing in certain turbulence theories.] Though the spectra for the two depths are quite similar, they are not indistinguishable, as in Rhine's (1971) diagram, perhaps partly because T's data was from surface moorings, but at least partly because the resolution and stability of the estimates are higher here.

Perhaps the most important feature of Fig. 5 is the lack of energy at periods between a week and a day. This is important, because theory suggests no mechanism for motions between tidal frequency and the highest bottom-trapped frequency $\alpha N$, except possibly very long Rossby waves, or the decaying tails of continental shelf waves. This spectral gap was also observed by Webster (1972), at other sites in the ocean.

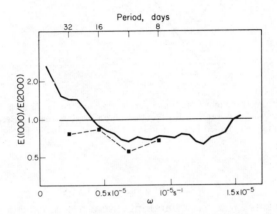

Fig. 6. The solid line is ratio of kinetic energy spectrum at 1000 m depth to that at 2000 m vs frequency, for the Site D array, on log–linear plot. The dots and dashed line represent predictions from theory and the wave-number estimates of section 5.

A point stressed above was that the theory implies the energy must increase downward, so the ratio of the spectrum for 1000 m to that for 2000 m depth should be less than one. The ratio is plotted in Fig. 6. The theory should be valid from about a week to a month. The observed ratio is less than 1 between 5 days and 18 days. The higher energy at 1000 m for periods of three weeks or longer suggests that some other dynamics than the linear theory of Rossby waves must be becoming dominant, at least at 1000 m. Inspection of the original records, XBTs and satellite photographs shows the presence of comparatively strong, surface-intensified 'Gulf Stream eddies'; in fact, three in one 6-month period. These are clearly nonlinear and better described as blobs of warm water drifting with the current than the waves of any linear theory.

The momentum flux, $\overline{uv}$, can also be decomposed into frequency components, as the real part of the cross-spectrum, re$[S_{uv}(\omega)]$. This is presented in Fig. 7, with about 35 degrees of freedom for each point. Not only is the total momentum flux negative, but the momentum flux associated with each period longer than a week is negative. This fits with the linear theory proposed, if the energy source for the waves is downslope. The correlation nicely drops to insignificance at the frequency limits of the theory.

142　　　　　　　　　　　RORY O. R. Y. THOMPSON

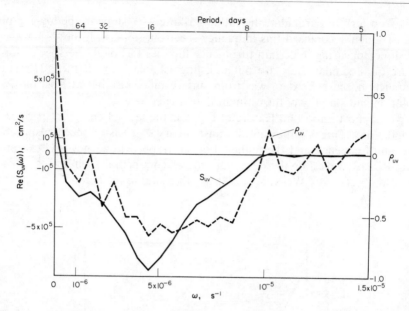

FIG. 7.　The hanned spectrum of the momentum flux per unit mass in the Site D area, $\mathrm{Re}[S_{uv}(\omega)]$, vs frequency on linear plot, in $(\mathrm{cm}^2/\mathrm{sec}^2)/\mathrm{sec}^{-1}$ vs $\mathrm{sec}^{-1}$. Also the correlation between (band-passed) $u$ and $v$, $\rho = \mathrm{Re}(S_{uv})/(S_{uu}S_{vv})^{\frac{1}{2}}$. The correlations were computed before hanning. There are nominally 35 degrees of freedom for each point.

## 4. CROSS-SPECTRA AND LAGS

Since the replacements for a set of moorings were not in exactly the same position, and records are of varying quality, it was decided to use the records indicated in Fig. 3 for two depths in four pieces, each about 8 months long. The first piece was D,W,S from 2 February to 10 December 1972, and was a merger of shorter records. This piece is like earlier Site D data (see T) in being patched together from short pieces and being partly on surface moorings. The remaining three pieces are all DWN, and much nicer, being all subsurface, and all being single records with no gaps. They were from 27 March to 14 October 1973; from 16 October 1973 to 9 April 1974; and from 9 April to 5 December 1974.

WEBSTER (1969) has discussed estimation of the mean velocity at Site D. Figure 8 presents the vector means over the four pieces, so some estimate of the variation can be gained. The estimates seem reasonably stable.

Since the interest here centers on periods of about a week to a month, it was convenient to break the records into 32-day subpieces and compute cross-spectra between the current meter records at 32, 16, 11 and 8-day periods. Currents have two components $(u, v)$, where $u$ is the eastward component of velocity and $v$ is northward, so there are four elements in a cross-spectral matrix. For simplicity, and following T, the records actually used in section 5 are only the component along an axis 30° south of east, namely,

$$\mu(t) = 0.833u(t) - 0.5v(t). \qquad (4.1)$$

Some 70% of the kinetic energy near Site D is in this component. It might be argued that using this component favors seeing the waves going upslope. This is exactly so; we

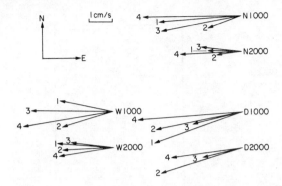

FIG. 8. Mean velocities for the four 8-month periods, for 1000 m and 2000 m depths at the three moorings D, N, and W. The arrows labeled 'l' are for the period extending until 26 March 1973, but with varying quality. For instance, at N, they only encompass 3 months, and at D, the data for 2000 m depth has been lost. The other arrows are all from subsurface moorings and close to 8 months each, extending to 16 October 1973, 9 April 1974, and 5 December 1974 respectively.

are looking for such waves. Under a null hypothesis of noise, it will not matter what component is used, so using this component does not bias the statistics.

In section 2, we noted that the waves should have no phase shift in the vertical. All 2178 days for which there were simultaneous records at 1000 m and 2000 m on the same subsurface mooring were used, and the resultant phase and coherence plotted in Figs. 9 and 10. According to standard formulas, there should be 272 degrees of freedom in each estimate, but the three moorings exhibit considerable coherence, so there could be as few as 100.

The upper dashed line in Fig. 9 is the 1% level of significance for 50 bands from AMOS and KOOPMANS (1963); the lower is the 5% level for 136 bands. The observed coherences are exceedingly significant. In fact, the approximate 95% confidence intervals are only $\pm 0.03$ from the observed. The observed coherences were used in the asymptotic formulas of HINICH and CLAY (1968) to construct 95% tests of zero phase shift. Under the hypothesis of zero phase shift in the vertical, the observed shifts should fall between

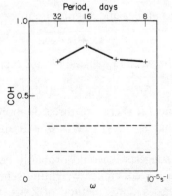

FIG. 9. Phase by which the 2000 m current meter observations led 1000 m, for the Site D array. The dashed lines are the 5% significance levels, using the sample coherences. There are nominally 272 degrees of freedom for each estimate. The lower dashed line is the 5% significance level therefore. The upper dashed line is the 1% significance level for 100 degrees of freedom.

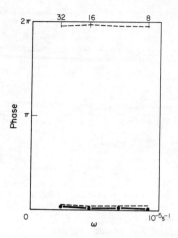

Fig. 10. Dot Coherence between 2000 m and 1000 m current meter observations for the Site D array. There are nominally 272 degrees of freedom for each estimate; the dashed lines are the 5% significance levels for testing zero phase.

the dashed lines in Fig. 10. All four observations do. We conclude there is strong coherence in the vertical and very little, if any, phase shift.

The cross-spectrum actually used in section 5 is

$$S_{12}(\omega) = \overline{\tilde{\mu}_1^* \tilde{\mu}_2},\tag{4.2}$$

where $\tilde{\mu}(\omega)$ is the Fourier transform of $\mu(t)$, and bar is the average over the subpieces; the coherence is $|S_{12}|(\overline{\tilde{\mu}_1^* \tilde{\mu}_1}\, \overline{\tilde{\mu}_2^* \tilde{\mu}_2})^{-\frac{1}{2}}$, and the lag at period $2\pi/\omega$ is

$$\tau_{ij} = \frac{2\pi}{\omega} \operatorname{Arg}[S_{ij}(\omega)].\tag{4.3}$$

## 5. WAVE NUMBERS

If we assume a single wave-number and angular frequency $\omega$ with phase

$$\phi = kx + ly - \omega t,\tag{5.1}$$

and measure phase-lags $\omega\tau_{ij}$ between $(x_i, y_i)$ and $(x_j, y_j)$, then we should get

$$\omega\tau_{ij} = k(x_j - x_i) + l(y_j - y_i),\tag{5.2}$$

and be able to solve for $k$ and $l$ if there are three or more moorings. Of course, since there is noise, it is unlikely that $\tau_{12} + \tau_{23} + \tau_{31} = 0$ as it should to allow two unknowns to satisfy three equations (5.2).

Experience with this data suggests it would probably be better just to throw out the least significant lag or to use the coherence weighting scheme of Hamon and Hannan (1974), but the unweighted least-squares method of Zalkan (1971) was used and found satisfactory. Using the known distances between the moorings and the lags from (4.3) in each piece, the wave-numbers were computed for both depths and four pieces. The results are displayed in Fig. 11 for 32, 16, 11, and 8-day periods. Each piece is about 8 months,

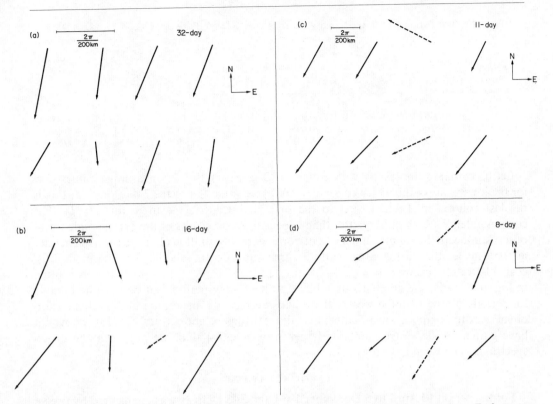

FIG. 11a.   Wave-numbers estimated from cross-spectra at 32-day periods of 32-day record pieces. Each arrow is independent of the others and represents approximately 8 months of data at three current meters; the upper row is for 1000 m depth, the lower for 2000 m depth (except S and 2500 and D partly at 2350 m in the first pair). The first pair are from DWS, the others DWN. The waves propagate in the direction of the arrow. An arrow of the length indicated would have a wavelength of 200 km; if half the length, 400 km. (b) As for (a), but 16-day period in 32-day record pieces. Dashed lines represent estimates based on cross-spectra with one coherence less than 0.35 and another less than 0.5. All solid lines have all three coherences greater than 0.35 and not more than one less than 0.5. (c) As for (b), but 11-day period. (32/3). (d) As for (b), but for 8-day period.

so each of the three current meter records contributes about 8 degrees of freedom to each estimate.

The data used to estimate one wave-number is independent of that in every other piece, yet there is reasonable consistency. On the basis of earlier data (not re-used here) from single moorings, and the energy flux expected, T hypothesized the wave-number vectors would point into the SW quadrant. This hypothesis was made before collecting the data, so a simple test can now be made to distinguish it from a null hypothesis that there is no definite wave-number, that is, the estimates are just noise. Seeing that there are four quadrants, under the null hypothesis, there is less than 5% chance that five or more out of eight vectors would be in the SW quadrant, and less than 1% chance of six or more. The actual numbers, from Fig. 11, are 7, 6, 7, 7 so the null hypothesis is rejected for all four periods, and at a significance level of 0.00038 for 32, 11, and 8-day periods. Further tests are given below, but this already indicates something real is being measured. If the reality is accepted, let's estimate the wave numbers. For this purpose, let us delete the doubtful estimates before averaging to get the estimates of Table 1.

TABLE 1.    BEST ESTIMATE OF WAVE-NUMBER VECTORS AT SITE D FOR PERIODS OF 32, 16, 11, AND 8 DAYS

| Period (days) | $k$ (km$^{-1}$) | $l$ (km$^{-1}$) | Wavelength (km) | $\theta$ | $\dfrac{-\sin\theta}{\tanh\left(\dfrac{NH\kappa}{f}\right)}$ | $\omega/(\alpha N)$ |
|---|---|---|---|---|---|---|
| 32 | $-0.007 \pm 0.002$ | $-0.026 \pm 0.002$ | 230 | $-15°$ | 0.34 | 0.22 |
| 16 | $-0.006 \pm 0.004$ | $-0.021 \pm 0.004$ | 290 | $-16°$ | 0.40 | 0.44 |
| 11 | $-0.026 \pm 0.003$ | $-0.035 \pm 0.003$ | 140 | $-37°$ | 0.64 | 0.66 |
| 8 | $-0.025 \pm 0.005$ | $-0.022 \pm 0.005$ | 190 | $-49°$ | 0.88 | 0.88 |

Let us compare the results with equation (2.4). From Fig. 2, an average value of $N$ for the lower part of the water column is $0.0014\,\sec^{-1}$, so $\omega/(\alpha N) \approx 7$ days/period; this is the last column in Table 1 next to the predicted value of $-\sin\theta\coth(NH\kappa/f)$. The correspondence seems rather better than expected. Another test of the correspondence of observation with the simple model was performed: from equation (2.5), the ratio of energies at 1000 m to that at 2000 m, with 2.7 km total depth, should be $[\cosh(30\kappa/2.7)/\cosh(30\kappa \times 2/2.7)]^2$. With $\kappa = (k^2 + l^2)^{\frac{1}{2}}$ inserted from Table 1, this prediction is plotted in Fig. 6. Clearly, the comparison is bad for the 32-day period, but perhaps acceptable for periods of one to three weeks. If we had reversed the procedure and used the ratio of energies to compute wave-numbers, as did THOMPSON and LUYTEN (1975), we would have gotten smaller wave-numbers (larger wave-lengths) than we got from the cross-spectral lags between moorings.

## 6. LARGE-ARRAY DATA

For the period 10 April to 5 December 1974, the Site D array was augmented by twelve additional moorings. These moorings were put in under the aegis of James R. Luyten for Gulf Stream studies, but are valuable for the present wave study as well. Figure 12 shows the positions of the moorings, the topography, and the mean flows; see also Table 2. The numbers beside the position dots are the Moored Array Program mooring numbers; the instruments on a mooring are numbered from the top, record 5246 being sixth from the top on mooring number 524. The isobaths are taken from a diagram by Elazar Uchupi (personal communication). The mean flows are simply the average over each low-passed record at 1000 m depth or deeper. Two arrows indicates different levels; the thinner arrow is the shallower. It is interesting to note that there are about 10 Sverdrups flowing to the west, apparently constituting one of the larger deep flows in the ocean. For the present purposes, one need only note that there appears to be two regimes—a rather homogeneous flow to the west in the northern part of the array, changing to an irregular pattern to the south of mooring 530. This change occurs near 4000 m depth, and will be seen in all of the statistics discussed here. Further discussion may be found in Luyten (1977).

## 7. NON-DIVERGENCE

Probably the most fundamental feature of potential vorticity dynamics is that the motion is nearly horizontally non-divergent. That is, we want to test if

$$u_x + v_y \approx 0. \tag{7.1}$$

If we are willing to assume the field of motion is dominated at a given frequency $\omega$ by a single wave $\{\exp[i(kx + ly)]\}$, then (7.1) becomes

$$(k, l)\cdot(u, v) \approx 0. \tag{7.2}$$

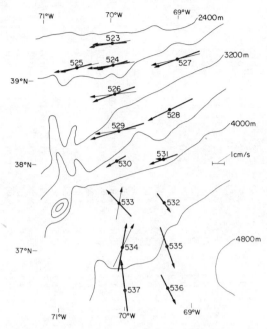

FIG. 12.    Locations and numbers of moorings in large array. The isobaths are taken from UCHUPI. The arrows show the mean flow over 8 months. The thick arrows are mostly 200 m above the bottom, and the thinner 1000 m.

This says the motion is transverse to the wave-number vector. We will check this in a later section by testing if the wave-number estimate at a given frequency is perpendicular to the principal axis of the velocities at that frequency, but here we will make a simpler test that can be done with a single current meter. Equation (7.2) says the velocities will be restricted to a line [perpendicular to $(k, l)$], so a scatter-diagram of the observed $v$ vs the observed $u$ should stretch out along one line and be comparatively squashed perpendicular to this line. Under a null hypothesis of noise (many waves, or divergence), there should be no particular preferred direction. So, we can compute the square of the ratio of the minor axis to the major axis. We can express this ratio ('flatness') by using formulae from FOFONOFF (1969) to get:

$$F = \frac{S_{uu}\cos^2\theta - 2\operatorname{Re}(S_{uv})\cos\theta\sin\theta + S_{vv}\sin^2\theta}{S_{uu}\cos^2\theta + 2\operatorname{Re}(S_{uv})\cos\theta\sin\theta + S_{vv}\sin^2\theta}, \tag{7.3}$$

where $S_{uv}$ is the cross-spectrum between $u$ and $v$, etc., and $\theta$ is the angle of the principle axis, given by

$$\tan(2\theta) = \frac{2\operatorname{Re}(S_{uv})}{S_{uu} - S_{vv}}. \tag{7.4}$$

The ratio (7.3) not only has a geometric interpretation in terms of the flatness, but has a dynamical interpretation as the ratio of the spectrum of the divergence to that of the vorticity. This interpretation shows that $F$ is a ratio of chi-square random variables which are independent under the null hypothesis, so $F$ has an $F$-distribution. This makes it easy to say when the scatter-diagram at a particular frequency is significantly flattened.

TABLE 2. STATISTICS

| | Lat. °N | Long. °W | C.m. depth km | Total depth km | $2\pi/N\alpha$ days | Isobath angle (deg) | $\bar{u}$ cm/sec | $\bar{v}$ cm/sec | KE cm²/sec² | $\sigma'_u$ cm/sec | $\sigma'_v$ cm/sec | $\overline{u'v'}$ | $\rho$ |
|---|---|---|---|---|---|---|---|---|---|---|---|---|---|
| 5232 | 39°26 | 70°00 | 1.00 | 2.50 | 8.3 | 0 | −4.2 | −0.1 | 9.7 | 3.82 | 2.18 | 0.5 | 0.06 |
| 5233 | 39°26 | 70°00 | 2.00 | 2.50 | 8.3 | 0 | −2.6 | −0.1 | 8.5 | 3.68 | 1.87 | −3.6 | −0.52 |
| 5245 | 39°08 | 70°00 | 1.00 | 2.50 | 8.7 | 0 | −4.3 | −0.4 | 10.2 | 3.50 | 2.84 | −3.7 | −0.37 |
| 5246 | 39°08 | 70°00 | 2.00 | 2.66 | 8.7 | 0 | −3.0 | −0.4 | 8.9 | 3.30 | 2.62 | −4.5 | −0.52 |
| 5247 | 39°08 | 70°00 | 2.50 | 2.66 | 8.7 | 0 | −2.5 | −0.7 | 14.4 | 3.58 | 4.00 | −8.0 | −0.56 |
| 5252 | 39°07 | 70°33 | 1.00 | 2.76 | 9.0 | 0 | −3.9 | −0.6 | 9.9 | 3.73 | 2.43 | −4.5 | −0.49 |
| 5253 | 39°07 | 70°33 | 2.00 | 2.76 | 9.0 | 0 | −2.4 | −0.4 | 11.3 | 4.21 | 2.20 | −4.2 | −0.45 |
| 5261 | 38°47 | 70°01 | 2.00 | 3.01 | 7.6 | 32 | −3.5 | 1.9 | 18.5 | 5.16 | 3.22 | −8.7 | −0.52 |
| 5262 | 38°47 | 70°01 | 2.80 | 3.01 | 7.6 | 32 | −3.7 | 1.1 | 32.5 | 7.47 | 3.05 | −7.1 | −0.31 |
| 5271 | 39°10 | 69°00 | 2.00 | 2.98 | 6.7 | 17 | −4.4 | 0.4 | 14.9 | 4.61 | 2.91 | −4.0 | −0.30 |
| 5272 | 39°10 | 69°00 | 2.78 | 2.98 | 6.7 | 17 | −4.3 | −0.0 | 21.0 | 5.88 | 2.93 | −10.8 | −0.63 |
| 5282 | 38°35 | 69°10 | 2.33 | 3.33 | 9.0 | 19 | −4.3 | −0.4 | 22.4 | 5.38 | 3.98 | −10.0 | −0.47 |
| 5291 | 38°21 | 70°00 | 2.48 | 3.48 | 6.6 | 30 | −3.1 | 1.5 | 30.0 | 7.10 | 3.09 | −8.0 | −0.36 |
| 5292 | 38°21 | 70°00 | 3.28 | 3.48 | 6.6 | 30 | −4.8 | 1.5 | 46.2 | 8.78 | 3.91 | −8.3 | −0.24 |
| 5302 | 38°01 | 70°01 | 2.82 | 3.82 | 9.5 | 25 | −1.8 | 0.0 | 39.3 | 7.86 | 4.09 | −10.7 | −0.33 |
| 5312 | 38°00 | 69°19 | 2.92 | 3.92 | 9.7 | 27 | −0.3 | 0.2 | 31.5 | 5.35 | 5.87 | −9.5 | −0.30 |
| 5313 | 38°00 | 69°19 | 3.72 | 3.92 | 9.7 | 27 | −1.8 | 1.8 | 48.4 | 6.24 | 7.61 | −14.7 | −0.31 |
| 5322 | 37°30 | 69°20 | 3.21 | 4.21 | 20 | 33 | −0.1 | −1.9 | 53.5 | 7.01 | 6.92 | +7.9 | +0.16 |
| 5331 | 37°30 | 70°00 | 3.18 | 4.18 | 18 | 16 | 1.4 | 2.3 | 56.1 | 8.31 | 6.55 | −4.0 | −0.07 |
| 5332 | 37°30 | 70°00 | 3.98 | 4.18 | 18 | 16 | −1.3 | 2.8 | 57.8 | 8.58 | 6.49 | −4.7 | −0.08 |
| 5341 | 37°00 | 70°00 | 3.34 | 4.34 | 32 | 18 | 2.9 | 3.0 | 67.1 | 8.04 | 8.25 | −2.7 | −0.04 |
| 5342 | 37°00 | 70°00 | 4.14 | 4.34 | 32 | 18 | 4.2 | 0.5 | 83.6 | 8.90 | 9.30 | 0.8 | 0.01 |
| 5352 | 36°59 | 69°20 | 3.45 | 4.45 | 33 | 70 | −2.9 | −2.5 | 59.7 | 8.88 | 6.35 | +18.7 | 0.33 |
| 5361 | 36°30 | 69°20 | 3.47 | 4.47 | 45 | 41 | −0.9 | −2.6 | 63.3 | 8.53 | 7.34 | 4.3 | 0.07 |
| 5362 | 36°30 | 69°20 | 4.27 | 4.47 | 45 | 41 | −0.8 | −2.6 | 74.8 | 9.16 | 8.11 | 5.5 | 0.07 |
| 5373 | 36°30 | 70°00 | 4.26 | 4.46 | 37 | 11 | 0.6 | 4.5 | 53.9 | 7.33 | 7.35 | −1.4 | −0.03 |

Here, and in all the following discussions, the records were broken into overlapping 32-day pieces, and the first, second, third, and fourth harmonics calculated by a complex Fourier routine. Thus the periods considered are centered at 32, 16, 11, and 8 days, which spans the range of most interest. The spectra were formed by averaging over pieces, then the ratio was calculated from (7.4) and (7.3), for each of the four frequencies at each current meter. The results are plotted for 16-day period in Fig. 13 vs total water depth at the mooring. These and some later results are plotted vs depth because of T's (1971a) argument that depth change from the region of generation (presumably the Gulf Stream) should have important effects on the statistics. Since each record is 8 months

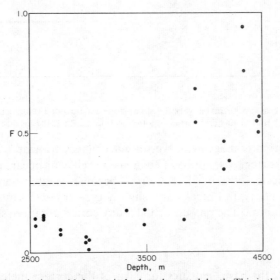

FIG. 13.   The 'flatness' of the velocity at 16-day period, plotted vs total depth. This is the square of the ratio of the minor axis to the major axis of the scatter-diagram. If one assumes a single wave-number, it is also the ratio of the spectrum of the divergence to the spectrum of the vorticity, at frequency $2\pi/16$ days.

long, and the pieces are a month long, the numerator and denominator in $F$ each have 8 degrees of freedom, so the 5% significance level for $F$ is 0.29; this is marked with a dashed line in Fig. 13. All of the ratios for the shallower depths are significant, with an abrupt change to insignificance at a depth of near 4000 m (that is, south of mooring 531). This change of regime coincides with that of the mean flows in Fig. 12. The results for the other three frequencies are similar.

Figure 13 should not necessarily be taken as evidence that the flows are divergent in the deeper water, since we introduced a secondary hypothesis much more doubtful than non-divergence when we assumed a single wave-number. The small ratios in the shallower water are indeed compatible with the theoretical requirement that the divergence should be small compared to the vorticity, but chiefly provide evidence that the motion at a given frequency tends to be dominated by a single wave, so to move back and forth at a particular angle.

A consequence of transverseness is that the velocity components should be either in or out of phase. Since there is expected to be a negative correlation between $u$ and $v$, they should be out of phase. Figure 14 shows the phase of the cross-spectrum $S_{uv}$ plotted vs depth for the 16 and 11-day periods. The phases are indeed near $\pm\pi$ in the shallower

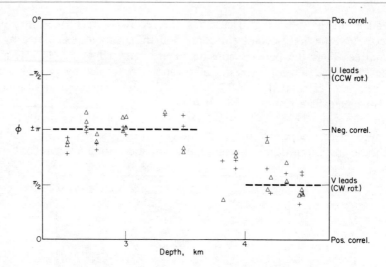

FIG. 14.   Phase of the cross-spectrum between the along-slope and up-slope velocity components for 16-day (+)
and 11-day (△) periods, plotted vs total depth at the mooring.

depths, so the motion is indeed transverse, but changes to near $\pi/2$ in the deep water, where the motion is clockwise rotary. The break in behavior occurs near the depth where the bottom slope decreases so much that $2\pi/\alpha N$ becomes greater than 16 days. The 32-day period behaves similarly, but breaks at a slightly greater depth (lesser slope) to counterclockwise rotary motion. The phases for the 8-day period are more scattered.

## 8. STATISTICS

Table 3 presents the basic statistics for each of the current meter records. The slopes and isobath angles with respect to East were estimated from Uchupi's chart, given in Fig. 12. The estimates are of moderately good quality in the northern part of the array, but become quite vague in the southern. In fact, the bottom topography is probably the greatest source of error in the present comparison of theory and observation. Anyway, taking the isobath angles as given by the topography, all statistics are with respect to coordinates rotated so $u$ is along the slope and $v$ is upslope.

The kinetic energy (without the means) is plotted vs depth in Fig. 15a, for the current meters which were well away (1000 m) from the bottom. The plot is log-linear to show the rather nicely exponential drop in energy upslope. This suggests an energy source in the deep (Gulf Stream) end, though one should remember the change in depth of the current meters. The energy drops by about an order of magnitude while the depth of the water is cut in half. Possibly this will turn out to be more usefully described as an e-folding decay distance of 100 km, since there is also a drop of energy south of the Stream, to judge from the single, earlier, 3-month record 4001. In Fig. 15b, the kinetic energy of records only 200 m off the bottom drops a little more slowly, possibly due to local generation of bottom-trapped disturbances, as suggested by Suarez (1971). The smoothness of fit of Schmitz's (1974) records 3451 and 3521 in these figures only accents the abrupt increase of energy reported by Schmitz as one starts up the much steeper continental slope. The dashed line marks the approximate boundary between slope and rise, estimated from Fig. 2 of Schmitz (1974).

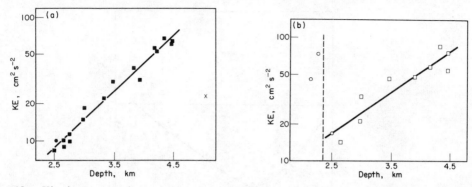

FIG. 15a. Kinetic energy on log scale plotted vs total depth, for records of intermediate depth. ■—1000 m off the bottom in this array. × —W.H.O.I. mooring 4001. ●—mooring 3451 from SCHMITZ (1973). Regression line is drawn.

FIG. 15b. Kinetic energy on log scale plotted vs total depth, for records closer to the bottom. □—200 m off the bottom in this array. ○—moorings 100 m off the bottom from SCHMITZ (1973).

The Reynolds stress, $\overline{u'v'}$, is plotted vs depth in Fig. 16. This is the cross-isobath flux, rather than the northward flux. This seemed more sensible, because the restoring forces are due to topography, because the isobaths close within the ocean, and because the Gulf Stream runs more closely along an isobath than due East. In Fig. 16, we see there is downslope momentum flux, strengthening to past 4 km depth, where it abruptly changes to upslope. An earlier, shorter, record (WHOI 4001) is included to extend the range across the Gulf Stream, to Site J. W. Sturges (personal communication) also observed positive Reynolds' stress at Site J (near 5 km depth) from longer, near-bottom U.R.I. records. There seems to be definite support for Thompson's (1971b) prediction of convergence of momentum into the Gulf Stream from both sides, caused by the flux of energy away.

The momentum flux can be decomposed by frequency as $\mathrm{Re}(S_{uv})$. It appears the momentum flux is dominated by motions with periods of a week to a month, and the 16-day period is a bellwether. Figure 17 presents the correlation $\mathrm{Re}(S_{uv})/(S_{uu}S_{vv})^{\frac{1}{2}}$ for 16-day period, vs depth. The one-tail 5% significance level is $-0.51$. The correlations are negative and statistically significant in the shallow end of the array, and possibly change sign near the deep end. There may be indication of reflection from the continental rise,

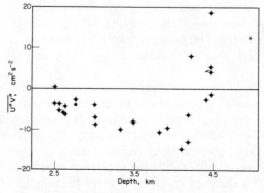

FIG. 16. Reynolds stress vs bottom depth in an 8-month array along 70°W. The * represents 4001, an earlier 3-month record.

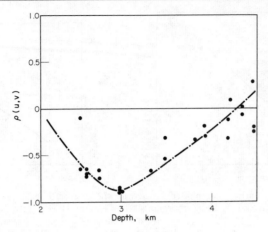

FIG. 17.  Correlation between up-slope and along-slope velocity components, for 16-day period. The dashed
line is just freehand.

in that the correlations may be smaller near it. Figure 5 of SCHMITZ (1974) suggests that
the principal axes approach the local isobaths, as the slope becomes steeper, so the
correlations go to zero.

Equation (2.4) can be written,

$$\frac{\omega}{\omega_0} = \sin\theta \coth(NH\kappa/f), \qquad (8.1)$$

where $\theta$ is the angle of the principal axis, given by (7.4), and $\omega_0 = N\alpha$ is the frequency
of internal waves moving straight up and down the slope $\alpha$ under the local Brunt–Väisälä
restoring frequency. If we expect the length scale to be that of a Rossby radius, so
$NH\kappa/f \approx 1$, then $\coth(NH\kappa/f) \approx 1.3$; in any case, it is greater than 1, so

$$-\frac{\omega}{\omega_0} \geqslant |\sin\theta|. \qquad (8.2)$$

The inequality (8.2) can be tested with individual current-meter records, so will be discussed
now; stronger tests of the complete dispersion relation are given below. Using values of $N$
averaged over the lowest 300 m, and interpolated to each mooring from CTD records
supplied by Gordon Volkmann, and using bottom slopes from Fig. 12, $|\omega/\omega_0|$ is plotted
vs $\theta$ in Fig. 18. The southern part of the arrays (depths more than 4 km) is indicated by
dots, and is not well behaved. The northern part is indicated by crossed dots, which do
seem to cluster near and above the curve $\omega/\omega_0 = \sin\theta$, and quite possibly with an
average factor near 1.3.

The phase lag in the vertical is important for theories of baroclinic instability. For the
neutral topographic waves of interest in this paper, there should be no phase lag in the
vertical, and indeed in the small array data above, the vertical phase lags were found
to be insignificant, and very small. For the large array, Fig. 19 shows the dot phase by
which the lower of pairs of current meter records leads the upper for each of the four
frequencies, plotted as angle, with radius the coherence between the records. The curve
$\theta = 1.645\,[(r^{-2}-1)/n]^{\frac{1}{2}}$ is sketched in, with $n = 16$ for the number of degrees of freedom.
This curve is the asymptotic 5% significance test for testing zero phase. Most of the
estimates lie within the figure, but several are significantly positive, so it is possible that

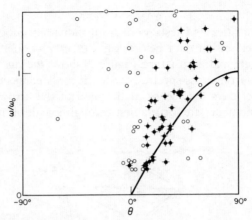

Fig. 18.   The relative frequency $\omega/N\alpha$ for each mooring and harmonics of 32 days, vs principal axis angle from the isobath. The theoretical lower limit, sin $\theta$, is drawn. ◆ represents water depth less than 4 km; ○ greater.

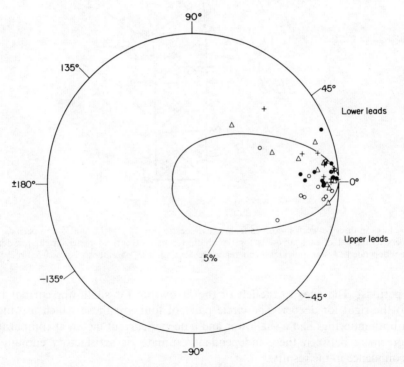

Fig. 19.   Phase between the lower (200 m off bottom) and the upper (1000 m off) of pairs of current meters on the same mooring, plotted with coherence as radius. The curve is the asymptotic 5% significance level for testing if the phase is different than zero. ●32, + 16, △11, ○8-day period.

the lower ones tend to lead. However, the large leads are all by 5342 vs 5341, so could conceivably be due to a time–base shift at some stage in the data processing. Perhaps we should decide to be more impressed by the coherences, some of which are quite high, showing almost columnar behavior through a thousand meters of water.

### 9. WAVE-NUMBERS

All of the above statistics and tests were from individual moorings, and have been approached from other sets of data before, e.g. Schmitz (1973). Now the entire array will be used to investigate spatial structure. Figure 20 shows the lags between the moorings. The lags are computed from cross-spectra between records connected by the arrows, and indicate the number of days the record at the base of the arrow leads the one at the arrowhead. The four numbers are for the first through fourth harmonics (32, 16, 11, and

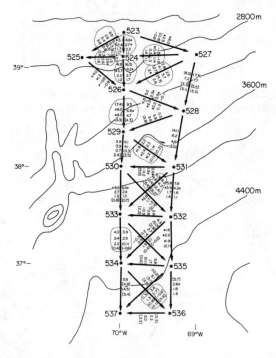

FIG. 20. Lags in days between moorings from cross-spectra, for 32, 16, 11, and 8-day periods. The quartet of numbers to the left of each arrow is for the shallower pair of current meters, right for deeper. Circled pairs are independent. A star indicates a coherence greater than 0.75, parenthesis below 0.5, bracket below 0.35.

8-day periods). The ones to the left of the arrow are for shallower current meters, the ones to the right for deeper. The circle pairs of four are those which are independent, in that both moorings had a shallower and a deeper current meter, at comparable depths. The agreement between these independent estimates is satisfactory enough to inspire some confidence in the results.

Figure 21 a–d shows the wave-numbers estimated from the lags of Fig. 20, by the least-square method. Figure 21a is for the first harmonic (32-day period); b, c, and d are for 16, 11, and 8-day periods. The wave-numbers point in the direction of wave phase-velocity, with lengths inversely proportional to the wavelength. The length corresponding to 250 km wavelength (40 km wave-scale) is indicated, and is seen to be typical of the 32-day waves. The shorter period waves have rather longer wave-number vectors (shorter wavelengths) in general. Notable features of Fig. 21: There is pretty good consistency, even between independent estimates. All the waves propagate southward. A little

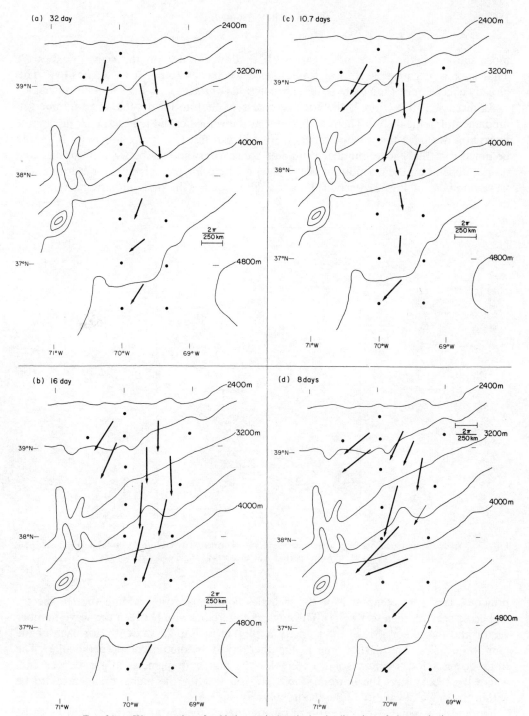

FIG. 21a.  Wave-numbers for 32-day period, pointing in direction of phase velocity.

FIG. 21b.  Wave-numbers for 16-day period.

FIG. 21c.  Wave-numbers for 10.7-day period.

FIG. 21d.  Wave-numbers for 8-day period.

155

more subtle, all the waves propagate with shallower water on the right ('westward'). That is, $k$ is negative in a local frame, since we have chosen $\omega$ to be positive. This already checks one of the more striking predictions of the theory. Let's look at others.

As discussed in section 7, the transverseness of the motion to the wave-numbers is fundamental to the theory. There we found that the velocities did tend to lie along definite lines; now we have the wave-numbers $(k, l)$, and can test (7.2) more precisely. It should be noted that these wave-number estimates come from the lags between moorings, and do not depend on the orientation of the $(u, v)$ to the principle axis at the moorings. The wave-numbers of Fig. 21 pertained to triangles (squares in the southern part), so the

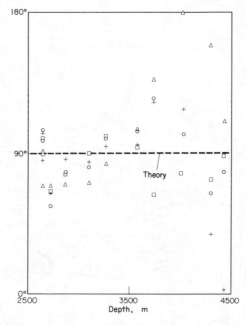

FIG. 22.    Angle counterclockwise from wave number vector to principal axis, plotted vs total depth at mooring.
□32, +16, △ 11, ○ 8-day period. The theoretical angle of 90° is dashed.

principal axes were computed for the triangles (and squares) by adding the $u,v$ spectra and cross-spectra, then using (7.4). The angle counterclockwise between the wave-number vector and the major axis of the velocity is plotted in Fig. 22 vs depth, for each of the four frequencies. The dashed line is the theoretical prediction of perpendicular. The transverseness looks pretty good, except for the higher frequencies at the deeper end, where the theory says the bottom slopes are too low for the higher frequencies to be within the range of validity of the theory anyway.

From (2.5), we may expect a ratio of kinetic energy spectra at depths $z_1$ and $z_2$ to be $\cosh^2(\kappa z_1 N/f)/\cosh^2(\kappa z_2 N/f)$. The array contained eight pairs of current meters with one 1000 m off the bottom and one 200 m off, plus four pairs of similar spacing from the small array near D. The ratios of the observed kinetic energy spectra from these pairs, at each of the four frequencies, are plotted against the predicted ratio in Fig. 23. The agreement seems to be within 'oceanographic accuracy'. The squares represent moorings

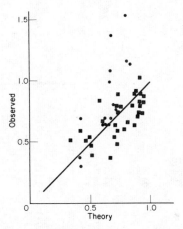

FIG. 23. Ratios of observed energy spectra from two current meter records on the same mooring, plotted vs theoretical ratio derived from the wavenumber estimate. The line is perfect prediction. ■ mooring entirely below the thermocline. ● mooring extends into or through the thermocline.

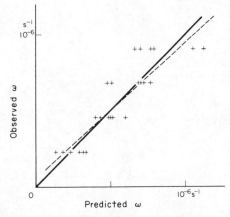

FIG. 24. Imposed frequencies plotted vs the theoretical frequency derived from the wavenumber estimates. The dashed line is the regression line. The solid line is perfect prediction and also the regression line through the origin.

entirely below the thermocline. It is notable that five out of the six ratios in the theoretically forbidden range above 1 are from moorings extending to 200 m depth. After noting also that five out of the six are at 32-day periods, one strongly suspects mooring motion caused by near-surface eddies is contaminating the observations. This was definitely the case with the temperature observations (not reported here). Incidentally, with the Gulf Stream and wind up above, one might well have expected the upper ones to be more energetic, if not guided by the theory. In fact, there is still a bit of a theoretical problem to explain how the energy does get down.

Having found that the basic assumptions of the theory hold well enough, and that there seems to be evidence that there are meaningful wave-number estimates, and something to the vertical structure prediction, let us test the dispersion relation (2.4). Table 3 gives

TABLE 3.

| Moorings | Units | 523 525 524 | 523 524 527 | 525 526 524 | 524 526 527 | 526 528 527 | 526 529 528 | 529 531 528 | 529 530 531 | 530 533 532 531 | 533 534 535 532 | 534 537 536 535 |
|---|---|---|---|---|---|---|---|---|---|---|---|---|
| Average depth $H$ | km | 2.6 | 2.7 | 2.8 | 2.8 | 3.2 | 3.3 | 3.5 | 3.7 | 4.0 | 4.3 | 4.4 |
| Brunt–Väisälä $N$ | $10^{-3} \text{ sec}^{-1}$ | 1.13 | 1.13 | 1.17 | 1.15 | 1.15 | 1.15 | 0.93 | 0.83 | 0.66 | 0.57 | 0.54 |
| Bottom slope $\alpha$ | | 0.007 | 0.011 | 0.007 | 0.009 | 0.009 | 0.009 | 0.009 | 0.011 | 0.008 | 0.004 | 0.004 |
| Rossby radius $NH/f$ | km | 30 | 31 | 33 | 32 | 37 | 38 | 34 | 31 | 27 | 26 | 25 |
| Isobath, from E | degree | 0 | 13 | 5 | 24 | 24 | 24 | 21 | 20 | 15 | 24 | 17 |
| **32 day**   Wavenumber angle from S, $\phi$ | degree | −8 | 9 | −9 | 12 | 13 | 14 | 6 | −21 | −22 | −52 | −35 |
| Cross-isobath angle, $\theta$ | degree | 8 | 4 | 14 | 12 | 11 | 10 | 15 | 41 | 37 | 76 | 52 |
| Wave number $\kappa$ | $\text{km}^{-1}$ | 0.026 | 0.028 | 0.022 | 0.025 | 0.021 | 0.029 | 0.012 | 0.022 | 0.022 | 0.018 | 0.026 |
| $N\alpha \sin\theta/\tanh(NH\kappa/f)$ | $10^{-6} \text{ sec}^{-1}$ | 1.8 | 1.3 | 3.2 | 3.3 | 2.9 | 2.3 | 5.9 | 10.3 | 6.1 | 5.2 | 2.7 |
| Period | days | 41 | 56 | 23 | 22 | 25 | 32 | 12 | 7 | 12 | 14 | 27 |
| **16 day**   $\phi$ | degree | −35 | −1 | −27 | 0 | 1 | −3 | −13 | −14 | −16 | −35 | −29 |
| $\theta$ | degree | 35 | 14 | 32 | 24 | 23 | 27 | 34 | 34 | 31 | 59 | 46 |
| $\kappa$ | $\text{km}^{-1}$ | 0.037 | 0.034 | 0.043 | 0.038 | 0.043 | 0.051 | 0.033 | 0.041 | 0.026 | 0.021 | 0.026 |
| Predicted $\omega$ | $10^{-6} \text{ sec}^{-1}$ | 6.0 | 4.0 | 4.9 | 5.1 | 4.3 | 5.0 | 5.9 | 6.0 | 4.6 | 4.1 | 2.4 |
| $2\pi/\omega$ | days | 12 | 18 | 15 | 14 | 17 | 15 | 12 | 12 | 16 | 18 | 30 |
| **10.7 day**   $\phi$ | degree | −29 | −8 | −45 | 1 | −9 | −15 | −21 | 13 | 9 | −3 | −42 |
| $\theta$ | degree | 29 | 21 | 50 | 23 | 33 | 39 | 42 | 7 | 6 | 27 | 59 |
| $\kappa$ | $\text{km}^{-1}$ | 0.036 | 0.025 | 0.045 | 0.040 | 0.024 | 0.050 | 0.043 | 0.013 | 0.017 | 0.021 | 0.030 |
| $\omega$ | $10^{-6} \text{ sec}^{-1}$ | 5.1 | 7.2 | 7.0 | 4.8 | 7.7 | 6.9 | 6.3 | 2.8 | 1.3 | 2.2 | 2.6 |
| $2\pi/\omega$ | days | 14 | 10 | 10 | 15 | 9 | 11 | 11 | 26 | 56 | 34 | 28 |
| **8 day**   $\phi$ | degree | −51 | −26 | −54 | −25 | −11 | −17 | −35 | −44 | −72 | −49 | −52 |
| $\theta$ | degree | 51 | 39 | 59 | 49 | 35 | 41 | 56 | 64 | 87 | 73 | 69 |
| $\kappa$ | $\text{km}^{-1}$ | 0.039 | 0.034 | 0.049 | 0.027 | 0.036 | 0.063 | 0.022 | 0.075 | 0.053 | 0.022 | 0.037 |
| $\omega$ | $10^{-6} \text{ sec}^{-1}$ | 7.9 | 10.5 | 7.7 | 11.3 | 6.6 | 7.0 | 11.2 | 8.4 | 6.0 | 4.3 | 2.5 |
| $2\pi/\omega$ | days | 9 | 7 | 9 | 6 | 11 | 10 | 7 | 9 | 12 | 17 | 29 |

the details of the comparison. The depth, bottom slopes, and isobath angles are averaged from Table 2 for the triangles corresponding to the wave-number estimates of Fig. 21. The Brunt–Väisälä frequencies are interpolated from root mean square values over the bottom 300 m of CTD soundings provided by Gordon Volkmann, and taken during the setting and retrieval of the array. The predicted frequencies are plotted in Fig. 24 vs the imposed frequencies $2\pi/32$ days, etc., but only for the upper part of the array (depths less than 4 km), since the lower part has consistently seemed to be a different regime. The dashed line is the regression line for the theoretical values predicting the actual, and has a correlation coefficient of 0.88. The solid line is a perfect prediction. It is also the regression line, with a correlation of 0.98, if we remember that transverse oscillations (as we have seen these to be) are composed of negative as well as positive frequencies, so the reality of the cross-spectral lags makes the figure odd-symmetric. While the scatter is not small, it is only of the order of the imposed frequency resolution, and there does seem to be considerable predictive ability in the dispersion relation (2.4), at least below the thermocline on the slope north of the Gulf Stream.

## 10. VORTICITY AND DIVERGENCE

In section 7, an assumption of a single wave-number was used to estimate the ratio of the spectra of the divergence to that of vorticity from individual moorings. For quasi-geostrophic flow, the ratio should be small, but we found in Fig. 13 that it was only small in the upper part of the array. Here, we will use direct estimates of the divergence and vorticity to estimate their spectra.

BRYDEN (1975) used a very nice way to show that data from the Mid-Ocean Dynamics Experiment was non-divergent, in what appears to have been the first direct oceanic test. Instead of comparing the spectra, he simply plotted $u_x$ vs $v_y$ to show a strong negative correlation.

In the present model, an upslope velocity $v$ causes a column of fluid to shorten, so to diverge. The divergence, combined with conservation of angular momentum, causes the water column to start to rotate more slowly than the Earth. Thus, the vorticity should be in quadrature with the divergence and upslope velocity. The divergence is small, so cannot be reliably estimated. The vorticity should lead the upslope velocity, since positive upslope velocity leads to negative vorticity. This will be checked.

The divergence, vorticity, and upslope velocity will be estimated as averaged over areas formed by connecting $n$ moorings at positions $(x_j, y_j)$. Using Stokes' theorem, the average vorticity is

$$\zeta = \frac{1}{A} \iint (\nabla \times \mathbf{u}) dA = \frac{1}{A} \oint \mathbf{u} \cdot d\mathbf{x}, \tag{5.4}$$

where the area $A$ is given by

$$A = \iint 1 \, dA = \iint \nabla \cdot (\tfrac{1}{2}\mathbf{x}) dA = \tfrac{1}{2} \oint \mathbf{x} \times d\mathbf{x} = \tfrac{1}{2} \sum (x_j \Delta y_j - y_j \Delta x_j), \tag{10.3}$$

where $\Delta y_j = y_{j+1} - y_{j-1}$, with $y_{n+1} = y_1$, etc. If $\mathbf{u}$ is linearly interpolated between $\mathbf{x}_j$ and $\mathbf{x}_{j+1}$, (10.2) becomes

$$\zeta = \tfrac{1}{2} \sum (u_j \Delta x_j + v_j \Delta y_j)/A, \tag{10.4}$$

the divergence is

$$D = \tfrac{1}{2} \sum (u_j \Delta y_j - v_j \Delta x_j)/A, \tag{10.5}$$

and the mean upslope velocity forcing is

$$v = \frac{1}{n} \sum (\alpha_j v_j / H_j). \tag{10.6}$$

It is necessary to use areas which are not small in each direction compared to a dominant wave scale, else the signal will not change compared to the noise in (10.4), so will cancel. It is also necessary that the dimensions be not large compared to the dominant length scale, else again the signal will cancel (i.e. truncation error). So, we must use dimensions comparable to, but preferably a little smaller than, a wave scale, that is, about 50 km. From Fig. 12, we can see that the areas are thus restricted to the smallest ones,

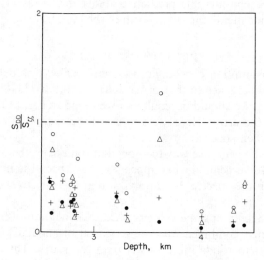

FIG. 25.   Ratio of spectral estimate for the divergence over the triangles to that for vorticity, plotted vs average water depth of the triangle. ● 32, + 16, △ 11, ○ 8-day period.

triangles in the northern part of the array, squares in the southern. Three points are not many for estimating an integral, but let's try.

The time series were formed from (10.4, 5, 6), and Fourier transformed by 32-day blocks, as previously. Figure 25 shows the ratio of the spectral estimates for the divergence to those for the vorticity. For the four frequencies, it can be seen that the ratio is smaller than 1, except for a single point. The ratios are not very small, but not much can be expected from three points. Notice that there is no tendency for greater relative divergence in deeper water, so we can conclude that the single wave-number hypothesis broke down in Fig. 13.

From the Fourier transforms we can also test whether $v$ and $\zeta$ are in quadrature. The normalized cross-spectral estimates for the four frequencies are plotted in Fig. 26. (This is the same as a plot with coherence plotted as radius and phase lag as angle.) The agreement is not overwhelming, but certainly satisfactory for 3-point integrals. Theoretically, the vorticity should have a 90° lead over upslope velocity. Of the fifty-six estimates,

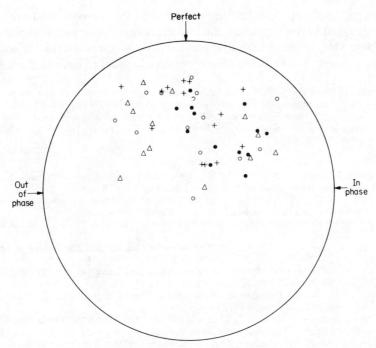

FIG. 26. Imaginary vs real part of the normalized cross-spectrum between upslope velocity and vorticity estimate over triangles. The radius is coherence, the angle from the real axis is the phase. 'Perfect' is the theoretical quadrature. ●32, +16, △11, ○8-day period.

all but one have the vorticity leading. The 5% significance level for the coherence (radius) is about 0.43, so most of the coherences are significant. Most of the phases are significantly different than 0°, but not from 90°.

## 11. CONCLUSIONS

An experiment has been carried out, essentially as proposed five years ago, to test whether the model of RHINES (1970) and T (1971a,b) actually describes the low-frequency motions north of the Gulf Stream. The results appear very positive: the non-divergence, transverseness, momentum flux, and horizontal and vertical energy distributions are as expected, and the dispersion relation checks well. Even the places where the model breaks down, at the Gulf Stream, and at the continental slope, are as expected. This part of the ocean seems to behave as a linear, rotating, stratified fluid on a moderate bottom slope.

Since this experiment was started, RHINES (1975) has shown that even strongly nonlinear flow may be expected to behave much as if it were linear. He argues that nonlinear quasigeostrophic flow will adjust itself so the particle velocities are the same order as the phase velocities of the linear theory. This may well be so in the southern part of the large array here, but it appears that the northern part has the same horizontal and temporal scales as the southern part (presumably imposed thereby), but much weaker energies, so the particle velocities are slower than the phase velocities. For instance, the slowest phase velocity considered is that for a 32-day period, with phase velocity about

250 km/32 days, or about 8 km/day perpendicular to the wave-fronts. Even if all of the kinetic energy is put at this frequency, the low-frequency motions around Site D are only 4 or 5 km/day. Also, we have seen that most of this is transverse to the wave-number. The dynamics around Site D, below the thermocline, appear to be indeed linear.

*Acknowledgements*—I thank Ferris Webster for essential encouragement over this long-term work, which was supported by Office of Naval Research contracts N00014-66-C0241 NR 083-004 and N00014-74-0262 NR 083-004. Of course, this work could not have been done without the instrument work, deployment, and recovery, and the data processing of the rest of the WHOI Buoy Project. This paper was written while visiting the University of Western Australia.

## REFERENCES

AMOS D. E. and L. H. KOOPMANS (1963) Tables of the distribution of the coefficient of coherence for stationary bivariate gaussian processes. Sandia Corp. Monograph SCR-483.

BRYDEN, HARRY LEONARD (1975) *Momentum, mass, heat, and vorticity balances from oceanic measurements of current and temperature.* Ph.D. thesis, M.I.T.-W.H.O.I. Earth Sciences, 130 pp.

FOFONOFF N. P. (1969). Spectral characteristics of internal waves in the ocean. *Deep-Sea Research*, supplement, **16**, 59–71.

GOULD W. J. and E. SAMBUCO (1975). The effect of mooring type on measured values of ocean currents. *Deep-Sea Research*, **22**, 55–62.

HAMON B. V. and E. J. HANNAN (1974) Spectral estimation of time delay for dispersive and non-dispersive systems. *Journal of the Royal Statistical Society, Series C*, **23**, 134–142.

HEINMILLER, ROBERT H., JR. and DONALD A. MOLLER (1974) Failure of a moored array in a Gulf-Stream eddy. *Marine Technical Society Journal*, **8**, 35–38.

HEINMILLER, ROBERT H. and R. WALDEN (1973) Details of Woods Hole Moorings. Woods Hole Oceanographic Institution Internal Report 73-71 (unpublished manuscript).

HELD, ISAAC M. (1975) Momentum transport by quasi-geostrophic eddies. *Journal of the Atmospheric Sciences*, **32**, 1494–1497.

HINICH M. J. and C. S. CLAY (1968) The application of the discrete Fourier transform in the estimation of power spectra, coherence, and bispectra of geophysical data. *Review of Geophysics*, **6**, 347–363.

LUYTEN JAMES R. (1977) Scales of motion in the deep Gulf Stream and across the continental rise. *Journal of Marine Research, Richardson Memorial Volume*, (in press).

RHINES, PETER B. (1970) Edge-, bottom-, and Rossby waves in a rotating stratified fluid. *Geophysical Fluid Dynamics*, **1**, 273–302.

RHINES, PETER B. (1971) A note on long-period motions at Site D. *Deep-Sea Research* **18**, 21–26.

RHINES, PETER B. (1975) Waves and turbulence on a beta-plane. *Journal of Fluid Mechanics*, **69**, 417–443.

SCHMITZ, WILLIAM J., JR. (1974) Observations of low-frequency current fluctuations on the continental slope and rise near Site D. *Journal of Marine Research*, **32**, 233–251.

STARR, VICTOR P., JOSE P. PEIXOTO and NORMAN E. GAUT (1970) Momentum and zonal kinetic energy balance of the atmosphere from five years of hemispheric data. *Tellus*, **XXII**, 251–274.

SUAREZ, ALFREDO A. (1971) *The propagation and generation of topographic oscillations in the ocean.* Ph.D. thesis, M.I.T.-W.H.O.I. Earth Sciences, 130 pp.

THOMPSON, RORY O. R. Y. (1971a) Topographic Rossby waves at a site north of the Gulf Stream. *Deep-Sea Research*, **18**, 1–19.

THOMPSON, RORY O. R. Y. (1971b) Why there is an intense eastward current in the North Atlantic but not in the South Atlantic. *Journal of Physical Oceanography*, **1**, 235–237.

THOMPSON, RORY O. R. Y. and JAMES R. LUTYEN (1976) Evidence for bottom-trapped topographic Rossby waves from single moorings. *Deep-Sea Research*, **23**, 629–635.

UCHUPI, ELAZAR (1965) *Map showing relation of land and submarine topography, Nova Scotia to Florida.* U.S. Geological Survey Washington, D.C. G65040.

WEBSTER, FERRIS (1965) Measurements of eddy fluxes of momentum in the surface layer of the Gulf Stream. *Tellus*, **XVII**, 239–245.

WEBSTER, FERRIS (1968) Observations of inertial-period motions in the deep sea. *Reviews of Geophysics*, **6**, 473–490.

WEBSTER, FERRIS (1969) Vertical profiles of horizontal ocean currents. *Deep-Sea Research*, **16**, 85–98.

Prog. Oceanog., Vol. 7, pp. 163–239.
© Pergamon Press Ltd. 1978.   Printed in Great Britain

0079–6611/78/0701–0163 $05.00/0

# Growth conditions of manganese nodules
# Comparative studies of growth rate, magnetization,
# chemical composition and internal structure

DIETRICH HEYE

*Bundesanstalt für Geowissenschaften und Rohstoffe (B.G.R.)*
*(Federal Institute for Geosciences and Natural Resources)*

(*Received* 11 *September* 1975; *in revised form* 28 *April* 1976; *accepted* 25 *May* 1977)

CONTENTS

**Abstract**—Twenty-four manganese nodules from the surface of the sea floor and fifteen buried nodules were studied. With three exceptions, the nodules were collected from the area covered by *Valdivia* Cruise VA 04 some 1200 nautical miles southeast of Hawaii.

Age determinations were made using the ionium method. In order to get a true reproduction of the activity distribution in the nodules, they were cut in half and placed for one month on nuclear emulsion plates to determine the $\alpha$-activity of the ionium and its daughter products. Special methods of counting the $\alpha$-tracks allowed resolution to depth intervals of 0.125 mm. For the first time it was possible to resolve zones of rapid growth (impulse growth) with growth rates, $s > 50 \, mm/10^6$ yr and interruptions in growth. With few exceptions the average rate of growth of all nodules was surprisingly uniform at $4$–$9 \, mm/10^6$ yr. No growth could be recognized radioactively in the buried nodules. One exceptional nodule has had recent impulse growth and, in the material formed, the ionium is not yet in equilibrium with its daughter products. Individual layers in one nodule from the Indian Ocean could be dated and an average time interval of $t = 2600 \pm 400$ yr was necessary to form one layer. The alternation between iron and manganese-rich parts of the nodules was made visible by colour differences resulting from special treatment of cut surfaces with HCl vapour. The zones of slow growth of one nodule are relatively enriched in iron.

Earlier attempts to find paleomagnetic reversals in manganese nodules have been continued. Despite considerable improvement in areal resolution, reversals were not detected in the nodules studied. Comparisons of the surface structure, microstructure in section and the radiometric dating show that there are erosion surfaces and growth surfaces on the outer surfaces of the manganese nodules. The formation of cracks in the nodules was studied in particular. The model of age-dependent nodule shrinkage and cracking surprisingly indicates that the nodules break after exceeding a certain age and/or size. Consequently, the breaking apart of manganese nodules is a continuous process not of catastrophic or discontinuous origin. The microstructure of the nodules exhibits differences in the mechanism of accretion and accretion rate of material, shortly referred to as accretion form. Thus non-directional growth inside the nodules as well as a directional growth may be observed. Those nodules with large accretion forms have grown faster than smaller ones. Consequently, parallel layers indicate slow growth. The upper surfaces of the nodules, protruding into the bottom water appear to be more prone to growth disturbances than the lower surfaces, immersed in the sediment. Features of some nodules show, that as they develop, they neither turned nor rolled. Yet unknown is the mechanism that keeps the nodules at the surface during continuous sedimantation. All in all, the nodules remain the objects of their own distinctive problems. The hope of using them as a kind of history book still seems to be very remote.

## INTRODUCTION

LARGE AREAS of the deep ocean floor are covered with manganese nodules although they are also found in shallower regions. Nodules similar to those found on the deep ocean floor but markedly higher in iron are even found on the shelf and in inland lakes. The formation of ferromanganese nodules seems to be a phenomenon common to different kinds of depositional environments, the conditions controlling this process still unknown.

The nodules are complicated aggregate structures, which are built up predominantly from manganese and iron compounds, and with minor or trace amounts of a wide range of other elements. The contents of copper, nickel and cobalt in some regions rise to sufficiently high levels to give the nodules a potential commercial value. Normally their size varies between a few centimeters to decimeters. Cross-sections show a concentric arrangement of layers.

Their origin and their material supply are still under discussion. One of the hypotheses assumes that the nodules draw their growth material from submarine volcanic exhalations or from the decomposition of underwater volcanic rocks, and therefore would be particularly abundant in active volcanic regions. According to another hypothesis material is exclusively supplied from seawater. The sediments below the nodules are also considered as a source of supply material. In this case it is assumed that the elements are brought to the surface by diffusion. Yet on closer investigation nodules are abundant in regions where no volcanic activity is known to exist and proof of supply from sediment has not been substantiated by measurements of dissolved manganese. At present the seawater is favoured as the source of the elements, whereby it is itself fed from all conceivable sources such as rivers, underwater decomposition and diffusion from sediments.

During the nodule formation the accretion of material at the nodule surface causes similar difficulties in comprehension. Besides the conception of a purely inorganic accretion under specific Eh- and pH-conditions, several possibilities of an organic deposition by bacteria or other organisms are under discussion. Here also the most contradictory possibilities stand side by side.

It is difficult to answer the question as to what mechanism holds the nodules at the sediment surface, the growth of a sediment being at a rate of the order $10^2$–$10^3$ times higher than that of nodules. If a mechanism which held them at the surface did not exist, the majority of the nodules should have long been buried, since their bulk density is too high to permit them to float on the sediment. Various ideas have been discussed, for instance that organisms would move the nodules during their search for food in the sediment, and in doing so, constantly lift them to the sediment surface. However, this question remains unanswered, since for the large nodules, such as the 'Horizon-Nodule' for example, which is about 1 m in diameter and weighs 62 kg, this mechanism must be excluded. This report also expresses an opinion on this problem. Since excellent summaries of the different formation theories already exist, as for example BONATTI and NAYUDU (1965), HERING (1971), SEIBOLD (1973) and HALBACH (1974), specific references have been omitted for the sake of brevity.

Perhaps it will also be clear from the description given here, how many questions still remain unanswered and also how many different explanations have been put forward. At times the number of theories may have exceeded the number of nodules examined, and probably there is hardly a credible possibility left, which has not been speculatively considered. Either several mechanisms which have been previously discussed are active in different ways in specific regions, with the result that a uniform description of the nodule formation is impossible, or some additional insight is still missing.

During the course of the investigations, which deal with the growth rate, the chemistry and the internal structure, the author tried to work independently of any of these theories. Some of the results are discussed in connection with hypotheses in cases where they clearly support or abolish current views. Most of the methods used or developed in the present investigations are non-destructive. This made it possible to carry out different measurements, for instance, age determination, chemical and microstructural analysis on identical material which, in the case of age determination, had not been possible previously.

## 2. PREPARATION OF NODULES AND SEQUENCE OF INVESTIGATIONS

Since the nodules are composed of very brittle material and there is a danger of irregular breakage during sectioning, they were first encased in epoxy resin (araldite) in vacuum. After 10 min, normal pressure was established in the chamber and the epoxy resin penetrated into the finer pores of the nodules. After the resin had hardened, the embedded nodules were cut in two, and one of the faces was ground and polished. This process not only prevented the whole nodule from shattering, but also protected the nodule surface in particular, from damage. Diffusion of the radioactive, inert gas radon can occur in the porous material of the nodules and with it the danger of errors in the age-determination. In order to prevent such a migration, the sectional surfaces are also sealed. Following the process mentioned above, these surfaces were covered with a thin layer of epoxy resin, which, when it had hardened was ground off until the nodule material appeared, although the inner pores near the surface areas still remained filled. Then a circle of holes (1 mm in diameter) was drilled in the resin surrounding one of the cut section surfaces. These served as fiducial marks in the age determination and detail-photographs. Thereafter the sequence of investigation on the nodules was as follows:

(a) One of the prepared nodule sections was chosen for the investigation and laid on nuclear emulsion plates for 1–2 months for radioactive age-determination. At the end of this exposure time, each hole around the nodule was pricked through with a sharp needle until it reached the underlying emulsion layer. After the development the pricks in the emulsion layer were macroscopically visible and a layer reconstruction of the nodule on the emulsion plate was possible at any time.

(b) The structure of the nodules was macroscopically and microscopically examined. On the base of these investigations suitable regions were chosen for age determinations.

(c) Then, for the magnetic investigations, provided the nodule was large enough, a 5 mm thick slice was sawn off from the nodule half. This investigation of the slice was again non-destructive and the examined face once again preserved.

(d) Finally the section face was treated with acid to gain a qualitative view of the chemical properties of the nodule. Naturally the section face was changed by this treatment, so that subsequent investigations, for example of its micro-structure, were now only possible with limitations. Therefore sections which yielded the most interesting results during the course of non-destructive investigations were stored as proofs, whilst the opposite face was used for etching.

## 3. RADIOACTIVE INVESTIGATIONS (AGE DETERMINATION AND GROWTH RATE)

### 3.1. *Review of previous radioactive age-determination studies on manganese nodules*

The first research on the radioactivity of deep-sea manganese nodules was carried out in about 1950. It was found that the activity of the outer layers is greater, and towards the interior, this activity decreases exponentially within a few millimetres. From this it was concluded that a daughter product of the uranium dissolved in seawater is incorporated into the outer layer of the nodule as it forms, and the concentration of this daughter product decreases through time towards the nodule centre) due to radioactive decay. At the time, however, the radioactive isotope was not properly identified and it was erroneously assumed to be radium ($Ra^{226}$). As a result of more detailed investigations, we know now that the isotope of the $U^{238}$ series incorporated in excess in the nodules, is not radium, but ionium (Io or $Th^{230}$). The half-life of ionium is significantly longer (the half-life of $Ra^{226} = 1617$ years and that of $Th^{230} = 75,200$ years). As a result of this mistake, the results given in the first publications on the growth rates of manganese nodules are too high by a factor of 50. Even today average growth rates given in the general literature vary considerably, and in part, are too high. Consequently it is worthwhile to make a critical examination of the actual growth rates of deep-sea manganese nodules. It should be kept in mind, however, that nodules in shallow water and in inland lakes might grow more quickly.

If the measurements of these first investigations, for example, BUTTLAR and HOUTERMANS (1950), are correspondingly evaluated, the previous results agree with the new results, (for example KU and BROECKER (1967 and 1969) and a very small growth rate of a few millimetres per million years is found. This very low growth rate was first viewed sceptically, and therefore, the results of the Io decay were checked with other radioactive age-determinations. Comparable results were obtained by examining protactinium ($Pa^{231}$), an isotope of the $U^{235}$ series, KU and BROECKER (1967). Datings on two nodules using the painstaking $Be^{10}$ method also confirmed the previous results of the Io age-determination within a marginal error, SOMAYAJULU (1967); KRISHNASWAMI, SOMAYULU and MOORE (1972). The basalt core of a nodule was also examined using the potassium–argon method and here corresponding results were also obtained BARNES and DYMOND (1967), from which it was assumed that the igneous rocks solidified only shortly before the beginning of the nodule growth. Thus, it can be concluded that today scarcely any fundamental objections can be raised against an Io dating of manganese nodules and the derived low growth rates. The $Be^{10}$ and the K–Ar methods are unsuitable as standard methods, since the first requires greater amounts of material than are usually available to manganese nodules and is in addition a more painstaking method, and the K–Ar methods are only practicable in exceptional cases. Therefore, at the moment only the Io, respectively, Pa age-determination method can be considered as a standard process for radioactive dating, respectively, finding the growth rate of manganese nodules. This has one disadvantage in that only the outer nodule regions to an age of about $3 \times 10^5$ years can be examined, since at that stage the Io has decayed to such a degree that no further dating is possible. The total age of the nodule can, therefore, only be assessed by extrapolation of this rate up to the nodule centre, respectively, the core boundary. Nodules have often been measured which have lacked the marginal zone of raised activity with its typical exponential decrease, KU and BROECKER (1969). These results would seem to indicate, that either the nodules have stopped growing (stagnation of growth) at least $3 \times 10^5$ years ago (or even longer), or even that the nodules are decomposing through dissolution processes in the seawater or through erosion, (see section 3.4.4.). Naturally it cannot be established from the radioactive measurements alone, if this very logical conclusion is correct. Generally it is beyond dispute that new ionium is produced from the radioactive decay of the uranium dissolved in the sea and it is available for nodule formation. Nor is there any known or conceivable mechanism, which could have temporarily prevented the incorporation of ionium during nodule growth.

In order to give a full picture, several peculiarities of the nodule formation must be mentioned. KU and GLASBY (1972) have examined several nodules from shallow waters and found that no Io had been incorporated from the seawater. The authors suspect the cause 'to be slightly more reducing conditions than in the open sea'. Datings were carried out on the growing Io (produced by $U^{234}$), which showed a growth rate of $(30 \pm 10 \, \text{mm}/10^6)$.

Accordingly, the radiochemical conditions of shallow water nodules vary from those of deep-sea nodules. These datings were based on another model, which is different from the model used for dating deep-sea nodules and is not in contradiction to the one used here. KRISHNASWAMI and MOORE (1973) examined $Ra^{226}$ in inland lake nodules and found an exponential decrease from the surface inwards. Indeed in these nodules the $Ra^{226}$ could have been incorporated from the freshwater above. If this assumption is evaluated, growth rates of $2-3 \, \text{mm} \, 1000 \, \text{yr}^{-1}$ are obtained. If these insufficiently explained radioactive relationships of freshwater nodules prove to be correct, then three different nodule groups have to be considered.

(a) Deep-sea nodules with excess Io incorporation and datings with Io decay.
(b) Shallow water nodules without Io incorporation and a possibility of datings through Io growth from $U^{234}$ decay.
(c) Freshwater nodules, mainly incorporating $Ra^{226}$, the decay of which can be used for dating.

During the investigation of deep-sea nodules it was noticed, that firstly, the nodule surfaces showed varying degrees of initial activity, which is the activity measured from the outermost layer. Secondly, not all of the Io produced in the sea water above the nodules is incorporated into the nodules. KRISHNASWAMI, SOMAYAJULU and MOORE (1972) went into this question in detail and arrived at three possible conclusions:

The nodules only grow during 10% of their lifetime.

A continual change between growth and erosion takes place on their surfaces and their effective growth is only 10%.

The nodules do not incorporate all the Io (and $Be^{10}$), that is formed in the seawater.

In their publications the authors favour the first possibility, in the literature this possibility is often cited, as if it were an established fact, that nodules only grow for 10% of their lifetime.

Estimations, based on the balances of elements in the ocean–sediment system, are often unreliable and hardly allow distinct conclusions. This is because they are only based on a few available readings of the Io content of nodules and sediment. Moreover, according to SEIBOLD (1973), the manganese nodules are found predominantly in regions of low sedimentation rate and possibly related to this a reduced Io supply as a result of suspended materials in the regions of manganese nodule formation, which make it impossible to compare balances. In my opinion, it seems more probable, for these reasons, that not all the Io produced in the seawater is to be found in the nodules, and that even the rate of Io-incorporation can be different in each nodule. The third possibility of the above enumeration has therewith been stated, which in my opinion, is just as likely.

The so called 'hydration-rind-dating' is a new attempt at determining the age of manganese nodules. It is based on the palagonitization of volcanic glasses through hydration within the manganese nodule cores, which is supposed to occur at a constant rate. The overall relationship of this method is set out in three publications MORGENSTEIN (1971 and 1973) and MORGENSTEIN and FELSHER (1971). The extent of the usefulness of this method is yet to be shown. It would at least have the advantage of reaching further than the $3 \times 10^5$ years of the Io method. However, the authors have already drawn very far reaching conclusions which amount to a new hypothesis on manganese nodule formation. It has yet to be seen, whether the authors' datings or derived theories can be substantiated.

### 3.2. Measuring technique of the radioactive investigations

In order to determine the exceedingly slow growth rate of deep-sea manganese nodules, it is necessary to measure the activity at very short intervals. For this purpose, layers can be scraped off, or chemically dissolved and after the radiochemical separation processes the whole family of radioactive isotopes Io, $Pa^{231}$, $U^{238}$, $Th^{232}$ can be measured individually, KU and BROECKER (1969). Thanks to their comprehensive investigations, we now have a precise knowledge of the distribution of the listed radioactive isotopes in the nodules and it is established that the Io and not the Ra is responsible for the excess activity in the nodules. Based on this fundamental clarification of the behaviour of radioactive isotopes in the manganese nodules, it has in the meantime become possible to revert to a simpler measuring technique for the determination of the growth rate. Therefore, in the investigations conducted here through the use of nuclear emulsion plates on sectioned surfaces, only the total distribution of $\alpha$-active elements was determined. Although this means a simplication of the investigation techniques and has other advantages, there are, however, difficulties which have yet to be faced and discussed. The first investigations of this kind were carried out in 1950 by V. BUTTLAR and HOUTERMANS.

The nodules sections are placed on a nuclear emulsion plate for 30–60 days and suitable specimen count areas are selected in each instance from between two marking points (bearing in mind the different aspects of the internal structure). The $\alpha$-tracks are projected by means of a projection microscope and are copied on to prepared sheets in a continuous assembly of individual specimen count areas. By gluing these individual sheets together, a magnification of $250 \times$ the total specimen count area of its $\alpha$-tracks distribution is obtained. At least two marking points in each count area are included and reproduced. The specimen count area of the nodule sectional surface, with the internal structure and the pertinent marking points, is photographed with $10 \times$ magnification on to a diapositive. If this diapositive is projected on to the joined sheets of the transcribed $\alpha$-tracks, and the marking points are superimposed, a very good combination of the alpha tracks and nodule structure is achieved. The nodule surface and the internal structure can be copied on the sheet of $\alpha$-tracks, (see Fig. 2). In this combination the projection with $250 \times$ magnification can be accurate to approx 5.0 mm, which corresponds to a real co-ordination accuracy of approx 0.02 mm for the nodule. Then, in the joined sheets of the $\alpha$-tracks,

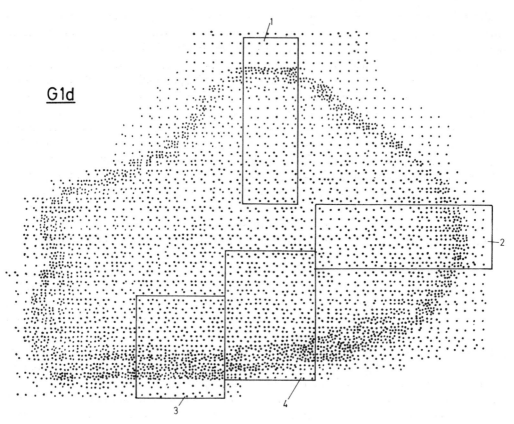

G1d

FIG. 1.   Distribution of α-tracks from a nodule. A dot is made for every 10 α-tracks. The nodule surface is delineated by its increased activity, caused by the incorporated ionium and its daughter products.

intervals of 0.125 mm are drawn parallel to the surface and to the internal growth lines of the nodules, and the α-tracks in the intervals are counted. Thus, in order to count the zones of the same age natural growth coordinates were used, and no longer the inflexible cartesian coordinates. Using this process, the age determination resolution could be substantially increased. Whereas the interval widths were previously 0.5 mm, HEYE and BEIERSDORF (1973), an interval width of only 0.125 mm could now be realised. Thus, it also seems that a limit has been reached which cannot be substantially minimised. For, firstly the α-tracks have a specific length and the width of the counting intervals must be large in relation to this, and secondly, the microstructure of individual nodules shows growth cusps, which are about 0.1 mm in size and it does not seem sensible, at the moment, to choose counting intervals which are smaller than these elementary areas of nodule formation.

In Fig. 1 of the frequency distribution of the tracks for a whole nodule is shown. The marginal zone of increased activity, which mainly originates through the Io and its daughter products, can be clearly seen. A small tracks frequency is also found outside the nodule, which shall be described here as $N_0$ background activity and is caused, for example, through the inherent activity of the photographic emulsion layer or through other α-rays. A larger track frequency is found in the centre than outside the nodule. This background in the

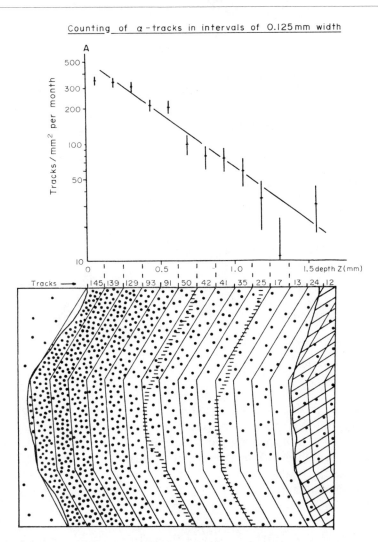

FIG. 2.   Number of α-tracks in intervals of 0.125 mm width. The counting area of a nodule with the reproduced α-tracks is given below. In the upper part the derived exponential decrease of the tracks is drawn in logarithmic scale as a function of depth.

track count stems from U and $Th^{232}$ in the nodule, which together with their daughter products, also leave α-tracks behind. According to KU and BROECKER (1969) between 3–150 ppm $Th^{232}$ and 6–10 ppm U are incorporated into deep-sea manganese nodules, and in the evaluation of their α-track contribution, must be subtracted from the tracks caused by Io and its daughter products.

## 3.3. *Determination in the growth rate*

The decrease in activity of the radioactive isotope $A$ whose half-life (respectively, decay constant $\lambda$) and the time $t$ which is interpreted as the age of the layer concerned, are connected as follows:

$$A = C_0 \exp^{-\lambda t}. \tag{1}$$

In which $C_0$ is the initial activity to time, $t = 0$.

If manganese nodules have a regular growth ($s$ = constant), then from the time function a linear function of depth $Z$ is derived, which is measured inward from the nodule surface:

$$Z = s \cdot t,$$

$$A = C_0 \exp\left(\frac{-\lambda Z}{t}\right). \tag{1a}$$

The equations for the Io, respectively, the Pa are as follows:

$$A(\text{Io}) = C_0 \exp\left(\frac{-\lambda(\text{Io})Z}{s}\right), \tag{2}$$

$$B(\text{Pa}) = D_0 \exp\left(\frac{-\lambda(\text{Pa})Z}{s}\right). \tag{3}$$

If one considers the background activity $N_0$ and $N_1$ (U, Th) caused by the U and Th within the nodule, the following total activity in the nodule depths $Z$, respectively, the trace frequency, is obtained:

$$A(Z,s) = A(\text{Io}) + B(\text{Pa}) + N_0 + N_1(\text{U, Th})$$

$$= C_0 \exp\left(\frac{-\lambda(\text{Io})Z}{s}\right) + D_0 \exp\left(\frac{-\lambda(\text{Pa})Z}{s}\right) + N_0 + N_1(\text{U, Th}). \tag{4}$$

The radioactive daughter products of Io and Pa are all extremely short-lived in comparison with the original isotopes and they are, therefore in radioactive equilibrium, if one ignores the short-lived radon losses during the nodule preparation. Their $\alpha$-rays do not appear separately, but as a multiplication of the activity of their mother products. The background activity $N_0$ can easily be determined by counting outside the nodule, and is immediately deducted from the determined track frequency $A_i^{(0)}$. The individual readings are as follows:

$$A_i = A_i^{(0)} - N_0,$$

$$A_i(s, Z) = C_0 \exp\left(\frac{-\lambda(\text{Io})Z_i}{s}\right) + D_0 \exp\left(\frac{-\lambda(\text{Pa})Z_i}{s}\right) + N_1(\text{U, Th}). \tag{5}$$

Protactinium is a daughter product of the rare isotope $U^{235}$ which only represents 0.7% of the total uranium. However, due to its short half-life it has 4.3% of the activity of uranium (activity ratio $U^{238}:U^{235} = 100:4.5$). This activity ratio is also applicable for the equilibrated U-daughter products Io and Pa. Again through their different half-lives and 15% excess of $U^{234}$, a theoretical ratio of Io/Pa = 10.8, SACKETT (1966), can be determined for the Io and Pa present in the sediment or in the nodules. In other words, this means at the moment of the Io and Pa accretion at the nodule surface, only one tenth of the activity is expected to come from Pa. Due to the quicker decay of Pa this ratio decreases with time and in equation (5), one can ignore the Pa activity compared to the Io activity. The last equation thus reduces to:

$$A_i(Z_i, s) = C_0 \exp\left(\frac{-\lambda(\text{Io})Z_i}{s}\right) + N_1(\text{U, Th}). \tag{6}$$

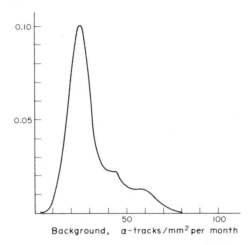

FIG. 3. Frequency distribution of counted background activity inside the nodule $N_1$ (U, Th) which has a mean value of $25 \pm 5$ α-tracks/mm²/month⁻¹ and showed high values for very few nodules.

The background activity, $N_1$(U, Th) must now be subtracted from the track frequency $A_i$. One obtains the value as α-track frequency from inside the nodule, from where the Io and its daughter products have decayed. The nodules taken from the investigation area of the Pacific registered a mean value, with a few exceptions, of $\sim (25 \pm 5)$ tracks/mm²/30 days. (See Fig. 3.)

After subtracting the background activity, equation (6) transforms to:

$$A_i(Z_i, s) = C_0 \exp\left(\frac{-\lambda(\text{Io})Z_i}{s}\right). \tag{7}$$

These track frequencies were drawn on a logarithmic scale as a function of depth for all nodules investigated. The results and the descriptions of the individual nodules are given in the appendix. In the parts where the results lie on a descending straight line, the weighted values were used to calculate the gradient, respectively, the growth rate and its error by computer. The growth rates are plotted as numerical numbers in the figures.

### 3.4. *Growth relations of the investigated nodules*

All previous publications on radioactive investigations of manganese nodules showed a continual growth. Within the investigations presented here, deviations could be proved through increased resolution. These possibilities are sketched in Fig. 4 and are interpreted as rapid growth rate and growth interruptions. These results represent a refinement of previous investigations and are not in contradiction to them. If one adds and averages the results obtained (depth intervals 0.125 mm) to the previously accessible interval widths (0.5, respectively, 1 mm), one almost always obtains a continual growth, as published by other authors. The result of the nodules 54(1) (specimen count area A–X) and 65 (specimen count area 3–4 and specimen count area 1–2) are recommended as examples for comparison to the reader. There are even single points scattering around the logarithmic straight line in the figures presented here as can be found in previous publications. Many deviations reported in earlier publications might be related to irregular growth processes, which hardly could be detected until recently because of poor resolution.

3.4.1. *Growth interruptions.* The evaluation of the nodules 77 (specimen count area C–D)

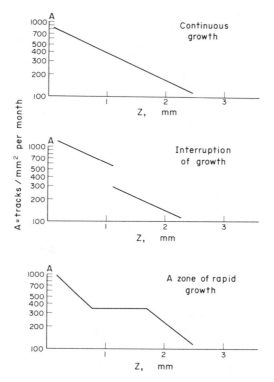

FIG. 4. Decrease of α-track distribution of different growth stages.

and nodule 84 (specimen count area 1–2) showed a dislocation of the logarithmic straight lines. This dislocation can only be interpreted as a growth interruption and the microscopic inspection at that depth showed a so called straight layer, as described later in more detail. Indications of an interim nodule erosion could not be recognised in the microstructure. Therefore, in these cases the growth interruptions were closely connected to a special internal structure. Even in the macrostructure of the nodule 77 (see description of nodule 77) the interruption with the discontinuation of a layer is particularly clear and evident.

However, the radioactive growth interruptions proved to be only isolated cases and they can provide no basis for the assumption, which describes nodule growth as a predominantly intermittent process, in which the interruption intervals are large and only relatively short growth periods determine the growth. According to the results obtained here, a continual growth seems to be the norm and interruptions are exceptions to the rule. However, the ratios are more involved in the case of the surface nodule 162(0). The α-track frequency at the surface indicates a sudden increase in radioactivity in a layer of 0.15 mm width. Microscopic investigations show that a bordering zone of 0.10 mm thickness is present. The differences in the radioactive and optical layer thicknesses can be explained firstly through the inaccuracy of the coordination, (approximately 0.02 mm) and secondly through the specific lengths of the α-tracks and the resulting blurring on both sides of the specimen count intervals. This indicates that a further diminution of the specimen count intervals is not advisable and that an error of 0.05 mm can be determined in this case for an interval width of 0.125 mm. Moreover, the microscopic inspection shows that the outer zone of 0.1 mm is not separated from the material lying underneath it by a straight layer. In this case, a different process might have led to a growth interruption.

3.4.2. *Rapid growth zones.* A relatively high number of nodules show a zone of constant activity, for example, the nodule 54(2). This constancy means that in this zone no proven age differentiation can be established with the $\alpha$-tracks and that this layer has been relatively quickly deposited. If this growth form is described as rapid growth, it is to be seen in relation to the half-life of the Io of 75,000 years. If the growth in such zone lasts $\sim$ 5000 years for instance, it would be too short to determine an age difference with the Io method and a rapid growth would have to be assumed. In the case of nodule 54(2) the differences in both specimen count areas can be clearly seen. A growth rate of $\sim$ 50 mm/$10^6$ yr has been determined for the specimen count area A–D, but the growth rate in the specimen count area F–X is quicker and can no longer be defined. Therefore two limits can be defined for rapid growth zones in general:

$$\text{age of layer} \leqslant 5000 \, \text{yr},$$

$$\text{growth rate} > 50 \, \text{mm}/10^6 \, \text{yr}.$$

On examining the microstructure it was found that rapid growth zones usually show large growth cusp patterns. In section 9 these relations are dealt with again.

At this point, the question is raised, whether it is conceivable that the nodule conditions could simulate the appearance of rapid growth. One could imagine, for example, an exponential decrease in density within a layer so as to result in a constant $\alpha$-track frequency. However, this possibility can be excluded as a general explanation for in such a case differences in density of up to factor $10\times$ would have to appear, which are hardly ever reached in nodules. In the polished sections one can distinguish between the resin-filled pores and existing nodule material, and one can also exclude much smaller differences in density. Therefore, differences in density are excluded in interpreting the zones of constant activity. Another possibility might be that a constant distribution of the radioactive isotope could have occurred through diffusion in the layers of rapid growth. At first this seems irrefutable for if one wants to exclude a falsification through diffusion, one is placed in a very bad position from the start. It is always easier to 'explain away' puzzling readings through diffusion, than to prove the opposite that diffusion has not occurred. But this counter evidence is still possible in the case of manganese nodules. If the constant radioactive values in rapid growth zones are caused by diffusion, the diffusion must be several orders of magnitude higher than in other zones. Such a significant change in the diffusion conditions must have its contributory factors—for instance a large increase in porosity, which should be optically visible and yet this is not the case.

In Figs 5 and 6 $\alpha$-tracks from the specimen count area of open nodule cracks near the surface are copied. In the crack system of these two nodules growth-cusp patterns were found, from which it can be ascertained that these cracks were rinsed by the seawater and that a transport of material into the nodule centre had occurred. If diffusion occurs the crack edges should experience either a decrease or an increase in radioactive isotopes. However, neither one nor the other occurs for the decrease in $\alpha$-track frequency from the surface is completely maintained near the crack. In my opinion these two figures prove that the Io is only incorporated during the surface growth and that neither the Io, nor one of the radioactive daughter products migrates. (See also the results description of the nodule 123.)

FIGURE 5.

FIGS. 5 and 6.  α-track distribution in area around open cracks near the surface. Growth activity
was optically visible in material accreted to the crack wells. This proves that the cracks were open
to the circulation of water. Near the cracks neither an increase nor a decrease in α-emitters can
be detected and the gradient is completely preserved. Therefore migration and diffusion processes
in the nodule material can be excluded.

3.4.3. *Special case of a recent rapid growth rate.* The cases described above were those
in which the α-track frequency decreased from the nodule surface towards the inner layers.
However, one often finds an increase in α-track frequency from the surface into the outermost layer. This is very pronounced in the nodule 115(2) and stretches over ∼ 1 mm.

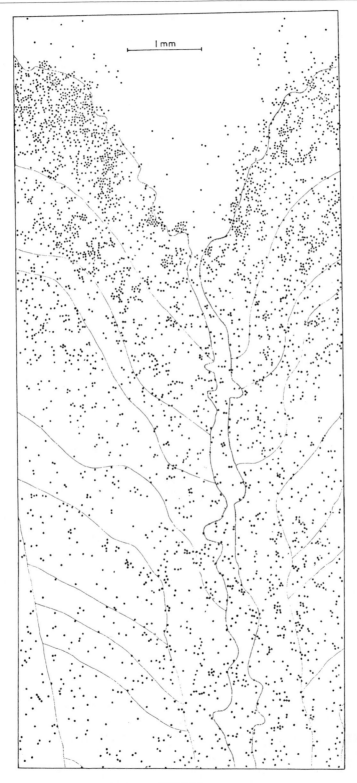

FIGURE 6.

Although, at first sight this result appears to be nonsensical, it proves to be an interesting special case, which is described in depth in the following pages.

In all the previous investigations it was presumed that Io is in equilibrium with its daughter products. This assumption is right for a regular growth of the usual order of magnitude (several millimetres per million years). However, conditions are different for recent rapid growth, where the $Ra^{226}$ and its daughter products must first be produced from the Io, and non-equilibrium exists for the first 8000 years.

The Ra-production is described by the following differential equation:

$$\frac{\partial(Ra)}{\partial t} = -\lambda(Ra)\,Ra + \lambda(Io)\,Io. \tag{8}$$

With the boundary condition $Io = (Io)\exp[-\lambda(Io)t]$, the solution reads:

$$Ra = \frac{Io\,\lambda(Io)}{\lambda(Ra)-\lambda(Io)}\{\exp[-\lambda(Io)t] - \exp[-\lambda(Ra)t]\}. \tag{9}$$

The number of $\alpha$-tracks $A$, which are measured after a fixed time $t$, are composed of one part of Io and five parts of Ra. (Five parts of Ra means 1 part $Ra + 4\,\alpha$-track emitting daughter products of Ra.)

$$A = Io\,\lambda(Io) + 5Ra\,\lambda(Ra). \tag{10}$$

If the solution (9) and the boundary condition are inserted into the equation, one gets:

$$A = (Io)\lambda(Io)\{6.11\exp[-\lambda(Io)t] - 5.11\exp[-\lambda(Ra)t]\}. \tag{11}$$

If, $(Io)\lambda(Io) = A_0$ as the initial activity and $Z = st$ are introduced, the final solution for fixed depths $Z_i$ is obtained:

$$A_i = A_0\left\{6.11\exp\left[\frac{-\lambda(Io)Z_i}{s}\right] - 5.11\exp\left[\frac{-\lambda(Ra)Z_i}{s}\right]\right\}. \tag{12}$$

If one returned to the original problem: the increasing $\alpha$-track frequency from the nodule surface the question is posed, whether equation (12) can adequately explain the given results for a certain growth rate. As a test, it was attempted to approximate the readings of nodule 115(2) in both measuring areas by a curve, according to equation (12). The readings and the curve can be compared in Fig. 7. The best fit to the data is found in the specimen count area Y–4, with a growth rate $S = 155 \pm 34$ mm/$10^6$ yr and in specimen count area X–1 with $S - 106 \pm 33$ mm/$10^6$ yr. It may be concluded from this close agreement that the explanation suggested above is correct. Thus, the possibility exists, for the special case of recent growth to determine a growth rate from the after-growth of $Ra^{226}$, which because of its size, could no longer be determined by the Io.

3.4.4. *Growth and erosion.* In the earlier publication (HEYE and BEIERSDORF, 1973), two nodules were investigated the surfaces of which showed no higher than background $\alpha$-track frequency. These results indicate either stop of growth during the last $3.10^5$ years or dissolution or erosion. The author does not exclude that erosion can happen during sampling procedure. But there are strong evidences for erosion such as fossil erosion surfaces over-grown during later growth stages (see Fig. 2), shiny surfaces of erosion planes as well as surface cutting the microstructures discordantly. This latter case was observed on the two nodules mentioned above (see Fig. 20 and description of the nodule M1). Such phenomena were missing on the nodules taken from the *Valdivia* and all the nodule surfaces show a state of growth.

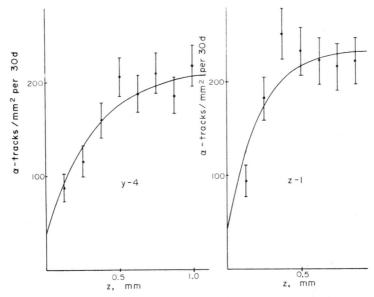

FIG. 7.   Nodule 115(2) special case of recent growth. The $\alpha$-track frequency increases from the surface, which can be explained by the increase of $Ra^{226}$ and its daughter products.

3.4.5. *Buried nodules.* In the cores 170 and 162, altogether 15 buried nodules were found at depths of more than 95 cm. They were all radioactively examined and only showed the mean background $\alpha$-track frequency of approximately 25 traces/mm$^2$/month$^{-1}$, which is usually found in the older part of surface nodules. No dissolution manifestations were microscopically visible on the surface of any of the nodules. It can only be confirmed that these nodules have not incorporated any Io in the last $3.10^5$ years. However, if buried nodules had still grown they could not have obtained an Io supply from the seawater above or the surrounding old sediment. Therefore, it cannot be concluded from these results that the buried nodules no longer grow, but only that it is impossible to prove a growth radioactively.

SUMAYAJULU, HEATH, MOORE and CRONAN (1971) published Io datings of nodules from the immediate surface vicinity on the subject of whether or not these nodules are still growing. The authors clearly come to the conclusion, in the abstract, that the nodules stop growing as soon as they are buried. The data can however also be interpreted in another way. These data would also fit the conception of nodules growing and incorporating Io from the surrounding sediment after having been buried.

Professor Beckmann from Clausthal University kindly gave me a core with three nodules from the surface vicinity taken from the position KG 92. With regard to this question no convincing results could be obtained from the investigations on these nodules. However, the investigations are not yet concluded, and are still being supplemented.

The present extent of knowledge seems to suggest, that the question of growth, respectively, a halt in the growth of buried nodules cannot be clearly answered by radioactivity.

## 3.5. *Discussion of the results obtained*

In Fig. 9 the frequency of the mean nodule growth rates is compared to the results of intermittent growth rates of individual zones. It is noticeable that the majority of growth rates for the whole nodules lie in a relatively narrow interval of $4-9\,mm/10^6\,yr$. There are seldom deviations and thus the growth conditions in the investigation area of the *Valdivia* in the Pacific are surprisingly uniform. The intermittent growth rate figure also shows the same mean interval ($4-9\,mm/10^6\,yr$). But the rapid growth (represented here as $S > 30$) and the growth interruptions (interval 0–1) often appear, but they hardly influence the average nodule growth rate.

If the results of the vertically sliced nodules are classified according to their sediment and water side, then no distinct differences are found. If the nodules grow, then they apparently grow at the same overall rate in all directions. This is even more surprising, if one considers that the material supply only comes from the water side and in addition any idea of the nodules being turned over by fish, or following on the sea-bed must be practically excluded for some nodules (as will be shown later).

The question arises at this point, of how the material supply of the Io gets to the under-neath side of the nodule. For age determination of deep-sea sediments one excludes the migration of Io through diffusion after deposition and yet Io is incorporated into the embedded halves of the nodules. This fact has not yet been adequately explained and rather surprisingly has also not been discussed at any point in previous publications. Therefore, at this point several mechanisms that could possibly work ought to be enumerated.

(a) While they grow the nodules are surrounded by a 'soft layer' in which an equal distribution of material around the nodule is largely possible and from which the growing nodule is supplied, for example with Io, on all sides. This soft layer could be an inorganic gel layer as well as a layer of biogenic settlement.

(b) The Io is mobilised (biogenically?) from the sediment by unknown mechanisms and incorporated into the nodules. This could also apply to other elements. In these cases it would even be possible that the buried nodules still grow, but their growth in deeper sediment cannot be radioactively proved, because the Io is missing in the deeper sediment.

Possibly the two processes function together. By mentioning these possibilities the author wants to point to the yet unsolved problem of Io supply. No conclusion is drawn whether buried nodules still grow or that Io is mobile in the sediment.

### 4. QUALITATIVE CHEMICAL INVESTIGATIONS OF THE NODULES

Precise investigations into the chemical relationships of the manganese nodules exist, for example CRONAN (1972), GLASBY (1972 and 1973), FRIEDRICH, ROSNER and DEMERSOY (1969) and HALBACH (1974). The chemistry of manganese nodules is shown accordingly: Fe and Mn are the main cation components of the nodules and their concentrations are inversely related. The elements of commercial interest Ni and Cu show a correlation with Mn and are therefore preferentially enriched in manganese-rich nodules. The relations with Co are not so clearly established, but a relationship with Fe has been found in the majority of cases. During experiments, which had a completely different objective, it was found, quite by chance, how to make the chemical relations of manganese nodule cross-sections visible. The polished sections are treated with HCl vapour and a layer of coloured chlorides forms at the surface (see Figs 10–15). There were discussions with Dr. GUNDLACH and it was noted that this reaction resulted in a separation into different colours. The reddish (respectively, the warm) colours had to come from Fe- and Co-chlorides and the colder colours (respectively, white, green, blue) had to form from Mn, Cu and Ni. Comparisons between results

(a) 95 cm depth

(b) 100 cm depth

(c) 324 cm depth

FIG. 8. Cross sections of buried nodule from core 162. (The boreholes around the nodules have 1mm diameter.)

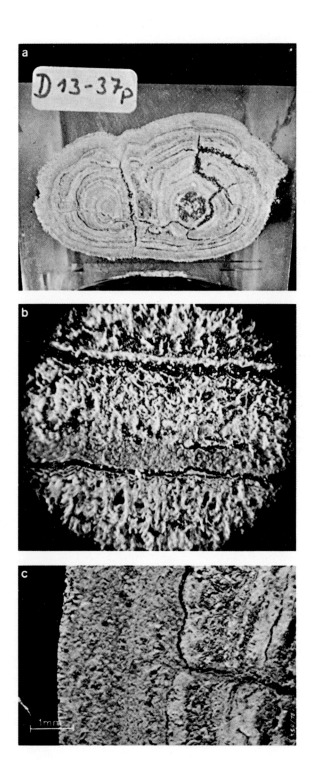

FIG. 10. Treatment of nodules in HCl vapour. (a) Nodule D13 37p. Largest nodule, diameter. 4.8 cm; (b) Nodule D13 37p. Diameter of circular photo 2.8 mm; (c) Nodule D13–37p. Sector in the vicinity of count area 3–4.

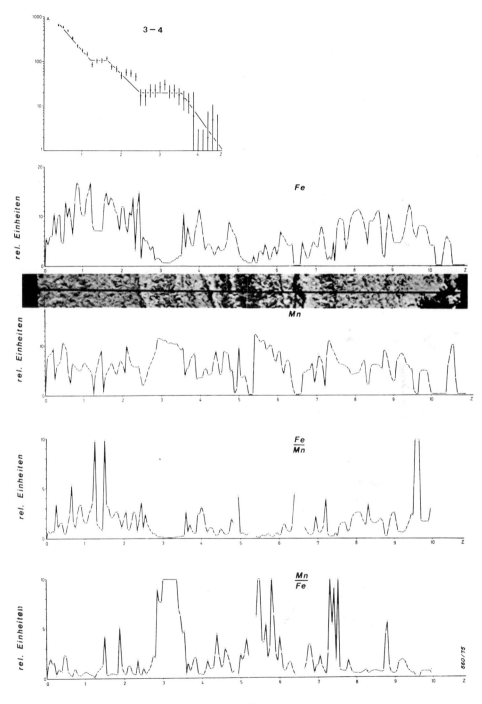

FIG. 11.

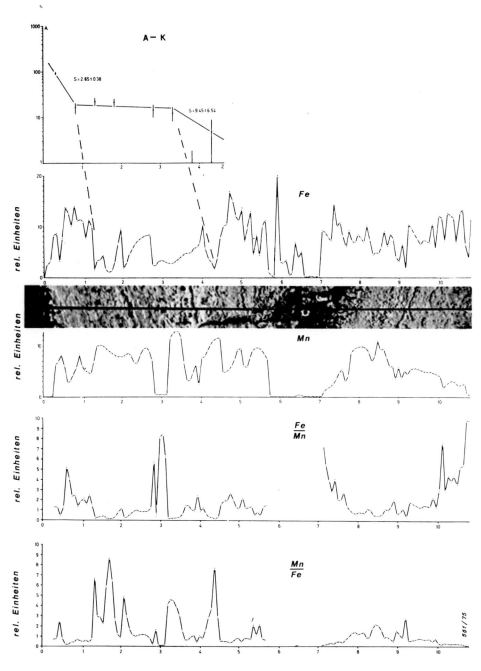

FIG. 12.

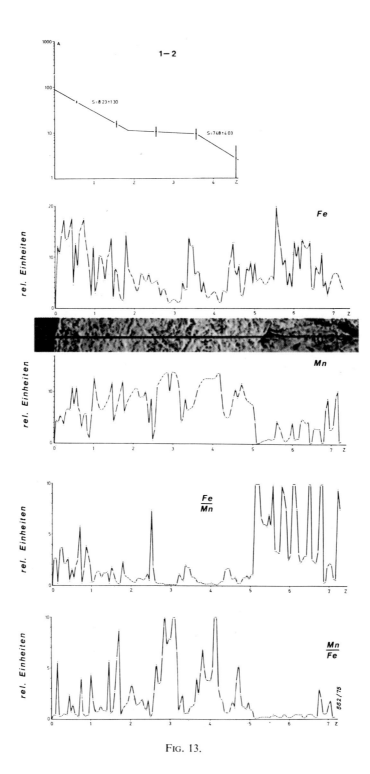

Fig. 13.

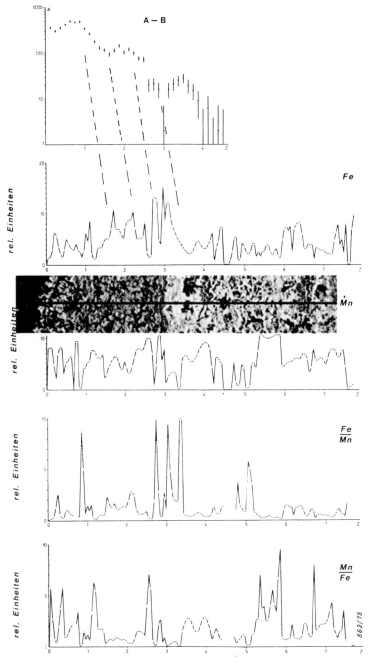

A – B

Fe

Mn

$\frac{Fe}{Mn}$

$\frac{Mn}{Fe}$

rel. Einheiten

563/75

FIG. 14.

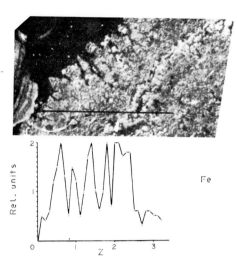

Nodule 32

Fe

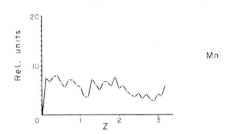

Mn

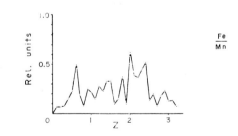

$\dfrac{Fe}{Mn}$

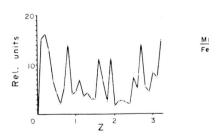

$\dfrac{Mn}{Fe}$

FIG. 15.

FIGS. 11–15. Comparison of the age determination (above, the relative Fe and Mn contents (microprobe) and the coloured chlorides after the nodule treatment in HCl vapour). The connections are discussed in the text.

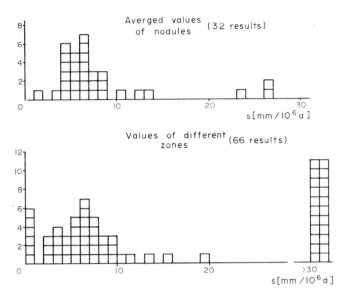

FIG. 9. Histogram of growth rates.

of the electro-microprobe analyser, with investigations by Dr WEISER, and Dr HALBACH, and the colour division of the nodules have in meantime confirmed these results (see Figs 11–15). With this HCl treatment it is not only possible to obtain macroscopically, in a very simple way, a qualitative view of the chemical composition of the cut nodules, but also to observe the chemical change in individual layers under the microscope and to make a comparison with this growth structure. Partially very sharp colour boundaries may be recognised, which clearly speak against a migration of the main chemical components within the nodules. Nodules behave differently during HCl treatment and require different exposure times. Therefore, comparisons of the change in the chemical composition can only be made with certainty within a single nodule. Comparisons from nodule to nodule are not so definite and are only allowed in the case of extreme differences.

The majority of nodules from the investigation region of the *Valdivia*, as can be seen from the description of the nodules, show a younger outer layer of a few millimetres in thickness. This young layer is attacked substantially more quickly by the HCl vapour than the older, inner layers of the nodule. Similar differences between the nodules' inner layers and the surface layers can be found in the cracking frequency of the nodules and are described in sections 7 and 8. A material shrinkage per unit time is responsible for the crack formation, and probably it required a longer treatment time of the HCl vapour on the material before the chlorides form. If this conception is correct, then qualitative age determinations could be read from these differences.

### 4.1. *Relations between the nodule chemistry and growth rate*

An essential aim of this investigation was to establish relationships between chemistry and growth rate. It is known that the slowly growing nodules from the Pacific have a high Mn contents with an accompanying higher Ni- and Cu-concentration, and that in the proven quickly-formed, millimetre-strong incrustations (for example, on cannonballs) a lot of iron has been found. These results suggest that slowly-growing nodules are enriched with more manganese and therefore more metals of commercial interest. The reverse was expected for the quickly-growing nodules which contain more iron and therefore are not of such interesting commercial value. In the literature, this opinion is at least hinted at and these investigations were started with these expectations in mind. In order to give a more detailed explanation, as many nodules as possible should be dated and compared with the chemical total analysis.

However, from the description of individual nodules it can be seen that they have an extremely varied formation-history and that in the majority of cases, the material comes from different growth periods. In each individual period, however, the growth rates and the chemical relations to the nodules can be different. The chemical total analysis then gives only average results from which material from the individual growth periods can differ. Up to now, however, the growth rate can only be determined in the outer zone of a nodule and does not have to be correct for the previous growth phases. It is clear from this enumeration that no results of scientific or commercial value can be expected from such general comparisons.

As far as the chemical composition of the nodules is concerned the relations in the different growth periods of the nodules should, in future, be examined in more detail. In my opinion chemical total analyses of nodules to examine scientific, respectively, genetic relations are of no use without a corresponding clarification. (This does not exclude such investigations on economic grounds.) As a result of the restriction of the radioactive dating methods to the surface part of the nodule, exact comparisons between growth rates and chemical composition can only be made in this area of the nodule. Whether such results can be generalised according to this and thus regarded as valid for older nodule parts, is another question. Only one of the investigated nodules from the Pacific showed a clear change in colour red–white in the surface area after the treatment with HCl. All the other nodules lacked colour-changes of this clarity and therefore the comparisons were particularly carried out on this one nodule. In the Figs 11–14, the relative Fe- and Mn-contents (results of a microprobe analysis by Dr WEISER and by Dr HALBACH, the distribution of the $\alpha$-track frequency and the colour photographs of the formed chlorides were placed together. The connection between the parts of reddish chlorides and the larger Fe-contents in different zones can be clearly seen. All in all, however, the Mn-content is almost constant. If the changes in Fe-content are compared with the distribution of the $\alpha$-track frequency, clear connections can be observed, which are also partly true for smaller areas. In the zones of small Fe-content the nodule grows quickly, and in areas of increased iron content the growth slows down and the reverse process occurs. During the quick growth little Fe is incorporated and during the slow growth more Fe is deposited. A connection with the concentration flow of the manganese is found to be less clear and only exists from time to time.

This result is also confirmed in a second nodule (Fig. 15). The polished section 32 was examined by Dr Halbach with the use of an electron-microprobe analyser and was kindly given to me later for dating and treatment. The area of the $\alpha$-track count and the traverse of the microprobe lay at an angle to each other with the result that the results cannot be shown together. The rest of the nodules also gave several indications towards this not unexpected connection. In my opinion it is still too early to generalise about these results and one should first wait for other results.

## 5. MAGNETIC INVESTIGATIONS OF THE NODULES

The Io age determination only reaches a maximum age of about $3.10^5$ years, and occasionally possible K–Ar and selected Be–Al datings are not suited generally to close the gap to the greater ages. Therefore, magnetic investigations were started in 1970, in order to try to find paleomagentic reversals and to use them for dating. However, these first investigations showed no satisfactory results, HEYE and BEIERSDORF (1973). In the interim, other magnetic investigations on manganese nodules have been published by CRECELIUS, CARPENTER and MERRILL (1973), where paleomagnetic reversals are supposed to have been found. These contradictory results need a fundamental clarification, and therefore, within the limits of these investigations, varying new experimental attempts were made to this end.

### 5.1. *Measurement with sound-heads* (*magnetic pick-up heads*)

Here sound-heads, as they are used in normal tape recorders, were pulled over polished sections of nodules and the output signal was amplified and recorded on a storage oscillograph. The investigations were based on the conception that enclosed small minerals of stronger magnetisations and of different directions in the nodule could be found and evaluated. The measuring techniques could be developed to its full functioning, but in doing so such an abundance of individual signals was recorded that an evaluation of individual particles was not possible.

### 5.2. *Measurement with spools*

In order to obtain measurements in this case two small spools, only several millimetres in size, were positioned in such a way that a nodule slice could be positioned between them. If the slicers were moved magnetisation induced a charge, which was also amplified and made visible on the storage oscillograph. These experiments, however, produced no satisfactory results.

### 5.3. *Measurements with flux-gate magnetometer*

Here it is a question of a further development of the process that was originally used. In the first investigations commercial sensors with an active length of 32 mm were used. They had a considerable integration effect and consequently a small resolution power of 11 mm, HEYE and BEIERSDORF (1973). In the new investigations, new sensors of only 7 mm in length were used, which caused a considerable increase in sensitivity and resolution. Changes in the direction of magnetisation can be precisely measured according to Fig. 17, which are only 3 mm wide and the limit of the resolution power should be 1–2 mm.

The measurements were carried out in the following way: the nodule slices that are to be measured are positioned between two sensors inside a magnetic shielding and the vertical field components are measured. A reproduction of the nodule with a pointer outside the shielding allows an exact orientation on to the nodule slice. The sensors and the pointer can be turned to various levels, so that the whole face can be measured point by point (see Fig. 16). Before and after each point measurement the probe is pushed out of the measuring area and the zero marker is checked. Using Helmholtz-coils inside the shielding, a demagnetisation with alternating fields up to 250 Oersted can be carried out. The two sensors of the flux-gate magnetometer are symmetrically positioned to the probe and are connected in such a way that a measurement of the field can be carried out in this position. In this arrangement only the field of the components of the magnetisation is measured, which lies vertically to the slice plane. A magnetisation component pointing in the direction

184                                   DIETRICH HEYE

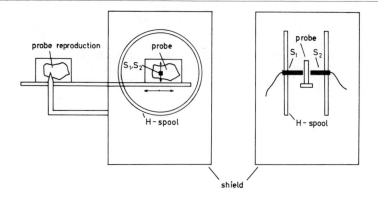

FIG. 16. Apparatus for measuring the magnetisation of slices.

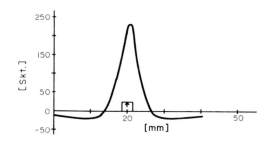

Artificial sample, 3mm width and very long

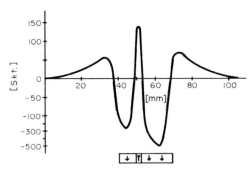

Simulation of a magnetic reversal

FIGURE 17.

of the slice plane is not measured, because in this case the sensors register a field of the same intensity, but with different signs. Even if inhomogeneities are present in this magnetisation component they only have an effect if they result in a disturbance of the mentioned symmetry. Even if the remanence of the slices does not lie exactly vertically to the slice plane, then in the main only the measured values are reduced, for only the field of vertical components is measured. The relationship between the vertical magnetisation

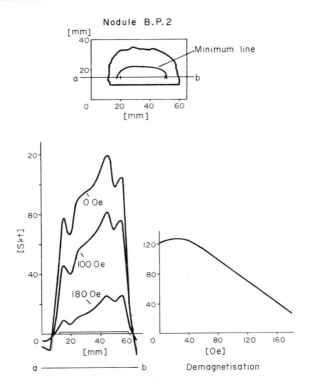

FIG. 18. Magnetic results of nodule slices.

components in the slice and the measured field is preserved, with the result that eventual layer-like changes in the arrangement of magnetic intensity or polarity also lead to the same changes in the measured magnetic field. If one disregards the resolution power or integration effect, a defined correlation between the measured changes of magnetic field intensity and the magnetization of the field, is given.

These conditions could be checked and confirmed by model experiments on artificial probes (see Fig. 17). Probes of 5 mm thickness were made from cast resin and fine iron powder, and then magnetised and measured.

Using this measuring technique about ten nodule slices in their original condition were measured at different stages of demagnetization. The majority of nodule slices gave no indications of paleomagnetic reversals. The results for half a nodule slice of the nodule BP2 from Blake Plateau were collated in Fig. 18. One can only determine a connecting line of minimum intensities. However, no ambiguous indications of paleomagnetic reversals were found. In the demagnetisation the remanence proved to be stable and decreased with increasing alternating field strength.

The results of nodule M1 are found in Fig. 19. No continuous sign-change of the magnetic field was found here either, although the maxima and minima could be joined with closed lines. These lines correspond to the growth structure of the nodule and to that extent the results are at least logical. It is unequivocally impossible to interpret these as paleomegnetic reversals and the maxima and minima should rather be regarded as areas

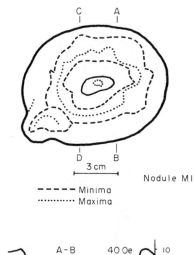

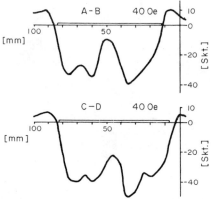

FIG. 19. Magnetic results of nodule slices.

of higher or lower magnetisation in the nodule. The remanence also proved to be sufficiently stable.

Despite the measuring accuracy and the resolution power of the further developed measuring process, which have in the interim become satisfactory, it can finally be stated that no paleomagnetic reversals were unequivocally proved from the measurements of the nodules examined. In the meantime, it can no longer be put down to the measuring techniques.

## 6. THE SURFACE CONDITION OF MANGANESE NODULES

Previously the surface colours of manganese nodules, which lie between medium brown and black, received little attention. Of the nodules examined here the nodules from the Pacific were all dark grey to black and only the nodule M1 from the Indian Ocean was dark brown. The origins of the different colours have not been examined in detail up to now. On the other hand, more attention has been paid to the structural surface condition and is described, for example by RAAB (1972). The author gives a graduated scale from smooth to macroscopical uneveness in the order 'smooth', 'gritty', 'goose bumps' and 'pisolitic or knobby'. He also points out that different surface forms appear on the same nodule and the

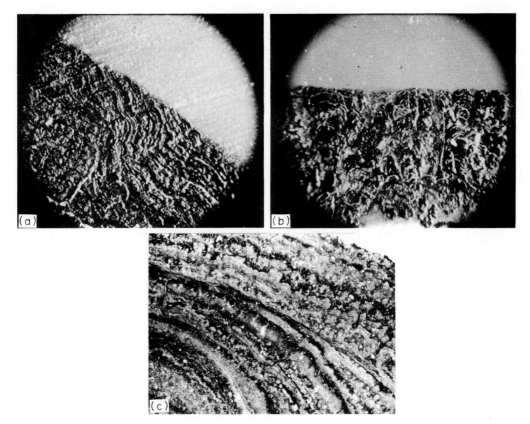

FIG. 20a.   Nodule BP1, nodule resolution in the microstructure at the surface (diameter of the photograph 2.8 mm).

FIG. 20b.   Nodule M1, nodule resolution and micro-structure at the nodule surface (diameter of photograph 2.8 mm).

FIG. 20c.   Nodule 114(1), nodule resolution in the micro-structure, which was surrounded by a new growth (diameter image sector 4 × 7 mm).

FIG. 21.  The microstructure of the polished manganese nodules. The holes bored around each nodule are 1 mm in diameter.

smooth surface originates on the waterside face. This graduation given by RAAB (1972) entirely corresponds to the visible appearance and to this extent, is complete. However, if one draws comparisons of the microstructural observations on ground sections and enquires about the origins of different surface forms, then in my opinion the graduation must be changed somewhat, to describe both observations. The following graduation is obtained:

### 6.1. *Smooth-shining surfaces* (*erosion surfaces*)

The nodule M1 (Indian Ocean) and the nodule BP1 (Blake Plateau) show this surface form. According to the radioactive age-determination no growth was found for the last $3.10^5$ years in the nodules (apart from the exceptions of small areas in M1). The microstructure of the two nodules show surface discordances (Fig. 20) and they were mechanically rubbed off, or chemically dissolved. Thus the smooth-shining surfaces point to a nodule erosion and the gloss seems to have originated from a kind of polishing. It is not certain whether a mechanical erosion through a rolling of the nodule on the sea bottom, or through a chemical dissolution, takes place. However, cavities are typical on these surfaces, which had graduated edges in the nodule BP1 and pointed to a mechanically caused peeling of individual layers. The nodule M1, on the other hand, showed circular cavities, which one can imagine occurring rather through chemical dissolution. These varied dissolution forms at the surface can also be predetermined just as well through the varying growth forms of the earlier nodule growth structure.

### 6.2. *Smooth surfaces without gloss 'unfavourable growth'*

At these parts of the polished nodule section surface one finds either a straight layer or very small, growth cusp patterns, which cannot be seen with the naked eye. Parallel layers, according to section 9, point to a growth under unfavourable, that is, adverse conditions. The nodule 156 had such a surface in its waterside part and the parallel layering can be recognised in the polished nodule section in the photograph in Fig. 31. Smooth surfaces with small growth cusp patterns, which are not visible to the naked eye, however, belong, from the microscopic observation, to the rough surfaces of this graduation.

### 6.3. *Rough surface* (*growth surface*)

These surfaces seem to be encrusted in a velvety way (SEIBOLD (1973), as if powdered) or show recognisable needles, small, rounded growth formations. Under the microscope these growth formations can be often identified as a kind of 'micro-nodule'. (However, the term 'micro-nodule' is unfortunately already used in another context and is therefore unavailable. Therefore, only the terms 'growth cusp patterns' or 'growth cusps' or 'buds' will be used in future.) These rough surfaces are found on the majority of nodules and the group contains, without any further subdivision, the surfaces from 'gritty' to 'knobby' according to the graduation by RAAB (1972).

Excellent scanning electron-microprobe photographs exist of the rough surface forms with their growth cusp patterns, for example HALBACH (1974). If such a surface form is found, then according to section 9 the nodule has grown to the last under favourable external conditions. 'To the last' means to the present surface, whereby the growth may well have suddenly ceased earlier, but where neither erosion nor an unfavourable growth has taken place (for example, if the nodule is suddenly buried and the conception should be correct that these nodules no longer grow). The graduation of the surface structure, as it is given here, is the first attempt to go beyond the purely superficial observation and to find the results of the nodule microstructure, the consideration of the datings and the causes of the formation of varying surface forms. A limited number of nodules formed the bases of this graduation and therefore it is surely necessary to have further supplements through other investigations. However, it should therefore be clear, that conclusions as to their last growth conditions can be drawn from the surface of a nodule, over and above a mere description.

## 7. INTERNAL MACROSTRUCTURE OF THE NODULES

### 7.1. *Nodule core*

Most manganese nodules of sectioning or X-raying show a central object around which they have grown. This can either be a fragment of an older nodule or a so-called foreign core. In the literature, these foreign cores have been identified as shark teeth or ear bones of whales, as basalt fragments, tuffite or palagonite. It can be presumed from this that the nodules grow around practically everything that can be found on the sea floor. According to the new hypthosis by HORN, HORN and DELACH (1973) the distribution of the potential cores is even the primary determining factor for the frequency of the manganese nodules. In the literature, transformations of core material are described, whereby volcanic glass is palagonitised and finally exists as zeolite for example phillipsite (see, for example, BONATTI and others (1965). In the work presented here no special investigations on the cores were carried out. Accordingly their description is cautiously formulated. At the beginning of growth the outer form of the manganese nodules is determined by the shape of the core. Its influence on the shape of the nodule decreases as growth proceeds.

### 7.2. *Erosion in the older parts of the nodule*

The majority of surface nodules from the working region of the *Valdivia* in the Pacific show macroscopically recognisable discordances, which point to an interim erosion of the nodules. In several nodules a multiple erosion can also be determined (see the details of the description of the nodules and as an example the photograph 20c). The relationships for a surface erosion of the nodules BP1 and M1 are already shown in detail in section 6. In the investigated nodules from the Pacific the erosion zones always lay in the older part of the nodules and never on the surface. They could all be recognised macroscopically and attempts to recognise further erosion zones of lower significance in the microstructure failed. (Although in the nodule D13–37p discordances were found in the microstructure of one spot in an inner layer they were only poorly defined and were not present $\sim 1/10$ mm away and therefore play no part in the total development of the nodule. It is impossible to infer a general erosion of the nodule from this, rather it will probably be a question of a small, local injury to the nodule.)

The outer layer of 3–5 mm thickness on the majority of the examined nodules from the Pacific, is worthy of note. This is formed as the last growth-phase of the nodules after an erosion. The extrapolated total age of this layer is different in the individual nodules and lies between $\sim 3.10^5$ yr–$10^6$ yr.

### 7.3. *Crack formation in manganese nodules*

If one examines the polished cross-sections of manganese nodules the majority show internal cracks, which are often connected over each other to a complete cracking system (Fig. 21). The majority of cracks have either a radial or tangential direction and up to now are generally explained by a volume shrinkage of the nodules, HALBACH (1974) and RAAB (1972). It seems necessary to examine in detail the processes occurring in this volume shrinkage. Some of the cracks will certainly have occurred during the drying-out subsequent to collection. However, under the microscope one can see cracks the edges of which are coated with nodule material and which have, therefore, certainly already originated during the nodule growth. It is also worthy of note that many radial cracks have not gone through to the nodule surface although some have ended at a specific layer. Therefore, in the main, the outer third of the nodule shows only few continuous radial cracks or in the extreme case, is free of them altogether (see nodule 114(1) in Fig. 21). In the main tangential cracks do not appear in the outer region. This is a generalisation from which exceptions exist. It seems necessary to examine in detail the processes occurring in this volume shrinkage. Therefore one should try, through the model calculation of different programmes of shrinkage processes, to describe the appearance more precisely, from which surprising and far-reaching conclusions can be drawn.

If one considers the concentric and layered build-up of the nodules, then one must proceed from an anisotropy of the mechanical sizes. The tangential and the radial shrinkage are not considered to be the same size. In the following model calculations of an irregular shrinkage results were obtained, which claim that the radial shrinkage part alone is sufficient for the formation of not only the tangential, but also the radial cracks. Therefore the question is posed, whether a tangential shrinkage is present at all, particularly since, according to the following model calculations, it is not cogently necessary for crack formation. The answer is supplied by the old, buried nodules, which have only a thin layer of nodule material grown around a basalt core. In this thin layer no cracks would have been allowed to form if a radial shrinkage only was present because the core of the nodule remains unchanged in size. However, since the nodules have several radial cracks a tangential shrinkage is also present. The relationship between the radial and tangential shrinkage means that the size of the anisotropy remains unclear. In the model calculations they are both given the same values.

7.4. *Model calculations for the shrinkage and crack formation of manganese nodules*

In a regular shrinkage in a plane the distance $P$ between two points, shortens by the amount $\Delta P$

$$\vert \longleftarrow \qquad P' \qquad \longrightarrow \vert \; \Delta P \vert$$

The new distance $P'$ is:

$$P' = P - \Delta P. \tag{13}$$

The contraction of the line $P$ by $\Delta P$ can also be described by a so-called shrinkage factor.

$$P' = P - a \cdot P. \tag{14}$$

If an area of a circle is now considered, which represents an idealised picture of a manganese nodule in cross-section, then $R$ is the distance of a point from the centre and $R_0$ the radius of the nodule. Because the nodules have many cracks in the inner layers and scarcely any in the outer layers, the shrinkage of the outer surface edge of the nodules is regarded as unchangeable (fixed).

The shrinkage occurs from the inner layer outwards and through this boundary condition of the unchanged surface, the distance of a point from the nodule centre increases.

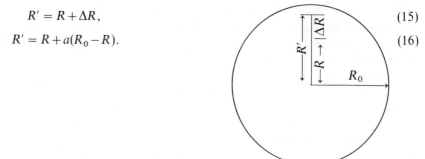

$$R' = R + \Delta R, \tag{15}$$

$$R' = R + a(R_0 - R). \tag{16}$$

If the shrinkage is allowed to be a changing function depending on the radius $R$, then the more general equation (15) is applied. If $\Delta R = F(R)$ is included and the equation is normalised, then for the radial shrinkage a more general expression is obtained:

$$\frac{R'}{R_0} = \frac{R}{R_0} + \int_{R|R_0}^{1} F\left(\frac{R}{R_0}\right) d\left(\frac{R}{R_0}\right). \tag{17}$$

In the case of tangential shrinkage of the circular plane a shortening of the different circumferences of the circle $U$ is obtained. ($U$ = crack width of the radial cracks.)

$$U' = U - \Delta U$$
$$= U - a \cdot U$$

or in a more general expression the width of radial cracks in arcs:

$$\alpha = 360 \cdot F\left(\frac{R}{R_0}\right). \tag{18}$$

7.4.1. *Constant shrinkage in the nodule.* The simplest model case is a constant shrinkage in the whole nodule, that is $F(R/R_0) = a$. If one includes that in equations (17) and (18), then one gets the solutions:

$$\frac{R'}{R_0} = \frac{R}{R_0} + a\left(1 - \frac{R}{R_0}\right) \text{radial shrinkage}, \tag{19}$$

$$\alpha = 360 \cdot a \quad \text{tangential shrinkage}. \tag{20}$$

In Fig. 22 (top circle), the cracks originating in this way from radial and tangential shrinkage are shown individually and jointly. A shrinkage of 10% is assumed ($a = 0.1$) and a circle of 120° is taken as a basis. From the comparison of the crack forms from the model calculation and real manganese nodules the case of a constant tangential shrinkage, whereby no tangential cracks can also originate, must certainly be excluded. One also gets nearer the observed conditions with the case of radial shrinkage alone. Radial and tangential cracks may occur at the same time, whereby the radial cracks go as far as the surface and the nodules would be completely cracked. This simple case is not sufficient to describe the nodule cracks.

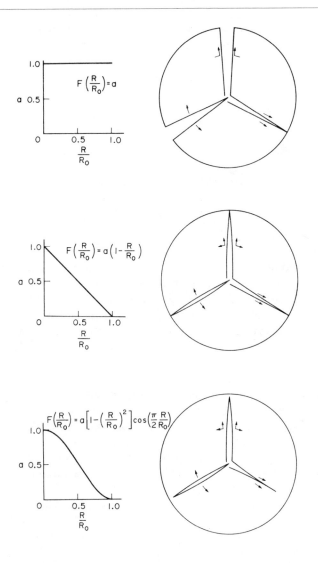

FIG. 22.   Model calculations for shrinkage and crack formation in the nodule.
At the top a case of a constant shrinkage with a fixed outer edge is produced. In the
middle a linear increase in shrinkage and below the somewhat more complicated case of the
shrinkage beginning gradually from the edge and reaching a constant final value in the nodules'
centre. In the last two cases the outer edge is also assumed not to vary. In the circle the cracks
always caused by radial shrinkage (below right), by tangential shrinkage (below left), are shown
individually and for both cases together (above), for a sector of 120°. In all three cases, only radial
cracks originated from a sole tangential shrinkage. In the radial shrinkage the circle sector is pulled
from the middle outwards (which can be divided into several tangential cracks) and additional radial
cracks are formed. If the radial and tangential shrinkage is taken together, then the radial cracks
spread even further. In order to understand this properly, the nodule 114(1) in Fig. 21 should also
be considered. On the right-hand side one can see radial cracks which go to the centre, in which
the individual sectors are detached from the middle. This photograph of the crack formation nearly
corresponds to the lowest case considered in Fig. 22. On the left-hand side of the nodule 114(1) it
may be seen how this process can also spread to several tangential cracks, whereby the radial cracks
remain narrow. Thus, the distribution of the shrinkage cracks into tangential and radial parts
remains a random process according to the model conceptions and can be different in every nodule.

7.4.2. *Linear shrinkage increase towards the nodule centre.* In this case, a linear shrinkage increase is assumed from the nodule edge towards the centre of the ideal model nodule.

$$F\left(\frac{R}{R_0}\right) = a\left(1 - \frac{R}{R_0}\right). \tag{21}$$

If these conditions are included in equations (17) and (18), then one gets the solutions:

$$\frac{R'}{R_0} = \frac{R}{R_0} + a\left[\frac{1}{2} - \frac{R}{R_0} + \left(\frac{R}{R_0}\right)^2\right] \text{radial shrinkage,} \tag{22}$$

$$\alpha = 360° \cdot a\left(1 - \frac{R}{R_0}\right) \text{ tangential shrinkage.} \tag{23}$$

In Fig. 22 (middle), the resulting cracks of this case are drawn for a circle sector of 120° and with a loss by shrinkage in the centre of 10% ($a = 0.1$). A tangential shrinkage alone does not quantatively describe the real crack-distribution in a nodule, for the crack goes as far as the edge, and tangential cracks are missing. Again in a radial shrinkage not only radial but also tangential cracks occur. A crack-distribution of a nodule cross-section is described quite well and the radial crack becomes narrow towards the nodule surface, but yet goes completely through to the surface. The nodule would still break, if no allowance was made for the elasticity of the outer edges. In this respect, the model has still to be improved.

7.4.3. *Non-linear shrinkage increase towards the nodule centre.* A function $F(R/R_0)$ is now looked for, which as in the previous case, increases towards the nodule centre, but in doing so fulfils two boundary conditions.
1) The shrinkage increases towards the nodule centre, but reaches a constant value in the nodule centre (in the centre $F(R/R_0) = a$).
2) In the most recent layers no shrinkage occurs on the nodule surface, i.e. the shrinkage and the crack formation starts later. (This could also imply that recent layers do indeed shrink, but that they are still too elastic to crack.)
One of these possible functions, which fulfils these conditions reads:

$$F\left(\frac{R}{R_0}\right) = a \cdot \left[1 - \left(\frac{R}{R_0}\right)^2\right] \cdot \cos\left(\frac{\pi}{2}\frac{R}{R_0}\right). \tag{24}$$

If this expression is included in the equations (17) and (18) the solution is obtained:

$$\frac{R'}{R_0} = \frac{R}{R_0} + a\left[\frac{16}{\pi^3} - \frac{2}{\pi}\sin\left(\frac{\pi}{2}\frac{R}{R_0}\right) + \frac{2}{\pi}\left(\frac{R}{R_0}\right)^2 \cdot \sin\left(\frac{\pi}{2}\frac{R}{R_0}\right) + \right.$$
$$\left. \frac{8}{\pi^2}\frac{R}{R_0} \cdot \cos\left(\frac{\pi}{2}\frac{R}{R_0}\right) - \frac{16}{\pi^3}\sin\left(\frac{\pi}{2}\frac{R}{R_0}\right)\right] \text{radial shrinkage,} \tag{25}$$

$$\alpha = 360° \cdot \alpha\left[1 - \left(\frac{R}{R}\right)^2\right] \cdot \cos\left(\frac{\pi}{2}\frac{R}{R_0}\right) \text{ tangential shrinkage.} \tag{26}$$

In Fig. 22 (lowest) the cracks in a radial and tangential shrinkage are shown individually and jointly for circle sectors of 120° and a final shrinkage of $a = 0.1$. In all three cases it can be seen that the radial cracks do not go as far as the nodule surface, but come to an end before they reach the edge of the section face. This model case comes closest in reality to the nodule and describes the crack formation in the nodules correctly in a qualitative manner.

7.4.4. *Age-dependent nodule shrinkage and crack formation.* Previously the model cases of the nodules were always regarded as if the shrinkage increasing towards the nodule centre were only a function of the given nodule radius. But since the nodules only grow on the outside through the addition of new layers, and the older layers always lie underneath, in reality a time function is hidden here. To make further observations and draw corresponding conclusions the function

$$F\left(\frac{R}{R_0}\right)$$

must be transcribed as a general time function $F(t)$, i.e. as a function of age. For this purpose one substitutes:

$$\frac{R}{R_0}+\frac{t}{\tau}=1 \rightarrow \frac{R}{R_0}=1-\frac{t}{\tau}.$$

For equation (24) of the third model case one obtains:

$$F\left(\frac{t}{\tau}\right)= a\left[\frac{2t}{\tau}-\left(\frac{t}{\tau}\right)^2\right]\cdot\cos\left[\frac{\pi}{2}\left(1-\frac{t}{\tau}\right)\right], \tag{27}$$

(where $0 \leqslant t \leqslant \tau$).

In this equation $\tau$ is the age from which the shrinkage processes of the nodules are finished. Indeed, the exact size of $\tau$ is unknown, but in further discussions it ought to be imagined as about $5\times 10^6$ yr. In equation (27) the size $t$ is the variable age of a nodule layer between $t=0$ and the final age $\tau$. Within the range of these model conceptions equation (27) has a general validity and different sections of the curve are valid for nodules of different ages.

## 8. TIME DEVELOPMENT OF NODULES (CRACK FORMATION, BREAKAGE OF NODULES)

The equation written as a time function at the end of the previous section is valid as a general model conception and nodules of different ages fill certain interval areas $\Delta t/\tau$ of the accompanying curve. Different cases ought now to be discussed.

### 8.1. *Continually growing nodules (Fig. 23)*

If a nodule grows continually, then it starts in the time diagram at $t/\tau = 0$ and widens to $\Delta t/\tau$ to the left. At first it has an age range $\Delta t/\tau < 0.1$ without shrinkage (Fig. 23, lowest case 1), or a range of sufficient elasticity of more recent layers in which case no cracks have appeared. If the nodules have grown out of the range $\Delta t/\tau < 0.1$ towards the left, then the shrinkage starts in the inner layers and cracks are formed inside at an age of $t/\tau \sim 0.2$–0.3 (Fig. 23 middle, case 1). Thus, it follows that a critical crack formation age is only reached after a certain time. This critical crack formation age is associated with a certain nodule size depending on the growth rate of the nodule. Or conversely, in a region of continually constant nodule growth all the nodules will show internal cracks after a certain size.

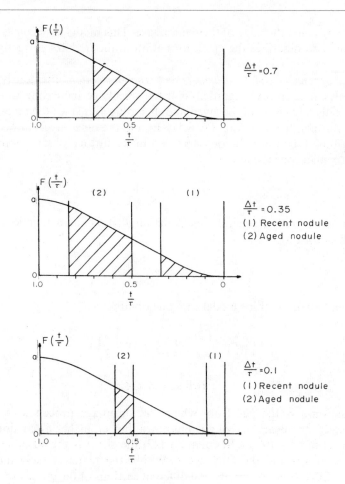

FIG. 23.   Model cases of time-dependent nodule shrinkage. Shrinkage of nodules as time depending processes.

## 8.2. *Nodules which have stopped growing (Fig. 23 middle)*

Consider the case of a nodule, which first reaches a certain age $\Delta t/\tau \sim 0.35$ and then proceeds to age without growing. In this case the age interval $\Delta t/\tau$ of the nodule no longer widens, although it wanders with constant widths to the left. In this way the nodule reaches the area of the greatest shrinkage difference between the inner and outer layers. In doing so the nodule can break into single segments. (If the elasticity of the more recent nodules plays a decisive part, then the layers age after a halt in growth, become brittle and the nodules crack as far as the surface.) If individual cracks go to the surface, then it is naturally questionable, whether the 3-dimensional nodule really breaks. A tilt of the 3-dimensional segments could play a decisive role and promote breaking. Moreover, after the growth has ended the ageing nodule also shrinks as a whole and the boundary condition for the surface (assumed not to vary), no longer holds. The individual segments shrink at a different rate because of their varying sizes and then do not fit together

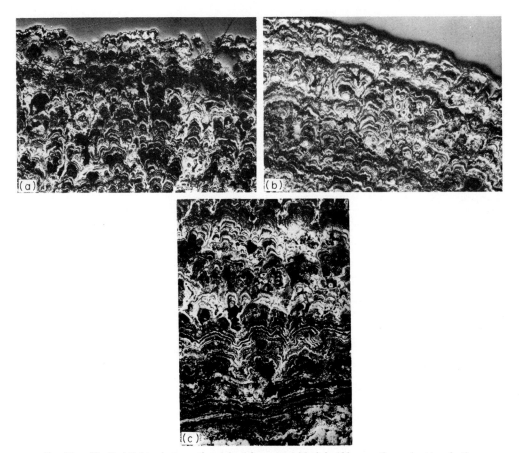

FIG. 24. (Vertical light, photo section 1.5 × 1.2 mm.) (a) Nodule 123, growth cusp pattern in the microstructure near the surface. (cuspate structure); (b) nodule 115(1), growth cusp pattern in the microstructure near the surface. (cuspate structure); (c) nodule 114(2), transition forms between straight layers and cuspate structure in the microstructure.

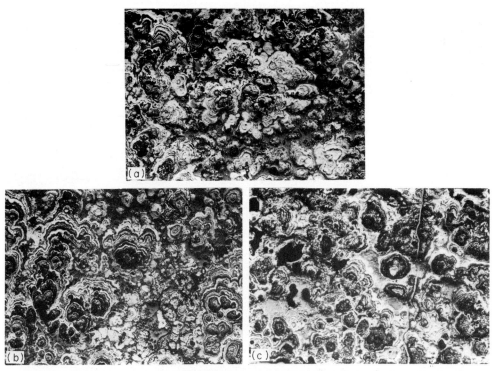

FIG. 25. (Vertical light, photo section 1.9 × 1.2 mm.) (a) Nodule 123. An example of non-directional growth in the microstructure of a cross-section; (b) nodule 123, an example of non-directional growth in the microstructure of a cross-section; (c) nodule 92, directional growth in an inclined section. Optical illusion of non-directional growth.

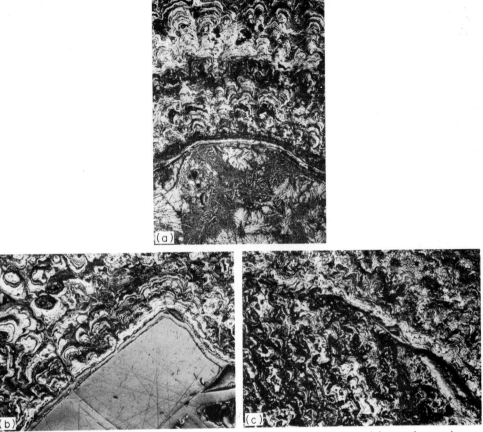

FIG. 26.   (Vertical light, photo section 1.9 × 1.2 mm.) (a) Nodule 162(0), the nodule growth around
a foreign core begins with a straight layer; (b) nodule 162(0), the nodule growth around an inter-
calation begins with a straight layer. After a small zone of directional growth the non-directional
growth, which is otherwise found in this part of the nodule, continues; (c) nodule 62, the nodule
growth around an old nodule fragment (below) begins with a layer, which approximately resembles
a straight layer.

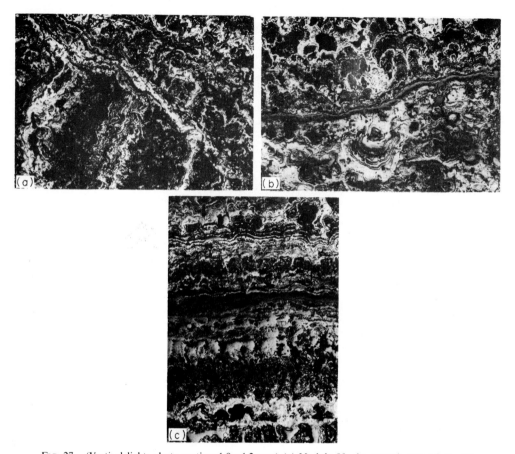

Fɪɢ. 27. (Vertical light, photo section 1.9 × 1.2 mm.) (a) Nodule 89, the growth around an old nodule fragment (below left) begins with straight layers, which are only slightly wavy; (b) nodule 123. The old nodule fragment is encrusted. At the inner breakage surface the growth cusp pattern is on both sides. Old and new growth directions are apposed; (c) nodule 114(2), straight layers near the surface nearest the water.

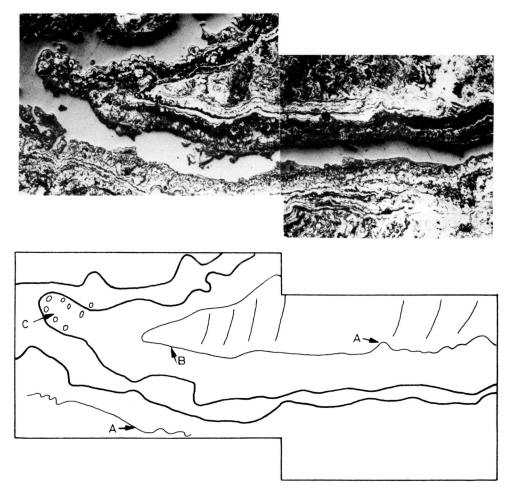

FIG. 28. Growth at crack edges of a manganese nodule. (Photosize = 3 mm, vertical light.)
A examples which show straight layers becoming wavy; B in the crack the first layer is a straight
layer; C example of non-directional growth in a crack.

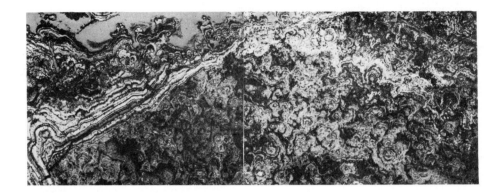

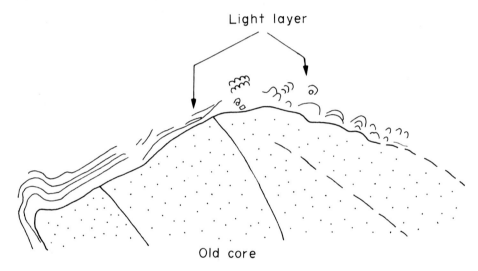

Light layer

Old core

FIG. 29. First growth around an old nodule fragment. At the break-face the growth begins with a light coloured straight layer and at the old nodule surface continues in the form of growth buds. (Photosize = 3 mm, vertical light.)

# Growth structure within a crack

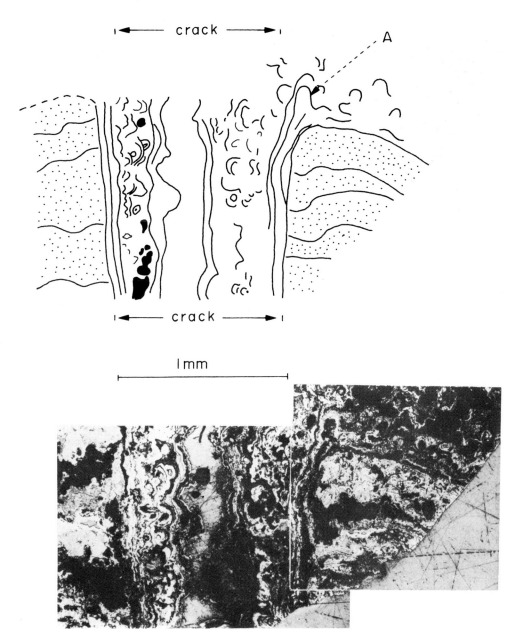

FIG. 30. Change in growth between straight layers—cuspate growth—straight layers in a crack. From point A a straight layer coming from a crack, reaches the old nodule surface and changes the growth form there.

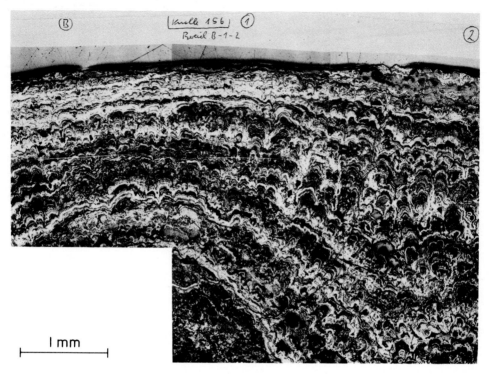

FIG. 31.  Change in the growth form of a manganese nodule at the water–sediment boundary
(Nodule 156, count area B–1–2, detailed discussion in the text).

3-dimensionally at all and the nodules finally break. The processes described here are generally valid for nodules of a certain growth duration $\Delta t/\tau$ and are independently valid of the growth rate for nodules of the most different sizes.

### 8.3. *Nodules of a shorter growth period (Fig. 23 lowest)*

The facts are different for nodules which are formed in a short growth phase. In Fig. 23 they only cover a small age interval of $\Delta t/\tau \sim 0.1$. During the growth no cracks are formed. If growth stops, then the relatively narrow interval moves to the left, but the shrinkage difference between inner and outer layers remains small. These nodules shrink altogether and when $\Delta t/\tau$ is sufficiently small no cracks are formed. If, on the other hand, $\Delta t/\tau$ is larger, then cracks can be subsequently formed during the ageing process without the nodule breaking. Depending on the growth rate the small $\Delta t/\tau$ are again valid for nodules of different sizes.

### 8.4. *Nodules of a longer growth period at critical breaking age*

If a nodule grows for a long time and reaches an age of $\Delta t/\tau \sim 1$, then it is probable that the outer layers, which otherwise hold the cracked nodule together are too small relative to the nodule diameter to maintain this function any longer. A continuing growth of nodule results in the nodule breaking into individual segments. This case is not included in the previous model conceptions, and yet some facts point in this direction.

(a) In the nodule 123 (see description) a fragment has been encrusted. Under the more recent incrustation the former growth of the old nodule could still be radioactively proved. The old nodule had therefore been broken up in the growth-stage and the fragment had become incrusted without a time lag.

(b) To my knowledge, very large nodules are never found in the regions of slow nodule growth in the Pacific. Therefore, a mechanism must exist, which prevents the nodules exceeding a certain size.

(c) Very large nodules, which for example had grown around deep-sea cables prove by this their young age and do not present any counter-evidence. They are rather to be included in the case of nodules of short growth phases and accordingly realised in arbitrary size.

The most far-reaching conclusion from the model conceptions developed about shrinkage and crack formation in nodules, is that, after they have reached a critical age the nodules break by themselves due to internal stresses. Thus, known observations are explained, for example, that nodules from regions with smaller growth rates only reach smaller sizes, than nodules from regions with larger growth rates.

### 8.5. *Nodules with surrounding zones of varying ages*

If a nodule has not grown continually, then the varying old parts of the nodule are settled in different interval areas of the age diagram. In these cases it is to be expected that the older parts have shrunk further and show more cracks than the younger areas. These relationships can be confirmed by observations of the so-called crack-distribution. Nodules 54(1), 62 and 65 are named as examples and can be referred to in the individual description. Due to the varying crack frequencies a time-lag between the individual growth phases can be inferred in these cases. In cases of older, encrusted fragments the different, old parts of the nodules are easily recognisable and if additional erosion has taken place the different, old areas of a nodule can be easily separated. On the other hand, it is more difficult if a nodule has continued growing after an interruption without an interim erosion. If by chance the radioactive investigations do not show a limit, then only the crack frequency is available for evaluation. In my opinion, the cracking frequency in nodules offers a good possibility of qualitatively judging the age differences in nodules since the exact times in which the processes occurred have been largely unknown up to now (e.g. the age $\tau$ can be not only $5 \times 10^6$ yr but also $10 \times 10^6$), in the meantime, it is only possible to estimate age differences.

### 8.6. *Comparisons between the model conceptions and the investigated nodules*

The model conceptions developed previously were developed on the basis of the investigated nodules. If such descriptions are to remain readable all the thought processes in the development of this conception can naturally not be reproduced. Therefore, at first the completed model conceptions were dealt with, and conclusions were then to be compared with the investigated nodules. But first some principal remarks must be made in advance.

Buried nodules can only be compared conditionally to the model conception. Due to the external pressure of the surrounding sediment at increasing depths the crack formation is made more difficult and the nodules are even prevented from breaking.

Nodules which lie on the sediment surface and are completely broken can remain together as fragments in their previous or changed order, as an accumulation of individual segments (e.g. nodule 84), or for example the lower fragments can stay there and the upper fragments can be transported away (e.g. nodule 156). These relationships can be easily evaluated on nodules, if the fragments of a more recent layer have been encrusted and the relationships fixed in this way.

The effect of erosion on the breaking of nodules must also be considered. For example, if a nodule is completely cracked inside and the external—holding—layers are dissolved, then the nodule of course also disintegrates through the effect of erosion.

Nodules from different regions not only grow at different speeds, but also show a different chemical composition. Therefore, in a similar way, the processes in the time-dependent volume shrinkage are also different. The tendency is for all nodules to act alike, but certain differences are yet to be expected and under certain circumstances already appear in nodules of the same region.

From these preliminary remarks, it appears that the relationships in the single case of a nodule can be quite different and involved. Therefore, one must reckon with exceptions and under the circumstances, with several possibilities of interpretation. Luckily, in addition to this, there are also simpler cases, which sufficiently correspond to the developed model conceptions.

8.6.1. *Small nodules without crack formation.* If nodules have only grown for a short time, according to the model conceptions of shrinkages, they have no cracks, but shrink all over. This time interval can now be estimated from the age determination of investigated nodules. Only the small nodules from the Pacific have no cracks, e.g. nodule 93, and also the smaller buried nodules 170(280), 162(33) and 163(246). The smaller daughter nodules also have no cracks, for example on surface nodules 92 and 115(1). They all show a maximum diameter of $\sim 8$ mm and according to the average growth rate of the region they represent an age of between $5 \times 10^5$ to $10^6$ yr. According to section 8.1.3. they correspond to the case of $\Delta t/\tau \sim 0.1$ and it can be concluded from this that the first crack formation in the nodules begins at the earliest at about $5 \times 10^5$ yr (critical age for crack formation). The generalisation would mean, for an age estimation of the nodules, that all nodules without cracks, independent of their size, are younger than $\sim 10^6$ yr (maximum age for uncracked nodules) and that all cracked nodules are older than $(5-6) \times 10^5$ yr (minimum age for cracked nodules).

8.6.2. *Nodules in which cracks have formed.* If the investigated nodules are sorted according to this point of view, then they must be divided into three groups.

The nodules 62, 123 and 86, which have all grown around an older nodule fragment, belong to the first group. The difference in the crack frequency between the older nodule part and younger encrustation is very clear, from which a time lag can be inferred between both parts. The same is also true for the poly-nodules 77, 81(1) and 81(2), many of which have encrusted old fragments and have, in doing so, grown together.

The nodules 54(2), 162 and 196 belong to the second group, which contain no recognisable old nodule parts and also show no traces of erosion in the inner layers. These nodules have probably grown in one period without any visible interruptions. In spite of different sizes they have only few cracks and may have just reached, or slightly exceeded the critical age for crack formation. Therefore, these three nodules are the only ones which are to be regarded as relatively young over all (about $10^6$ yr).

The nodules 87 and 156 with the excessively wide cracks in the older part belong to the third group. These were nodules which were already broken and then encrusted again, and to such an extent that it was less a question there of cracks but rather of cavities. The fragments have also been partly tipped against each other, as can be seen from the growth lines.

8.6.3. *Encrustation of a solid, foreign core.* If a fixed core of a constant size is encrusted, then the relationships are somewhat different. Here the smallest tangential shrinkage already leads to a formation of radial cracks even if the critical age for crack formation in the crust has not yet been reached. This case can easily be imagined and therefore further explanations may be foregone.

8.7. *General conclusions about the manganese nodule formation*

The model conceptions about shrinkage and crack formation, which were developed and discussed in the previous section, together with growth and erosion, seem completely to explain the mechanism for nodule formation. It was essential that a mechanism which functions without 'catastrophic' causes was found for the breaking of nodules into individual fragments. In the publication by RAAB (1972) external forces were still considered necessary to break the cracked nodules. Moreover, it is comprehensible why nodules from different regions have different sizes and why, for example nodules from the Pacific do not reach arbitrary sizes, and nodules of ~ 1 m in diameter are missing. At the same time the occurrence of large nodules from other regions, which have grown around deep-sea cables are explained by the same model with the result that there is no counter-evidence against regions of slower growth rate, for example the Pacific.

If one wants to compare nodules from different oceans one must take into consideration the differences in the composition mentioned previously, which can lead to gradual deviations. In addition there are other differences, for example the cracks in the nodules from the Blake Plateau filled with calcium carbonate. The cracks are glued together through this filling and the possible breaking apart of the completely cracked nodules is prevented for a filled crack grows together on the outside. In a Pacific nodule the cracks would have remained open and would, in the end, lead to a total disintegration of the nodule.

### 9. THE MICROSTRUCTURE OF MANGANESE NODULE CROSS-SECTIONS

Microscopical observations of manganese nodule cross-sections have only been carried out in increasing proportions in recent years. When Sorem and Foster began their investi-

gations several years ago they were amazed to find that very little existed in the literature
SOREM and FOSTER (1972). Other publications are CRONAN and TOOMS (1967), HALBACH
(1974), SOREM and FOSTER (1972) and MARGOLIS and GLASBY (1973). All these publications
contain excellent micro-photographs of the internal structure composition of the nodules.
The photographs resemble each other so closely that it can be assumed that all the authors
found approximately the same internal structures in the nodules. The terms used, however,
and the interpretations vary and can only be conditionally compared. Since the origin of
manganese nodules—whether inorganic or biogenic—has in no way been established, the
terms already in use imply certain interpretations and determine certain conceptions. The
graduation achieved within the frame-work of these investigations is based on purely optical
appearances and corresponds in certain aspects with the description by Sorem and Foster,
and with that by Margolis and Glasby. The terms which were used here are chosen
arbitrarily and were not influenced by model conception.

If a polished manganese nodule cross-section is placed under a microscope and examined
under a 10–60× magnification, with vertical or inclined lighting, then two structural
elements, both growth forms, are particularly noticeable. The first are straight layers and
the others are cusps or cuspate growth elements (Fig. 24, Fig. 27c), which enclose each
other. Transition forms between the two are seen in Fig. 24c, but they are seldom found
in the nodules investigated here. The straight layers or parallel layers enclose the former
growth surface. The nodule BP1 was completely constructed from these parallel layers (Fig.
20). They are called 'straight laminations' by MARGOLIS and GLASBY (1973) and they correspond approximately to the 'laminated zone' of SOREM and FOSTER (1972). The cuspate
growth patterns are a cross-section of the rounded growth elements, which are visible on
the nodule surface. In the following description they are referred to as *growth buds* or
*cuspate structure.*[1] Margolis and Glasby refer to them as 'arcuate cups' and I believe that
they correspond to the 'columnar zone' described by Sorem and Foster. In Pacific nodules
the size of the growth buds is usually less than 0.2 mm and they reach ∼ 10 single layers.
Then the growth is discontinued and continues in another bud. However, in the nodule M1,
from the Indian Ocean, they are substantially larger and they contain more layers. Margolis
and Glasby correctly call the straight layers and the cusps of the growth buds the smallest
elementary units of the nodule formation and measured their thickness as 0.25 to 10 μm.
The same authors estimated the age of these elementary units to be between 25 and 10,000
years. In this respect, the reader should refer to the dating of single layers of the nodule M1
(see nodule description), which gave an age of $t = 2600 \pm 400$ yr for a single layer.

If the microstructure of the nodules is examined, then it is noted that the structure can
be seen much better and more clearly at the surface layers than in the inner layers. Probably
the microscopic structure of the nodules becomes indistinct through the material shrinkage
or through the eventual changing mineralisation stage of the materials. Moreover, considerable differences between vertically and horizontally sliced nodules are established. In a
horizontal cross-section the growth lines can be followed relatively well around the whole
nodule and one gets a picture of approximate symmetry. In a vertical cross-section one can
hardly ever follow the lines around the whole nodule and the lines usually end at the
height of the estimated water–sediment border. RAAB (1972) has already referred to this
appearance and thus it should be concluded that the growth of nodules occurs differently
in the side towards the sediments and in the side exposed to the water.

---

[1] The German nickname of 'Zwiebeleuter' is translated as 'Onion-Udder'.

### 9.1. *Directional and non-directional growth*

From the asymmetrical form of the cuspate structure the growth direction, whereby the round cusps always point outwards (Fig. 24a–b), can be established. However, in a few places in the inner part of some nodules symmetric growth structures are also found, which infer a non-directional growth (Fig. 25a–b). At this stage probably no whole nodule had yet formed with a closed surface and the individual growth buds were formed separately but next to each other, whereby mixtures with the sediment can be observed. When the individual growth buds grew together a single nodule with a closed surface was formed and the directional growth began.

The non-directional growth is also very often found in nodule cracks, which were partially filled with light coloured silicate material (possibly phillipsite from the core transformation or possibly sediment). Perhaps, in the non-directional growth in the inner layers of nodules it is also sometimes a question of a kind of substitute growth in which a foreign core (defined in 7.4.) is dissolved and replaced by nodule material. However, similar round structural forms also appear in the tengential section of the nodules as can be easily conceived geometrically (Fig. 25c). Caution is therefore required in interpreting the symmetrical forms as non-directional growth. If the nodule is not cut precisely through the centre these forms appear automatically in the inner layer of a concentric nodule which has grown directionally. Yet it is certain that not all the structures, which appear to be symmetrical in the cross-section can be traced back to this geometrical falsification and a non-directional growth really exists. Sorem and Foster describe the so-called 'mottled zones' in the nodules. It cannot be determined from the study of the literature alone how far the 'mottled zones' correspond to the areas of non-directional growth described here. In order to do this comparisons from the same object would be required. On the basis of the observations carried out on the microstructure it can be concluded that both a directional and non-directional nodule growth may occur. Additionally the directional growth can be classified according to the structural form in growth buds and straight layers and their mixed forms.

### 9.2. *Nodule erosion in the microstructure*

The erosion of nodules in the microstructure at the surface can be clearly seen. Figures 20a–b are given as examples. In the inner layers of a nodule similar forms were observed, which were only present at one small spot, however, and therefore were of no significance for the whole nodule.

### 9.3. *Occurrence of straight layers and their possible significance*

If one would like to find the cause of growth in nodules in the form of straight layers or as growth buds, then one must first describe in more detail the local occurrence of both growth forms and examine the transition zones.

Obviously, the only nodules suitable for such comparisons are those in which both forms occur next to each other. Nodules of uniform growth are excluded from these comparisons. Thus only a few samples are left, as in the nodules from the Pacific mainly the cuspate structure and only sporadically occurring straight layers are found. It was noticeable in the datings (see nodule description) that the rapid growth and the large growth cusp patterns often coincided. Smaller structures gave smaller growth rates and in the vicinity of growth interruptions straight layers were often observed, which either had to be the last layer of the old growth or the first layer of the new growth. From this it can be concluded that straight layers are a sign of still slower growth. (MARGOLIS and GLASBY (1975) had expected the opposite connection.) Further cases will be subsequently enumerated in which straight layers occur as a rule or are at least frequently found. In the main, the growth around a foreign core begins with a straight layer, which changes into a cuspate growth (Figs 26a–b).

At first, old nodule fragments are enclosed at their *fragment-face* by a straight layer (Fig. 26c and Fig. 27a). It is noticeable here that the growth occurs as cuspate growth at the surface of a nodule fragment, while at the fragment-face it forms as straight layering (Fig. 29). If growth takes place again in an internal crack the first coat is a straight layer, which may change into a cuspate growth later on (Figs 28 and 30). Moreover, in Fig. 28 one observes cases which rarely occur, where straight layers of a short distance change sideways into the cuspate structure. In Fig. 30 a somewhat similar occurrence is noticeable. There a crack is coated with a first straight layering and at the same time a cuspate growth is indicated by a wavy line at the former nodule surface. Incidentally it should be noted that the change in this crack (Fig. 30) from parallel layering–cuspate structure–parallel layering is an interesting rarity. Before further cases of the occurrence of parallel layers are discussed, which will complicate the picture further, it would be wise at this point to consider an interim balance and a discussion of the possibilities of interpretations. According to the previous cases a straight layer is formed if the growth begins anew, irrespective of whether it is coating a foreign core, an old nodule fragment, an internal crack or initial growth after stagnations. In the surface vicinity of the nodule, a straight layer coming from a crack changes again into a waved form, which signifies a cuspate growth. There are three conceivable interpretations of this phenomenon.

(a) The straight layers are only formed on a mechanically consolidated material surface. In permanent nodule growth at the surface the material surface remains 'soft' and cuspate growth forms. Therefore it is also to be understood that after a straight layer cuspate growth begins. If the straight layers on foreign cores cannot be seen, it is not because they do not exist, but because they are 'accidentally' invisible.

(b) The straight layers represent growth forms, where growth is still possible (meagre growth, growth under unfavourable conditions). After a stagnation, which is caused by external conditions, the external factors slowly change and first of all growth begins under unfavourable conditions. Here a straight layer is formed and as the growth conditions improve growth buds occur. This process would not only explain the straight layer around the core, but also the straight layer in the cracks, where the conditions would be always slightly worse than at the surface. Foreign cores, which were set free in times of favourable growth conditions, would then be encrusted immediately without a straight layer and if they were formed in times of unfavourable conditions, then the growth began with a straight growth began with a straight layer.

(c) The nodule formation takes place biogenically and the first settlement (of organic bacteria, for example) forms a straight layer and only afterwards does a formation of clumps lead to bud growth. The process is repeated after an interruption.

If the observations are now extended to the rest of the straight layers, which are found in the Pacific nodules, then it can be established that they occur predominantly in the upper part of the nodules. Often the growth on the water-side surface finishes with a straight layer, e.g. nodule 156 (above), as is already mentioned in section 6 about the surface conditions of nodules. Cases, in which a cuspate structure in the lower part of the nodule can be clearly traced, as it changes into a straight layer at the upper side, are few. Normally the change is accomplished over longer distances and it is hardly possible to follow these

distances with a microscope. An exception in this respect is the nodule 156, where this change takes place within a distance of $\sim$ 2 mm (see Fig. 31). There one can clearly recognise a concentration of layers from right to left and in the upper part near the nodule surface the horizontal change from the cuspate structure to the straight layering (see in addition the dating of this area). Moreover, at the nodule surface a layer comes from the right, which ends in the right third of the figure and is missing on the left. It was reconstructed, before the nodule was prepared, that this was the depth at which the nodule had lain in the sediment.

The question is now raised as to the significance and the causes of this change in the growth of the nodule. Only one interpretation is possible here—after the straight layers are formed under unfavourable conditions. Thus, in the case of nodule 156 (Fig. 31), the development would be as follows. At first the nodule was in an environment of favourable growth and growth buds formed in the whole width of the section. But the growth was much greater on the right than on the left, as is also confirmed by the age-determination (see results in the nodule description). Afterwards the growth conditions at the water-side area of the nodule's surface deteriorated and only straight layers still grew there, while the growth in the sediment-side area continued unchanged. Finally, the growth conditions on the water-side deteriorated to such an extent, that the nodule growth in the water-side surface stopped, although an additional straight layer could still grow in the sediment. On the other hand an interpretation that the layer which is missing at the water-side surface originally encrusted the whole nodule and was then dissolved must be rejected for several reasons. (For example, the surface gloss which is formed during nodule dissolution is missing and moreover, in the inclined light one can see a final, slightly less defined straight layer over the whole transition zone, which can only be seen indistinctly in the vertical light.)

### 9.4. *Conclusions drawn from the investigation of the microstructure*

In summary it can be concluded that a formation of straight layers in Pacific nodules can be uniformly explained by the conception of a slow growth under unfavourable conditions. In the cuspate structure the growth rate was found to be dependent on the size of the growth cusp pattern. From the above conclusions the conceptions about the manganese nodule formation, are the following.

(a) The nodule growth is less disturbed in the sediment than at the water-side of the nodule, since straight layers form more frequently at the water-side.

(b) The disturbance of the nodule growth caused by the water does not reach the sediment side of the nodule, for, there the nodules are more protected.

(c) Since the straight layers usually occur only at the top of a nodule, that is at the water-side of the nodule, in some cases the conception can be excluded that the nodules are rolled or turned during their growth at the sediment-surface. Thus, another mechanism must be found, which keeps the nodules at the surface of the sea bed during constantly occurring sedimentation.

(d) The hypothesis that from time to time volcanic eruptions, with a rapid material supply of growth, are the cause of straight layers is also unacceptable.

### 10. FINAL REMARKS

In the investigations described here, where five to seven different methods have been applied to over 30 nodules, a few results about the growth of manganese nodules were found which were previously unknown. However, if one wants to compare 30 nodules in five to seven aspects, then an almost unbounded network of connections ensues, from which only the most important can be handled in detail here. Thus, it remains for the interested reader to supplement a few aspects through his own comparisons. Exceptions were also found which could not yet be explained. For example, if slow growth is related to an increased Fe-enrichment and parallel layers also mean slow growth, then strictly speaking, all straight layers must be rich in iron. However, this relationship has not yet been confirmed. Or, another example: at places with large growth rates, large growth cusp patterns could often (but not always) be found in the cross-sections. But the inverse relationship that large growth cusp patterns always mean fast growth is not always true. In a few places in the literature one is convinced that some aspects of geology—at least of the oceanic areas—can be explained from the investigations of the nodules and could serve as a kind of history book. However, this objective still seems to be very remote, because manganese nodules have too many problems of their own.

*Acknowledgements*—The investigations carried out here were promoted by the Bundesministerium für Forschung und Technologie. I would like to thank Professor Dürbaum B.G.R. for supporting the research application. I am grateful to Professor Beckmann and Dr. Halbach from T.U. Clausthal for kindly giving me some of the nodules which were investigated. I would also like to thank Dr. Weiser, Dr. Rösch, Dr. Gundlach and Dr. P. Müller, B.G.R., for carrying out investigations and, respectively, discussing special questions. I am indebted to Dr. Wendt, B.G.R., for helpful discussions about radioactive evaluations and reading this manuscript. I am especially grateful to Dipl. Phys. H. Meyer, B.G.R., who, during the investigations was a critical and therefore an especially valuable partner in discussing many particular questions. I am grateful to the technical laboratory assistant Mr. Lindner, and the temporarily employed students Miss Maronna, Miss Vollmer and Miss Klaus for taking the readings, counting the α-tracks and drawing the figures. I am grateful to Mr. P. Donnelly (Nottingham University) for helping translate the paper from German into English. Special thanks are due to Professor G. Arrhenius and the editor, Mrs. J. C. Swallow, who improved the manuscript considerably by helpful criticism and suggestions.

## 11. APPENDIX: NODULE DESCRIPTIONS AND RESULTS OF THE DATINGS

### LEGEND FOR THE DETAILED DESCRIPTION OF THE NODULES

*In the age-determination diagrams*

A = $\alpha$-tracks/mm$^2$/month
Z = Depth from nodule surface (mm)
S = Growth rate (mm/10$^6$ years)

*Symbols of the structure*

I, II, III numbered correlatable boundaries

ccccc    small growth cusp pattern

c c c    large growth cusp pattern

C C C    very large growth cusp pattern

straight layers

structureless boundary

structureless layer

other boundary

foreign core

boundary crust core

araldite-filled cavities

sediment part

ccc⊃⊃    direction of growth on different sides

*Sketches of the nodules:*

Str = sketched structure of the nodules at the examined section face
Ph = generations of nodule build up (growth phases or growth periods)

I    first growth phase

II    second growth phase

III    third growth phase (youngest part of the nodule)

– – –    cracks in the nodules

*Nodule M1*

The nodule comes from the Indian Ocean and was kindly given to me for investigation by Dr. D. Horn (Lamont-Doherty Observatory). The nodule had a shiny smooth surface with cavities, which infers a chemical dissolution of the nodules. In the cross-section one can see a foreign core with a first growth period. The nodule is encrusted on the outside with a thin, dark layer, in which a daughter nodule has formed. The first age-determination experiment showed no increased track frequency in the surface vicinity, from which it was

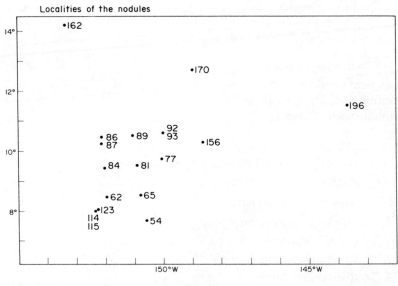

FIG. 32.

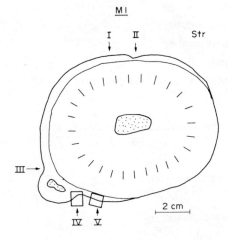

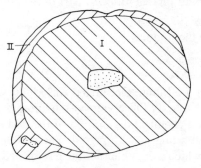

FIG. 33.

MI

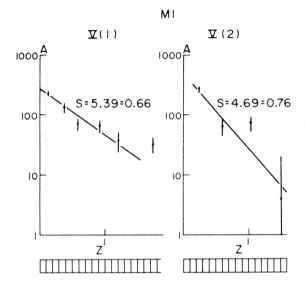

V(1)

S = 5.39 = 0.66

V(2)

S = 4.69 = 0.76

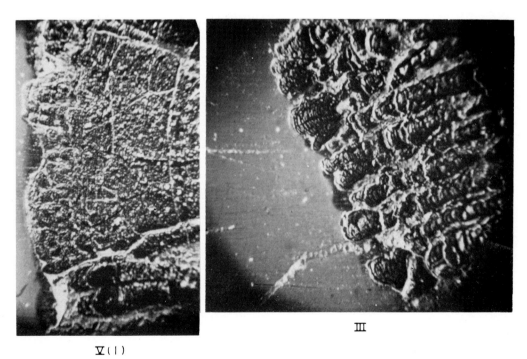

V(1)

III

FIG. 34.

concluded that this nodule was not at a growth stage, HEYE and BEIERSDORF (1973). These results were only modified as supplementary investigations of the microstructure of the nodule were carried out. Discordances were seen over almost all the surface, which confirmed the assumption of the nodule dissolution (see Fig. 20b from Area 1). The result published in 1973, about the missing growth, is thus in agreement. However, three small hollow-areas were surprisingly found in the microstructure, which show no surface discordances and are not dissolved. At these hollow regions the radioactivity ($\alpha$-tracks) is higher and shows an exponential decrease towards the inner layers. These hollow areas are either to

(a) be regarded as protected places, where the dissolution has no effect and that in the last $3 \times 10^5$ yr growing materials have been preserved there or

(b) to be regarded as places, in which growth still takes place, although nodule dissolution is predominant in the outer surface regions.

In Area V two specimen count areas with straight layering were found, in which the decrease in $\alpha$-tracks was evaluated. In Specimen Count Area 1 a growth rate $s = 5.39 \pm 0.66$ mm/$10^6$ yr was found. Under the microscope between 0 and 0.5 mm depth, $36 \pm 3$ straight layers were counted in addition. (According to the counting-line one followed, one obtained a different number of layers and therefore, a statistical error in this result.) From these two values one obtains a mean age for a single straight layer of

$$t = (2600 \pm 400) \, \text{yr}.$$

In Area 2 the evaluation showed a comparable growth rate of $s = 4.69 \pm 0.76$ mm/$10^6$ yr. But as a result of imperfections in the material the number of straight layers could not be consistently counted, so a check of the mean age of a layer was impossible.

*Nodule D 13–37p*

This Pacific nodule comes from a water-depth of 5000 m and was kindly placed at my disposal in 1972 by Dr. Fellerer (Preussag, Hannover). It was found during a trip by the *Prospekta*, SE of Hawaii.

Exceptionally clear zones in red–white colours showed when it was treated with HCl which infers fluctuating Fe and Mn contents (see Figs 10–15). In order to examine the nodules in detail individual segments of the nodule were prepared. These sections were examined with the electron micro-probe analyser, α-tracks were counted and finally they were treated again with HCl. In its internal build-up the nodule resembles the nodule 114(2) and since only special investigations were carried out here no detailed nodule description is given. The results of the datings are compared in section 4, with the internal chemistry.

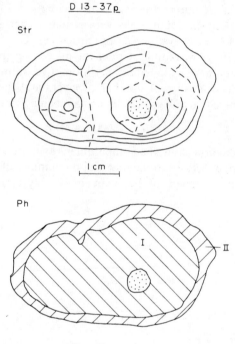

FIG. 35.

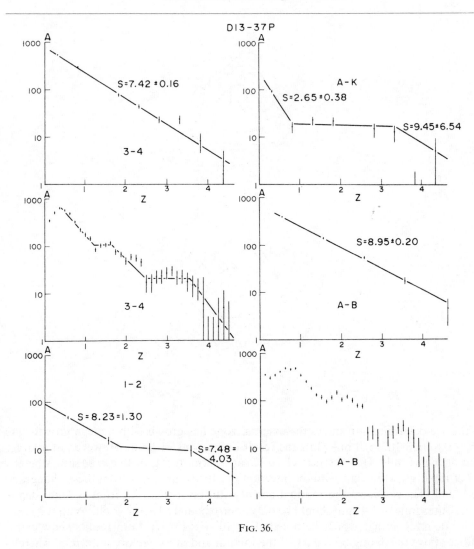

FIG. 36.

*Nodule Polished-Section* 32

The section was given to me for further investigation by Dr. Halbach (T.U. Clausthal). Investigations with the electron micro-probe analyser (Fe and Mn distribution) and age-determination were compared. Since the nodule was already enclosed in Araldite the surface condition, the position in the sediment and the nodule form can no longer be reconstructed. In the microstructure it can be seen that an old nodule fragment was encrusted by a 4–6 mm thick layer. Sharp-cornered fragment-faces can also be seen on the nodule fragment, so there are no indications of erosion (dissolution) after the break-up of the old fragment. In some places in the microstructure a straight layer is visible around the old nodule fragment, however it is only found on the fracture-faces. The growth on the old surface of the fragment continues with the formation of cuspate-structures. (The straight layers on the fracture-faces can also be coatings of an old nodule crack that have grown before the break-up of the old nodule along this crack.)

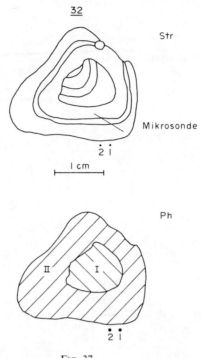

FIG. 37.

    As the age-determination shows the external zone has grown with a mean growth rate
of $8.59 \pm 0.61$ mm/$10^6$ yr, if one plots the result in intervals of 1 mm. However, at a larger
resolution in intervals of 0.5, respectively, 0.125 an intermittent growth can be seen, whereby
zones of rapid growth occur with an interruption. If one proceeds from these changes in
growth and also considers the residual activity still present in the Io at $\sim 4$ mm depth,
then the outer zone of 4–6 mm should be only approximately $4 \times 10^5$ yr old. With the mean
growth rate a higher age would be noted ($4.5 \times 10^5$–$7 \times 10^5$ yr). Both results show extra-
polations above the dating boundary of the ionium and are therefore uncertain, whereby
the extrapolation from the mean rate might be more uncertain, since this mean rate does
not reflect all the changes in growth. In the area of rapid growth below 1.5 mm in depth
the microprobe investigations showed a decrease in the iron enrichment. See section 4 for
these results.

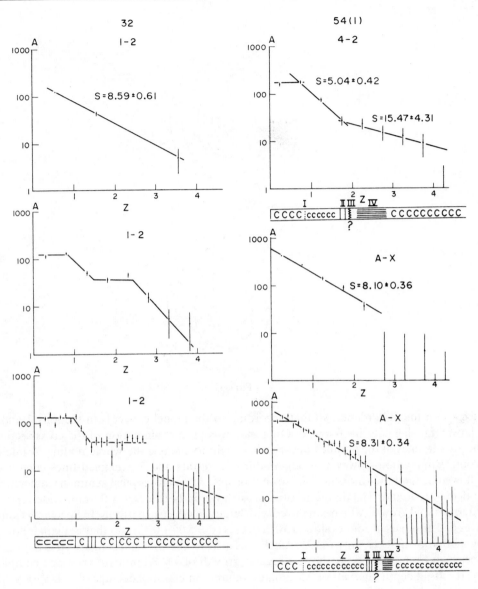

FIG. 38.

*Nodule 54(1)*

The round nodule comes from a water-depth of 5085 m. It has grown around a foreign core (probably volcanic rock). Growth cusp patterns were visible on the surface, which was rough on all sides, and the position of the nodule on the sediment remained unclear. Since it showed no adhering residue and also because of its symmetry, a specific position can not be reconstructed. Macroscopically visible discordances show in cross-section an interim erosion of the nodule. There is a two-phased growth and the last external layer is 5 mm thick. The crack frequency, respectively, the crack widths of the nodule show a sudden

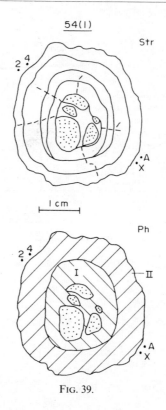

FIG. 39.

increase in the discordance, so that according to the model concepts in sections 7 and 8 a larger time-lag is to be assumed. On a macroscopical examination of the cross-section, the growth lines of the nodule appear at first sight to encircle the whole nodule. A microscopic examination, however, produces different results and less defined lines cannot be followed around the whole nodule, so that in spite of macroscopical symmetry differences in detail are found. The dating in two areas shows in both cases a 0.6 mm wide zone of external rapid growth, which concurred with large growth cusp patterns. In Specimen Count Area 4–2 an adjacent zone of slow growth occurs ($s = 5.04 \pm 0.42$) and that is again followed by a zone of quicker growth. As can be seen from the appended figure, the dating in Specimen Count Area A–X shows a regular growth of $s = 8.1\,\text{mm}/10^6\,\text{yr}$, if one compiles the results in depth intervals of 0.5 mm and obtains an extrapolated age of $\sim 6 \times 10^5\,\text{yr}$ for the external incrustation.

*Nodule 54(2)*

This nodule comes from the same location as the previous one and has also grown symmetrically around a foreign core, but is substantially larger. The rough surface is covered all over with small growth cusp patterns. For the same reasons it is not possible to reconstruct a location.

In the cross-section the growth lines are easily followed macroscopically around the whole nodule. However, indications of erosion are not visible. Hence, the nodule could have grown in a single growth phase. In relation to its size it only shows a few cracks and according to the model concepts in section 8, its total age might be relatively small. It is

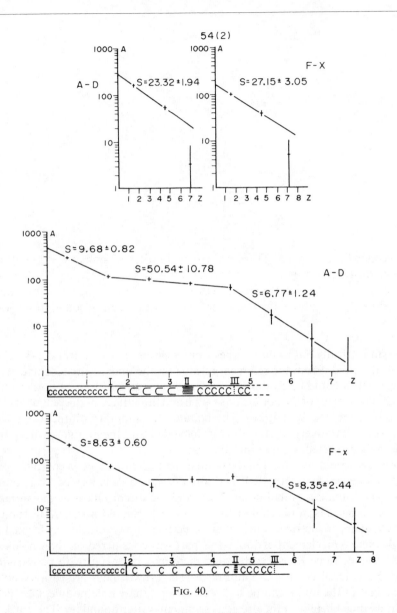

FIG. 40.

worthy of note that another $\sim 3\,mm$ thick compact zone has grown around the core, which consists of distinctly smaller growth cusp patterns, than the outer regions.

Under the microscope, one can see that the nodule is partly constructed from substantially larger growth cusp patterns than all the other examined nodules from the Pacific. The growth of the nodule around the core, however, begins with a straight layer. The evaluation of the $\alpha$-tracks at intervals of 0.125 mm showed no continual decrease of the values and only if the values are averaged in much larger intervals can a smooth curve be produced. An average evaluation in intervals of 3 mm showed a remarkably high growth rate of $\sim 25\,mm/10^6$ yr for both areas. In the evaluation with depth intervals of 1 mm, one obtains, within a continual growth of $\sim 8\,mm/10^6$ yr, a 3 mm broad zone of quicker growth.

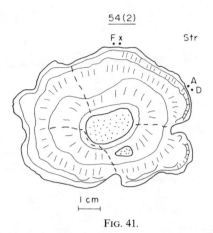

FIG. 41.

In the Specimen Count Area A–D this growth could still be found at $s = 50 \pm 10 \, \mathrm{mm}/10^6 \, \mathrm{yr}$. It is also interesting to note the clear correlation between the large growth cusp patterns and these zones of rapid growth. Through extrapolation a total age of $\sim 10^6 \, \mathrm{yr}$ is obtained for the nodule. The results of the magnetic investigations can be found in section 5.

*Nodule 62*

This asymmetrically flat nodule comes from location 65 ($\sim 5180 \, \mathrm{m}$ depth of water) and was vertically sliced. From the sediment residue on the underneath side and the shield on the reverse side, it could be established with sufficient certainty, as to how the nodule was positioned. The surface of the nodule was rough and the surface of the sediment-side showed clearly visible growth cusp patterns. The nodule was smoother on the water-side. As this smoother spot is relatively small it can be concluded that the nodule lay deep in the sediment and only projected $\sim 5 \, \mathrm{mm}$ into the water.

Under macroscopic examination the nodule cross-section shows a complicated build-up, which cannot be clearly interpreted. The core of the nodule is formed from an old nodule fragment. The rounding of its fracture-faces and the discordances at the surface indicate erosion, which had taken place after the break-up of the old nodule. This nodule is surrounded by an unsymmetrical zone, which could have, however, already formed on the old nodule. As there is an element of doubt, this part was considered on the whole as the older core and assumed to be Phase 1, although there may be two growth periods around this core and the external crust is 2–3 mm thick. The growth lines are easily visible with the naked eye and in the main, can be followed around the whole nodule. Less defined lines end at the assumed height of the previous sediment–water boundary. The crack formation of the nodule (i.e. frequency and width) shows a clear difference between the old nodule and the symmetrical surrounding growth, so the core, according to the model concepts in sections 7 and 8, is considered to be substantially older. The growth on the old nodule core began with a straight layer. However, no further straight layers are found in the nodule. One specimen count area from the water-side (T–1–2) and one from the sediment-side surface (x–2) were selected for the dating. The mean growth rates are much the same, only the initial activity in the sediment-side count area x-2 is lower by more than half.

An age of $\sim 4 \times 10^5 \, \mathrm{yr}$ for the last growth phase can be extrapolated from the two equal growth rates and can also be read from the still recognisable residual Io-activity in the layers beneath. If one excludes growth interruptions between period II and III, then an age of $\sim 10^6 \, \mathrm{yr}$ could be extrapolated for the initial growth of period II.

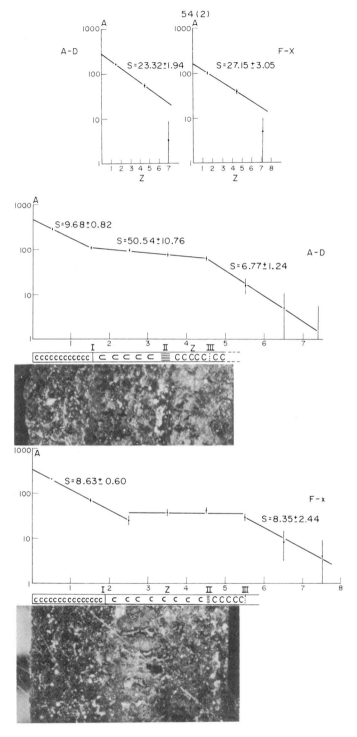

54(2)

A-D      S=23.32±1.94

F-X      S=27.15±3.05

S=9.68±0.82

S=50.54±10.76

A-D

S=6.77±1.24

cccccccccccc | C C C C ⫶ CCCC⫶CC

S=8.63±0.60

F-x

S=8.35±2.44

cccccccccccccc | C C C C C C C ⫶CCCCC⫶

Fig. 42.

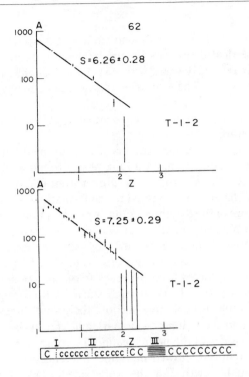

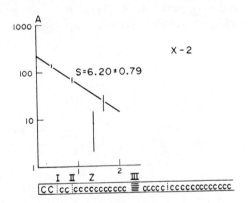

FIG. 43.

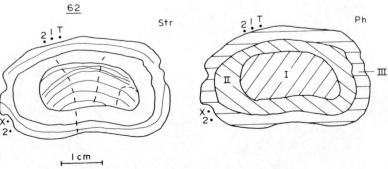

FIG. 44.

*Nodule from core 65*

This nodule was taken from the sediment surface of the piston core 65 (5050 m depth of water). Its unsymmetrical figure can already be seen from an external examination. The water-side surface is relatively small with just a little roughness from small growth cusp formations (smooth shield). The sediment-side surface is uneven, very rough and covered with formations of large growth buds. At first sight, it seems from the vertical cross-section, as if a dissolution took place at the sediment-side surface, but this is not confirmed by microscopic examination.

In the cross-section it can be macroscopically determined that the nodule has grown in two periods. The first nodule formation took place around a foreign core and a 2–3.5 mm thick outer layer grew after an erosion. A splinter from a supposedly different nodule has grown into the outer layer, which has led to the formation of a daughter nodule. After the erosion period the core of the daughter nodule settled on the main nodule and the daughter nodule has also grown during the whole growth period II. (However, the contact-line of the two nodules was not visible in the examined cross-section, because it lay in another plane.) According to the model concepts in sections 7 and 8, a larger time-lag between growth periods I and II can be inferred from the crack frequency. It can be seen under the microscope that the growth period I and II each started with straight layers (as well as the daughter nodule). In the vertical cross-section, the growth lines can only be followed in both the water-side and the sediment-side part, and end in the transition area. Therefore, the lower- and upper-sides of the nodule cannot be compared.

The age-determinations showed the same differences and rapid growth zones alternated with a continual growth. For the water-side surface (Count Area 1–2), using an internal width of 0.5 mm, one obtains a growth rate of approximately $5 \, \text{mm}/10^6$ yr and for the sediment-side surface a growth rate of $7.2 \, \text{mm}/10^6$ yr. From these varying values a crust age of $7 \times 10^5$, respectively, $4 \times 10^5$ yr can be extrapolated.

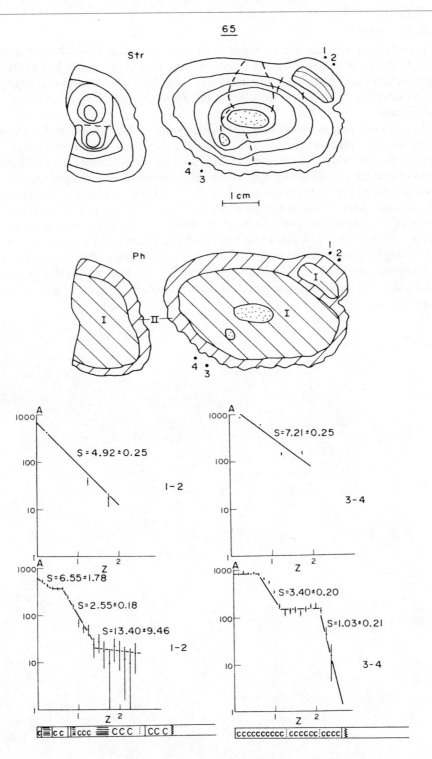

Fig. 45.

*Nodule from core 77*

This nodule comes from the sediment surface of the piston core 77 (5120 m depth of water). This nodule has an irregular shape, because it contains many fragments of older nodules, which were encrusted. The surface was smooth on all sides and only small growth cusp formations were visible. There were insufficient indications for a reconstruction of the nodule location.

In the cross-section the older nodule fragments are visible, which are surrounded and held together by younger material. The old nodule fragments seem to be broken in themselves, for they show several complete cracks. So it can be assumed that the fragments of one old nodule have remained here in a heap after the break-up and were encrusted again. No indications of a possible erosion period are found. The evaluation of $\alpha$-tracks in two areas showed a very similar mean growth rate of $s = 4.0 \pm 0.3$ mm/$10^6$ yr. If the $\alpha$-tracks are plotted at intervals of $\frac{1}{8}$ mm, then an interruption with a time-lag of about 7000 yr is found in Count Area C–D. This result agrees with the macrostructure of the nodule growth. The time-lag is caused by a wide layer (which is visible in the nodule) becoming narrower and narrower, and finally ending in the Specimen Count Area (C–D). To this extent, the age-determination, which shows an interruption, is in complete agreement with the nodule structure.

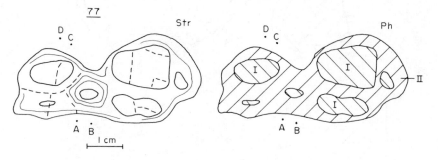

FIG. 46.

77

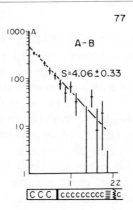

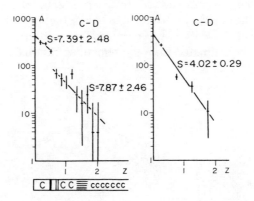

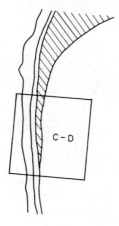

Fig. 47.

*Nodule* 1 *from core* 81

This nodule was found on the sediment surface of piston core 81 (depth of water 5185 m) and already shows externally an intergrowth of many single parts. Its surface was uniformly smooth and only covered with small growth cusp formations. It was not possible to reconstruct its position on the sediment. In a cross-section several old nodule fragments can be seen, which are surrounded and held together by an external layer. The older nodule fragments show many cracks and according to the model conceptions in sections 7 and 8 would be considerably older. Some cracks run completely through the old fragments. It is therefore conceivable that the fragments of a single nodule have remained together at one place and were again encrusted. The growth began again with straight layers and straight layers are also found elsewhere in the external crust. No clear indications of an erosion period were found, which could have caused the old nodule to break. In all the previously enumerated points, there is a great similarity to the nodule from core 77.

The evaluation of the $\alpha$-tracks showed an external zone of rapid growth of $\sim 0.5$ mm, which is followed in the inner layers by an area of constant growth at different growth rates. An extrapolated age of $\sim 4 \times 10^5$ yr was obtained for the external, surrounding layer of this nodule and can also be estimated from the remaining residual activity in the deeper part of this layer in Count Area M–N.

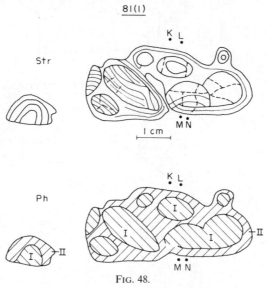

FIG. 48.

*Nodule* 2 *from core* 81

This nodule also comes from core 81, and in its surface condition and internal structure resembles nodule 81(1) and also nodule 77. However, clear traces of erosion are found on the old fragments in the inner layers of the nodule. After the break-up the edges of one fragment have been rounded off.

In the microstructure it can be seen that the growth of the external crust began with a straight layer. Several straight layers have also formed elsewhere in the external area. The evaluation of the $\alpha$-tracks shows no similarity in both areas. In one case (Count Area J–H), one finds an extremely slow growth with $s = 1.73 \pm 0.08$ mm/$10^6$ yr and yet in the other area

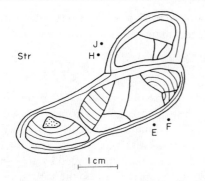

81 (2)

FIG. 49.

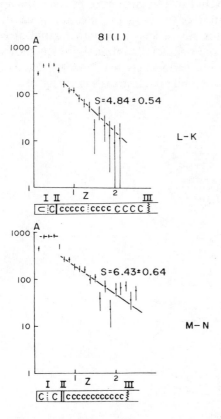

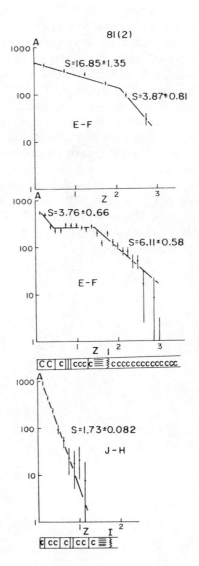

FIG. 50.

values up to $s = 16.85\,\text{mm}/10^6$ yr. Moreover, in Count Area E–F, the decrease in $\alpha$-tracks still reaches the old nodule fragments, so this nodule has many contradictions. The reasons for this could not be explained.

### Nodule from core 84

This nodule was taken from the sediment surface of piston core 84 (5220 m depth of water). The nodule was scraped on one side by the piston core, so this could be used to find the upper-side and the position of the nodule in the sediment very clearly. The surface of the nodule shows medium-sized growth cusp formations and was slightly rough. In a vertical cross-section it can be seen that an old eroded nodule was surrounded by a 2–3 mm thick younger crust. In the old nodule, one can see in the growth lines, the formation of an equatorial bulge, a growth form, which is missing in the last layer. The old nodule is completely cracked and in part shows very wide cracks. According to the model concepts in sections 7 and 8, the inner nodule, because of the cracks, might be considerably older.

The growth lines can only be microscopically traced in small distances and they all end at the height at which the nodule in the sediment might have laid. The upper and lower growth structures of this nodule vary. Therefore, in the case of this nodule it can be definitely determined that the growth of nodules in the sediment and in the seawater are independent processes. Since the transition zone always lies at the same height in the external crust, a change in position of this nodule (due for example to rolling or a turn on the seabed) during the formation time of the crust, can be excluded. The differences between the upper and lower parts can also be seen in the evaluation of the $\alpha$-tracks. In Count Area 10–11 (water-side of the nodule) a continual growth is stopped by an external rapid growth, whereas in Count Area 1–2 (sediment-side of the nodule) a continual growth shows an interruption. The growth interruptions fall in the period between $214{,}000 \pm 20{,}000$ and $1{,}220{,}000 \pm 12{,}000$ yr B.P. From the growth rates of both nodule areas a crust age of $4.10^5$ yr can again be extrapolated.

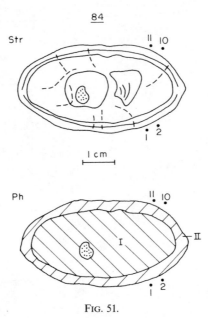

Fig. 51.

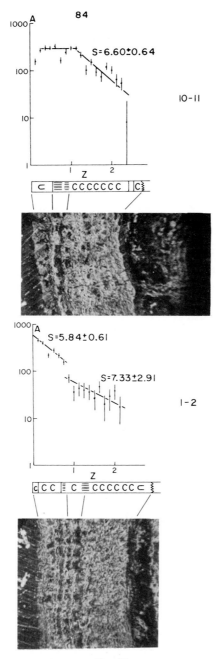

Fig. 52.

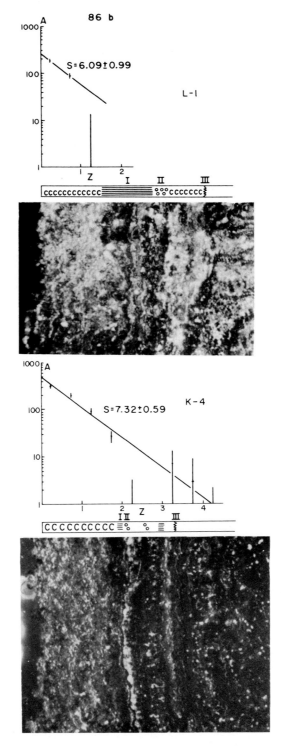

FIG. 54.

*Nodule 86b*

The surface of this nodule (5202 m depth of water) was rough and showed clearly recognisable growth cusp formations. A clear reconstruction of the nodule in the sediment was not possible, since all the sides showed sediment residue. In the cross-section one can see an old nodule-half in the inner layers, which was surrounded by an external 3–4 mm thick layer. From the microscopic examination, it can be clearly concluded that an older nodule has broken up here and that the older nodule is not a typical case of an asymmetrical growth. Cracks were present in the old part of the nodule. However, according to the model concepts of sections 7 and 8, these do not indicate a large time-lag between the old nodule and the external layer. Indications of erosion were missing on the old nodule-half, so that in this case the break-up of the old nodule was only caused by the crack formation and not aided by nodule dissolution. The start of the growth in the external crust began again with a straight layer. In the external crust, growth lines cannot be followed around the whole nodule, and yet the two evaluated count areas L–1 and K–4 are similar in many respects, although the single zones are formed at different widths. Therefore, a direct com-

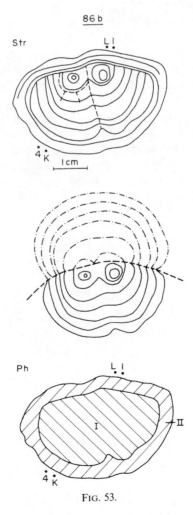

FIG. 53.

parison is impossible and thus the evaluation of the α-tracks also shows differences in detail. An age of between $(4–7) \times 10^5$ yr can be extrapolated for the external crust.

*Nodule from core 87*

This flat nodule comes from the sediment surface of piston core 87 (5195 m depth of water) and the surface is rough on all sides due to growth cusp formations. Its position in the sediment could be reconstructed with reasonable certainty from the adhering sediment and accordingly the nodule remained in its original position when it was raised from the seabed.

In the vertical cross-section, it can be seen that an external layer of 2–3 mm had grown around a heap of old nodule fragments, which probably belonged to a single nodule. Judging from the growth structure of the single parts the fragments were partly mixed up. Fragments now lying directly beside each other were seldom adjoining in the old nodule before it broke up. Therefore, the gaps between the fragments are mainly a question of 'pore spaces caused by the piling up' of single parts, and are only in a few instances a question of shrinkage cracks in the material. However, judging by the small number of existing shrinkage-cracks, the old nodule map have been considerably older than the incrustation. The single fragments still show sharp fragment faces and the corners are not rounded off. Thus, after the break-up of the nodule no erosion can be proved. On the other hand, erosion took effect on the old nodule before the break-up. The details of all these reconstructions may seem somewhat fanciful to the reader, but surprisingly enough in the case of this nodule, they can be clearly recognised.

In the microstructure the growth lines of the crust cannot be followed around the whole nodule and end at the height on the nodule at which the sediment–water boundary may have laid. This nodule also shows differences in its water and sediment area. The evaluation of the α-tracks in the water-side area (14–15) yields less growth than in the sediment-side area (29–30). A total age of more than $4 \times 10^5$ can be extrapolated for the external incrustation, but this could also extend to $10^6$ yr.

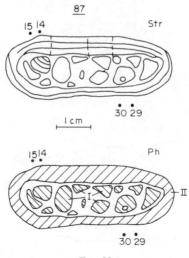

FIG. 55.

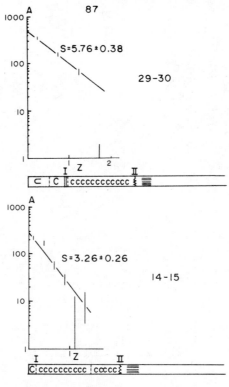

*Nodule 89*

All the sides of this twin-nodule were covered with growth cusp formations and also loaded with sediment (5090 m depth of water). It was not possible to reconstruct a position in the sediment. In the cross-section two nodule fragments can be seen, which were encrusted, and thereby grew together to form a twin-nodule. Both the fragments show no correlatable line, but the structure layers are curved in approximately the same way, so they could have come from the same nodule. (In one of the fragments, an even older fragment is encrusted, so here parts of nodules from three generations can be found.) There are few (also uncertain) indications of erosion of the old fragments after the break-up. Cracks are missing in the old fragments, however these are so small that the shrinkage case for small $At/\tau$ could be applicable, whereby, according to the model concepts in sections 7 and 8, a shrinkage without cracks would be possible. At one place it seems as if an old fragment has detached itself from the crust, which supports the view for this shrinkage case. (The crack visible there, is filled with sediment, so it cannot be decided with certainty, whether it is a question of a filled crack or whether it is a zone with the original sediment additive.)

In parts of the microstructure, it can be seen that the growth of the external layer began with a straight layer. The growth lines can only be followed for a certain distance, so no direct correlation can be found between the two specimen count areas. In the evaluation of the $\alpha$-tracks, comparable growth rates of $s = 6.7$ and $s = 6.4$ mm/$10^6$ yr are obtained. The youngest incrustation is older than the Io dating boundary and can be estimated as $t \leqslant 4 \times 10^5$ yr.

89

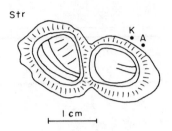
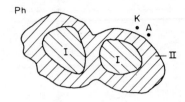

FIG. 57.

89

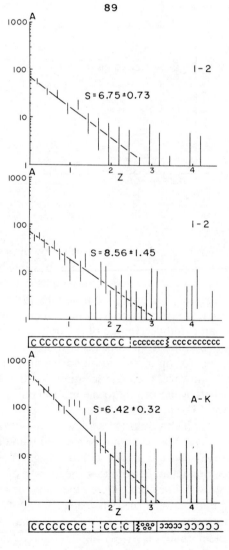

FIG. 58.

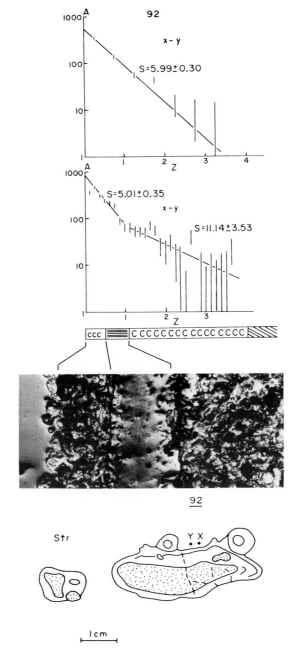

FIG. 59.

*Nodule 92*

This nodule comes from position 92 (5135 m depth of water). On the surface it was slightly rough and had an irregular shape. It was not possible to reconstruct a position in the sediment with any degree of certainty. In the cross-section one can see a foreign core, which was encrusted and which has small daughter nodules. The incrustation is of varying thickness and lies between 1–4 mm and the growth lines cannot be followed around the entire nodule. Only one area was evaluated as it can be already seen from the structure that the growth conditions for the single parts of the nodule must have been very different. At intervals of 0.125 mm depth, plotted points showed a change in the growth rate from $s = 5 \pm 0.3$ mm/$10^6$ yr in the outer layer to $s = 11.1 \pm 3.5$ mm/$10^6$ yr in the more interior layers. If one takes the average value at intervals of 0.5 mm depth an overall rate of $s = 6.0 \pm 0.3$ mm/$10^6$ is obtained. This is a good example to illustrate how interesting details can disappear when measurements are taken in large depth-intervals (here the average). The age of the crust cannot be specified, although the minimum age must be larger than $4 \times 10^5$ yr. At the beginning of the growth around the foreign core a straight layer was again formed. The area with the larger growth rates also shows the larger growth cusp formations again.

*Nodule from core 93*

This small twin-nodule comes from the sediment surface of piston core 93 (5142 m depth of water). Its surface is also rough on all sides because of growth cusp formations. In the cross-section, one can see nodule cores in both parts, which partially consist of foreign material. It remains uncertain whether these are sediment or partially dissolved volcanic material. In the nodule-half, in which the evaluations of the $\alpha$-tracks took place, the growth lines could be followed. First the Count Areas 1–2 and 3–4 were evaluated and the distribution of the $\alpha$-tracks showed contradictory growth relationship. Afterwards, the intermediate Count Areas 5–6 and 6–7 were evaluated, and the contradictions were even larger, and the results were inconsistent with one another. The causes of this remain uncertain and if, just by chance, only this nodule had been examined, the Io-dating of manganese nodules would have been proved to be useless because of these contradictions. However, if the results of this nodule are seen in relation to the results of other nodules, then it seems that such a far-reaching conclusion is in no way justifiable. Moreover, one occasionally obtains completely inexplicable results in any investigation and often an explanation is only found much later, or not at all.

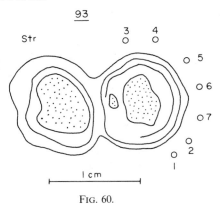

Fig. 60.

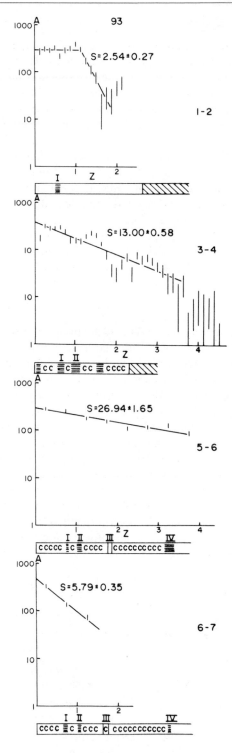

FIG. 61.

At this juncture, lengthy discussions could have been held about the varying obvious and possible cause (for instance, the microstructure of the nodule), but since none of these theories would have been conclusive, such dubious interpretations were deliberately omitted.

*Nodule* 114(1)

This lens-shaped nodule (5160 m depth of water) was rough on all sides and covered with larger growth bud patterns. In the horizontal cross-section two foreign cores and indications of an older and a younger part of the nodule could be seen. In an additional vertical cross-section the older part of the nodule could be clearly seen, but no signs of an erosion period could be found in both sections. On account of the crack frequency in both parts of the nodule, according to the model concepts in sections 7 and 8, a larger time-lag must exist between the older layers of the nodule and the younger incrustation. The lens-shaped appearance of the nodule is determined by the two foreign cores and the old part of the nodule. A so-called equatorial bulge with a faster growth to the sides is only hinted at in the outer nodule crust, but is not decisive for the flat nodule form. (In addition, see nodule 114(2).)

All the detailed investigations were carried out in the horizontal cross-section plane and the growth lines could be followed relatively easily in this plane. Here it was clearly shown that the nodule structure is more uniform in horizontal cross-sections than in vertical sections. In two Count Areas the evaluation of the $\alpha$-tracks showed a fast growth rate of 10–12 mm/$10^6$ yr. However, in order to obtain a reasonable decrease the values had to be averaged at intervals of 1 mm. The inhomogeneities of the nodule are already so large

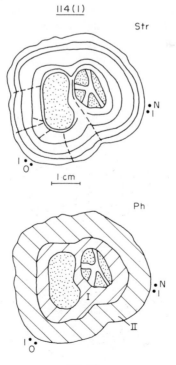

Fig. 62.

during this fast growth that plotting in smaller intervals only reflects variations of these inhomogeneities. (Also see nodule 54(2).) One obtains an extrapolated total age of $(4-5) \times 10^5$ yr for the age of the crust.

In this nodule it was also interesting to observe the formation of the cracks and their coatings. One can see many growth cusp patterns in the crack system, from which it can be concluded that the crack system was open for a long time during the formation of the nodule. It can be concluded, from the material supply of the growth formations within the cracks that the cracks were rinsed by water. It can be seen from the structure of cracks reaching as far as the surface that they remained open for the whole of the final growth phase. It offered a good opportunity to determine whether or not any diffusion of the Io (or its daughter products) was possible. In Fig. 6 the $\alpha$-tracks are individually sketched in the count area of this crack and it can be seen that a diffusion of $\alpha$-trace-emitting isotopes from the open crack into the older parts of the nodule (or vice versa) was not detectable. The growth formations within the cracks also have no detectable increase of $\alpha$-tracks from which it can be concluded that they are older than the Io-dating boundary.

*Nodule* 114(2)

This nodule comes from the same position as nodule 114(1) and the bottom-part showed a surface roughness caused by growth cusp patterns. But the top of the nodule was smooth. All the investigations on this nodule were carried out on the faces of a vertical cross-section. It also shows two foreign cores with an initial growth period, which, after a period of erosion, was enclosed by a second growth phase. It was possible to reconstruct the position in the sediment as a result of the adhering sediment residue. As in nodule 114(1) the large crack frequency of the older nodule indicates a greater time-lag between both growth periods.

As is customary in vertical cross-sections the growth lines could only be followed for a limited distance. The two cores and the older nodule are responsible for the shape of the nodule. The visible formation of an equatorial bulge due to larger growth at the sides (for example, larger growth cusp patterns at the surface), only plays a small part in the total shape of the nodule.

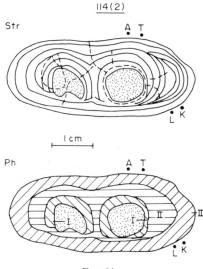

FIG. 64.

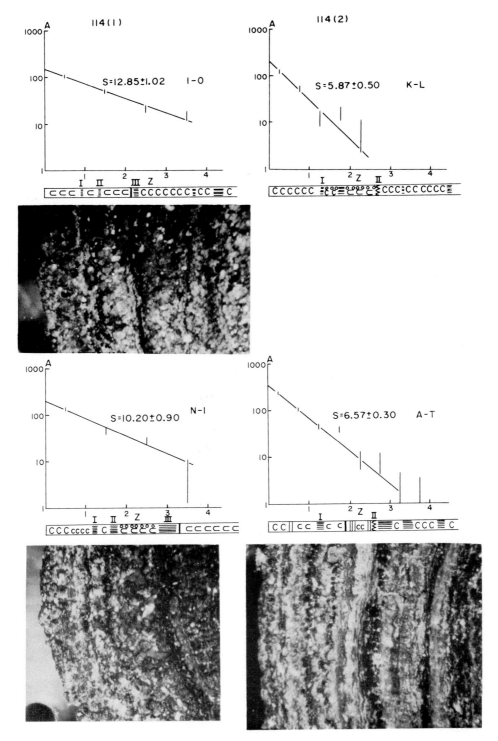

Fig. 63.

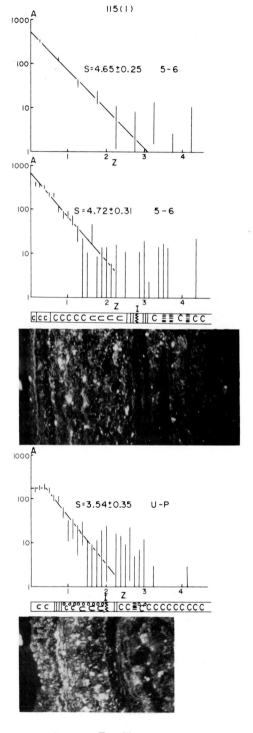

Fig. 66.

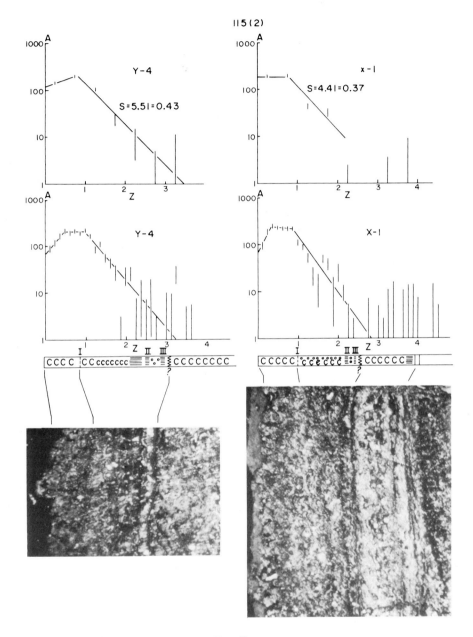

FIG. 67.

The same growth rate of $\sim 6\,\mathrm{mm}/10^6$ yr was obtained for the sediment-side and water-side surface but in the micro-structure the top and bottom sides varied considerably. Previously, it was noted that many straight layers grew in the water-side part of manganese nodules, which according to the concepts in section 9, point to a growth under unfavourable conditions. These layers are missing altogether in the sediment-side part of the nodule. An age of $\sim 5 \times 10^5$ yr was extrapolated for the outer layer.

*Nodule 115(1)*

This nodule was rough because it was covered with growth cusp patterns and only the top was relatively smooth (water depth 5180 m). Because of the adhering sediment residue it was easy to reconstruct the nodule position in relation to the sediment. The nodule was vertically sliced and had a foreign core. This foreign core was enclosed (apparently together with an older nodule fragment) during an initial growth period and after an erosion period, it was again encrusted. The daughter nodule settled in the last growth period. The inner part of the nodule is completely cracked, so according to sections 7 and 8, a time-lag to the last incrustation can be assumed. However, the internal construction of the nodule is so involved that four generations of nodule material could be counted between the embedded nodule splinter and the outer crust. It could also be the case that the older nodule was

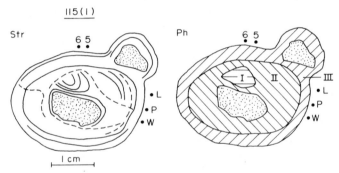

Fig. 65.

completely broken up and that the fragments were encrusted again in the same position. There are certain allusions to this in the excessive width of internal cracks, which could then have to be regarded as pore spaces between the fragments. The evaluated $\alpha$-tracks showed comparable growth rates for the upper Count Area (5–6) and the side Count Area (U–P–L). In Count Area U–P–L one can again see the connection between a zone of rapid growth and very large growth cusp patterns. Straight layers are found on both sides of this nodule. A total age of $\sim 4 \times 10^5$ yr can be extrapolated for the external crust.

*Nodule 115(2)*

This slightly flattened nodule comes from the same position as the nodule 115(1). It was rough on all sides because of the growth cusp patterns and was horizontally sliced. It was possible to reconstruct the position of the nodule in relation to the sediment with the aid of the adhering sediment. One can see two foreign cores and encircling them an old eroded nodule with many cracks. The latest incrustation is 2–3 mm thick. In the cracks lighter coloured coatings (presumably minerals) and also growth cusp patterns can be seen. The growth around the foreign core began again with a straight layer. In the outer area the

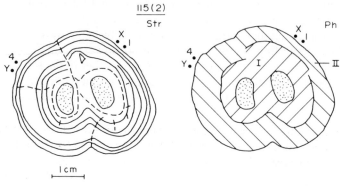

FIG. 68.

growth lines can be followed around the whole nodule and the two evaluated α-track areas can be correlated. Using intervals of 0.5 mm an external rapid growth is obtained in both cases, as was similarly obtained in nodule 115(1). In the inner layers a constant growth of 4.4, respectively, 5.5 mm/$10^6$ yr was obtained. However, if the results are plotted in intervals of 0.125 mm, then a decrease is shown towards the surface. The radioactive interpretation of this decrease is that the rapid growth took place only very recently and that the daughter products of the Io are not yet in equilibrium, whereby the $Ra^{226}$ with its half-life of 1617 yr determines the increase. The special case of this recent rapid growth is already described in more detail in section 3.4.3.

*Nodule 123 (Fig. 27b, (section 9.3))*

The surface of this flat nodule (5168 m depth of water) was rough on all sides because of growth cusp patterns, but a reconstruction of the position in the sediment was not possible. In the cross-section, the residue of an old nodule could be seen below a 2–2.5 mm recent incrustation. This older nodule part shows only a few small cracks, and according to the model concepts in section 7 and 8, might well be relatively young, and indications of an erosion phase cannot be seen. It can be reconstructed from the fragment that a small segment has been enclosed by a considerably larger nodule.

The growth of the external crust again began here and there with a straight layer, which

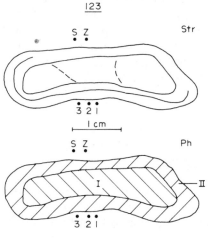

FIG. 69.

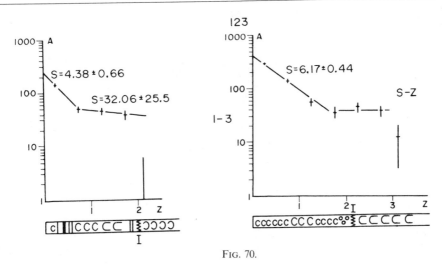

FIG. 70.

can also be found once again in the Count Area 1–3 in the outer layers. Under the microscope it is only possible to follow the growthlines over small distances. Therefore, there is no chance of showing a detailed correlation between the two counting areas. From the α-tracks one obtains a change of growth in the Count Area 1–3 from $s = 4.4 \pm 0.6$ to $s = 32 \pm 25$ mm/$10^6$ yr. However, at the edge of the core of the nodule there is a sudden decrease in the α-track frequency in this count area to the value of the usual background effect. In Count Area S–Z there remains a distinctly higher α-track frequency in the old nodule part. If one observes the growth direction in the old nodule fragments, then one can see that the older part of the nodule is enclosed in Count Area 1–3. The former surface of the broken nodule is encrusted at the side of Count Area S–Z. (See in addition the structure-sketch under the α-track diagrams and the structure photograph, Fig. 27b in section 8.) The incrustation is so recent, that its growth can still be followed through the whole thickness with the Io-method and it began about $(2.80–3.20) \times 10^5$ years ago. In Count Area S–Z a residual activity in the surface region of the old fragment is detectable. It can be concluded from this residual activity that the old fragment was immediately encrusted after the break up of the old nodule (without a time-lag). The boundary is only marked by a small layer of sediment deposit. In Count Area 1–3 the change in the growth rate can be correlated well with the change in size of the growth cusp patterns. The result of the residual activity present at the old nodule surface (Count Area S–Z) and its sudden decrease at the core boundary (Count Area 1–3) again gives a clear indication of the absence of any diffusion processes. If the ionium was mobile and would diffuse through into the nodule, then it would have also had to reach the older part of the fragment.

### Nodule from core 156

This nodule comes from the surface of piston core 156 (5300 m depth of water) and is probably the most informative nodule in the series of investigations. It was clearly possible to reconstruct a location and its surface was mainly rough and covered with growth cusp patterns. Only on its water-side surface did it have a smooth spot with a sharp edge. At first sight, it looked as if a material dissolution had taken place on this spot and the edge would be a depth interval. But the typical gloss and holes were missing for a material

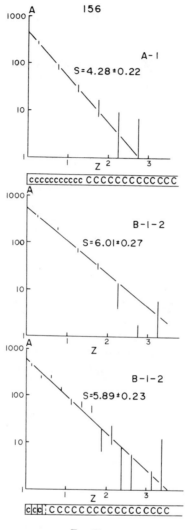

FIG. 71.

dissolution. This area of the nodule is considered in more detail in the cross-section (see section 9, Fig. 31).

In the cross-section one sees a foreign core in an asymmetrical position and several growth periods of the nodule. In the inner layers the fragments of an old nodule can be seen. This nodule was completely broken up, for the fragments were slightly tilted against each other, and were a further distance apart than could be explained by the crack formation. From the position of these fragments in relation to the core of the nodule it can be concluded that the upper fragments were transported away after the break-up and are now missing. At this point there were no indications of erosion. Then a second growth period with subsequent erosion followed, the material of which is preserved only in the lower part of the nodule. The erosion affected the upper fragments of the older nodule. (It could also be the case, that only then were the upper fragments of the first nodule dissolved

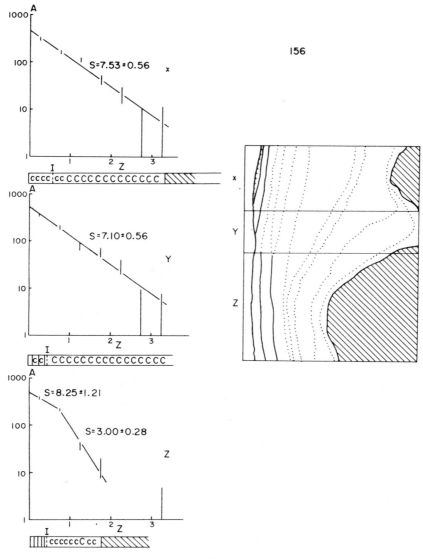

FIG. 72.

or transported away.) The surface of this nodule has then been encircled by another layer of irregular growth, which is 4 mm at the bottom and at the top only 2 mm thick.

The growth lines can be followed macroscopically around a large part of the nodule. It can be seen that the outer layer began to grow earlier in the sediment than in the water and it can be concluded that the growth conditions became favourable earlier in the sediment than in the water. In section 9, there are more detailed explanations of aspects of the micro-structural observations.

In the sediment-side Count Area A–1 the evaluation of the $\alpha$-tracks showed a continual growth of 4.3 mm/$10^6$ yr, which amounts to an extrapolated total age of $\sim 10^6$ yr for the outer incrustation. The other Specimen Count Area B–1–2 lay in the transition zone water-

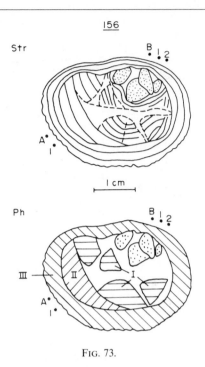

FIG. 73.

sediment and corresponds to the section discussed in detail in section 9 (see photograph Fig. 31). First of all the number of the α-tracks in the total Area B–1–2 was evaluated (growth rate ∼ 6 mm/10⁶ yr). In addition to this the specimen count area was divided into the sections x, y, z. These individual sections were evaluated and compared with the structure lines. The α-track diagrams contain a sketch of the structure and one can conclude the following.

(a) In all three areas the initial activity at the surface is practically the same. It cannot be radioactively determined, whether the terminating upper layer has additionally grown in section x, or later peeled off in section x.

(b) In section x one obtains a continual growth ($s \sim 7.5 \pm 0.6$), like that in section y ($s \sim 7.1 \pm 0.6$). The somewhat lower growth rate in section y corresponds to the small concentration of the growth lines in this transition area. The section z, on the other hand, has a significant change in the growth rate. Compared with section x, the growth lines with a depth > 1 mm are considerably concentrated and the growth rate falls to $\sim 3$ mm/10⁶ yr.

(c) Here the extrapolated crust age is about $\sim 4 \times 10^5$ yr and is therefore lower than on the bottom of the nodule, which agrees with the macroscopic observation.

Straight layers are again found around the nodule core and only in the water-side of the nodule.

*Nodules from core* 162

In piston core 162 (6100 m depth of water) nodules were found not only at the surface, but also at different depths. The surfaces of all these nodules, even of the buried ones, were relatively smooth, but had small, recognisable growth cusp patterns. Both an external and a microscopical examination of the cross-section revealed no indications of a nodule

162(O)

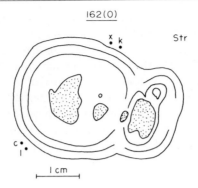

FIG. 74.

Core 162 with buried nodule

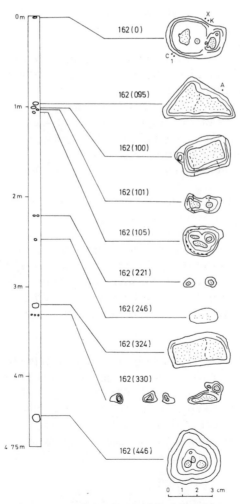

FIG. 75.   Core with buried nodule.

dissolution in the case of buried nodules. The theory, that buried nodules were *always* dissolved and that this dissolution in the sediment produced a supply of material for the build-up of surface nodules, can now be positively refuted. All the buried nodules of this core had surface growth cusp patterns that were extremely well-preserved.

In the cross-section one can see foreign cores in the 14 nodules of this core, which according to their external appearance might be weathered basalt. In the case of smaller cores the mineral transformation is also further advanced. In three nodules it was only a question of relatively thin incrustations of a larger core. For the buried nodules, the $\alpha$-track frequency only showed the value of the usual background effect. Thus, there were no radio-active indications of a growth in the last $3 \times 10^5$ yr.

### The surface nodule 162(0)

Shows no indications of erosion between the nodule core and the surface, and therefore could have grown without interruption. The relatively large growth cusp patterns are particularly noticeable in the hollows in the surface. A reconstruction of the nodule position in the sediment could not be clearly established. In the cross-section the growth lines can be followed macroscopically around the whole nodule. But under a microscope these lines cannot be followed so clearly.

The evaluation of the $\alpha$-tracks in two areas showed in both cases an external zone of rapidly decreasing activity and a constant growth of $\sim 7\,\text{mm}/10^6$ in the inner layers. If one evaluates the rapidly decreasing activity at the surface, the very small growth rates of $\sim 1\,\text{mm}/10^6$ yr are obtained. If, one also takes the microstructure of the nodules into consideration, then an outer layer of material of 0.1 mm can be seen, which is clearly defined and which could also have been a rapid growth. Therefore, there are two possible evaluations of the $\alpha$-track frequency for this outer layer.

(a)  Constant growth of $\sim 1\,\text{mm}/10^6$ yr in the surface area and a purely accidental boundary in this section.

(b)  Rapid growth of an outer layer of $\sim 0.1$ mm and a simulated rapid decrease of the $\alpha$-tracks due to insufficient dissolution. (In addition, see section 3.4.1.)

This last possibility should accurately describe the relationship of this nodule.

*Remark.* The photograph of the microstructure under the $\alpha$-track diagrams is not of one of the specimen count areas. Instead, it was taken at a different position, where the outer layer was particularly easy to recognise, but where the growth cusp patterns of this layer happened to be smaller than usual. In other parts of the nodule the size of the growth cusp patterns in the outer layer are the same as those in the zones lying underneath it.

### Nodule from core 170

In core 170 (depth of water 5130 m) two buried nodules were found at different levels. One of the nodules had a diameter of only 6–7 mm and a concentric build-up. The other had only a thin incrustation of a zeolite. (Zeolite Investigation, Dr. Rösch, B.G.R.) Both nodules had only the normal background activity. To interpret this phenomenon one should refer to the buried nodules of core 162. These buried nodules from core 170 also showed no signs of erosion or dissolution at the surface and the conclusions drawn from this correspond with the buried nodules of core 162.

### Nodule from core 196

This nodule comes from the piston core 196 (5250 m depth of water). As an experiment

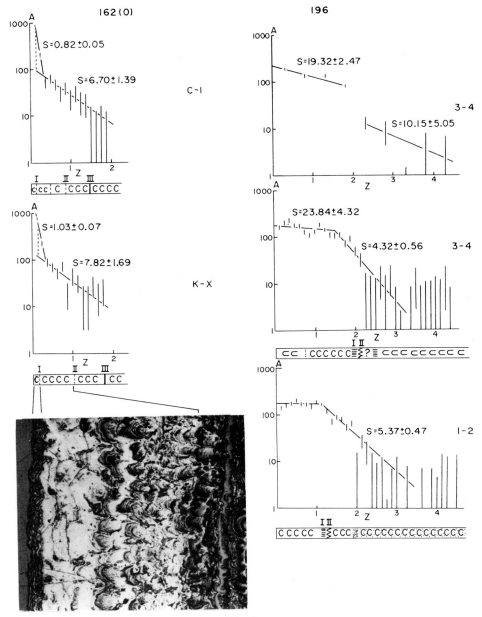

Fig. 76.

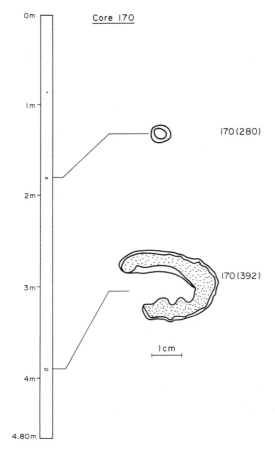

FIG. 77.

the nodule together with the adhering sediment was prepared in order to investigate possible radioactive relationships between the nodule and the adhering sediment. Therefore, a description of the surface and the attempt to reconstruct the position of the nodule in relation to the surface had to be foregone. In the cross-section one can see an inner part of the nodule without macroscopically visible cracks and a surface zone of 1–2 mm. There are no indications of an erosion period. In the cross-section growth cusp patterns are visible at the surface. Therefore, judging from the internal structure the nodule should be in the growth-stage and the surface should have been rough. The intended nodule-sediment investigations could not be carried out in the desired form, since the drying process and the casting of the nodules in epoxy resin caused a peeling of the sediment. If the $\alpha$-track frequency is plotted at depth intervals of 0.125 m, then one obtains a rapid growth for the outer layer and a constant growth in the inner layers of the nodule.

The results of Specimen Count Area 3–4 are given again in depth intervals of 0.5 mm and by taking such mean values a completely different result can be obtained and a growth interruption is simulated. Both count areas are situated on different nodule halves and therefore cannot be directly correlated by growth lines. Afterwards, the Count Area 1–2

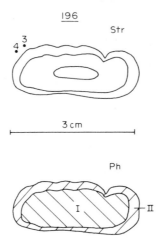

FIG. 78.

might appear to be an unfortunate choice and scarcely representative of the nodule. If one interprets the result of Count Area 3–4, then the outer layer is to be regarded as recent and to have grown quickly. However, the inner layers of the nodule can still be measured with the ionium dating-method. A relatively young total-age of between (0.5–1) $\times 10^6$ yr is obtained for the whole nodule.

## REFERENCES

BONATTI E. and Y. R. NAYUDU (1965) The origin of manganese nodules on the ocean floor. *American Journal of Science* **263**, 17–39.

BARNESS S. S. and J. R. DYMOND (1967) Rates of accumulation of ferro-manganese nodules. *Nature, London* **213**, 1218–1219.

v. BUTTLAR H. and F. G. HOUTERMANS (1950) Photographische Bestimmung der Aktivitätsverteilung in einer Manganknolle der Tiefsee. *Die Naturw.*, **37**, 400–401.

CRECELIUS E. A., R. CARPENTER and R. T. MERRILL (1971) Magnetism and magnetic reversals in ferromanganese nodules. *Earth and Planetary Science Letters* **17**, 391–396.

CRONAN D. S. and J. S. TOOMS (1968) A microscopic and electron probe investigation of manganese nodules from the northwest Indian Ocean. *Deep-Sea Research* **15**, 215–223.

CRONAN D. S. (1972) Regional geochemistry of ferromanganese nodules in the world oceans. In: Horn, 19–30.

FRIEDRICH G., B. ROSNER and S. DEMISOY (1969) Erzmikroskopische und mikroanalytische Untersuchungen an Manganerzkonkretionen aus dem pazifischen Ozean. *Mineral. Deposita* **4**, 298–307.

GLASBY G. P. (1972) The geochemistry of manganese nodules from the northwest Indian Ocean. In: Horn, 93–104.

GLASBY G. P. Manganese deposits in the southwest Pacific. *Phase I Report*, 137–169.

HALBACH P. (1974) Vergleiche stofflicher Eigenschaften limnischer und mariner Mangankollen. *Erzmetall*, **27**, 161–168.

HERING N. (1971) Metalle aus Tiefsee-Erzen. *Stahl und Eisen*, **91**, 452–459.

HEYE D. and H. BEIERSDORF (1973) Radioaktive und magnetische Untersuchungen an Manganknollen zur Ermittlung der Wachstusmgeschwindigkeit bzw. zur Altersbestimmung. *Zeitschrift für Geophysik*, **39**, 703–726.

HORN D. R. (Herausgeber) (1972) Papers from a conference on ferro-manganese deposits on the ocean floor. 293 S.

HORN D. R., B. M. HORN and M. N. DELACH (1973) Factors which control the distribution of ferromanganese nodules. *Technical Report No. 8, NSFGX* 33616.

KRISHNASWAMI S., B. L. K. SOMAYAJULU and W. S. MOORE (1972) Dating of manganese nodules using beryllium-10. In: Horn, 117–122.

KRISHNASWAMI S., and W. S. MOORE (1973) Accretion rates of freshwater deposits. *Nature Phys. Sci. Ed.*, **243**, 114–116.

KU, T. L. and W. S. BROECKER (1967) Uranium, thorium and protactinium in a manganese nodule. *Earth and Planetary Science Letters*, **2**, 317–320.

Ku T. W. and W. S. Broecker (1969) Radiochemical studies on manganese nodules of deep-sea origin. *Deep-Sea Research*, **16**, 625–637.

Ku T. L. and G. P. Glasby (1972) Radiometric evidence for the rapid growth rate of shallow-water continental margin manganese nodules. *Geochimica et Cosmochimica Acta*, **36**, 699–703.

Margolis S. V. and G. P. Glasby (1973) Micro-laminations in marine manganese nodules as revealed by scanning electron microscopy. *Phase I Report*, 77–83.

Morgenstein M. (1971) A study of growth morphologies of two deep-sea manganese mega nodules. *Pacific Science*, **25**, 308–312.

Morgenstein M. (1973) Sedimentary diagenesis and rates of manganese accretion of the Waho Shelf, Kauai Channel, Hawaii. *Phase I Report*, 121–135.

Morgenstein M. and M. Felsher (1971) The origin of manganese nodules: a combined theorie with special reference to palagonization. *Pacific Science*, **25**, 301–307.

*Phase I Report* (1973) Inter-university program of research on ferro-manganese deposits of the ocean floor, April 1973, sponsored by the sea bed assesment program international decade of ocean exploration. *National Science Foundation*, Washington, 358 S.

Raab W. (1972) Physical and chemical features of pacific deep-sea menganese nodules and their implications to the genesis of nodules. In Horn, 31–49.

Sackett W. M. (1966) Manganese nodules: $Th^{230}$: Pa ratios. *Science*, **154**, 646–647.

Seibold E. (1973) Rezente submarine Metallogenese. *Geol. Rundschau*, **62**, 641–648.

Somayajulu B. L. K. (1967) Beryllium-10 in a manganese nodule. *Science*, **156**, 1219–1220.

Somayajulu B. L. K., G. R. Heath, T. C. Moore Jr. and D. S. Cronan (1971) Rates of accumulation of manganese nodules and associated sediment from the equatorial Pacific. *Geochimica et Cosmochimica Acta*, **35**, 621–624.

Sorem R. K. and A. R. Foster (1972) Internal structure of manganese nodules and implications in benefication. In: Horn, 167–180.

Sorem R. K. and A. R. Foster (1973) Mineralogical, chemical and optical procedures and standards for study of growth features and economic potential of manganese nodules. *Phase I Report*, 23–38.

# AUTHOR INDEX

Gucker, F. T.   97, 98, **132**
Gurney, R. W.   109, **132**

Haedrich, R. L.   5, 53, **54**
Haffner, R. E.   5, 40, **54**
Haggis, G. H.   118, **132**
Halbach, P.   166, 180, 187, 189, 198, **238**
Hamon, B. V.   144, **162**
Hannan, E. J.   144, **162**
Harned, H. S.   97, 98, 106, 123, **132**
Harris, M. J.   4, 6, **53**
Harrisson, C. M. N.   10, **54**
Hart, J. E.   63, 72, **89**
Hasted, J. B.   118, **132**
Heath, G. R.   179, **239**
Heinmiller, R. H.   138, **162**
Held, I. M.   138, **162**
Hempel, G.   52, **54**
Hering, N.   166, **238**
Herring, P. J.   46, **54**
Heye, D.   163, 170, 178, 183, 205, **238**
Hindman, J. C.   111, 114, 118, **132**
Hinich, M. J.   143, **162**
Hjort, J.   4, 14, 20, 22, 25, **54**
Hoff, E. V.   107, 108, 110, 111, **132**
Hogg, N. G.   61, 72, **89**
Holtzer, A.   113, **132**
Horn, B. M.   188, **238**
Horn, D. R.   188, **238**
Horne, R. A.   91, 117, 118, 123, **132**
Hoskins, B. J.   69, 72, **89**, 90
Houtermans, F. G.   167, 169, **238**
Humboldt, A. V.   60, **90**

Jespersen, P.   14, 16, 19, 20, 36, 50, **54**
John, H. C.   47, **53**, **54**
Johnson, D. R.   87, **89**
Johnson, M. W.   5, 11, **55**
Jolicoeur, C.   113, 119, 124, **131**

Kampa, E. M.   4, 5, **53**, **54**
Karyakin, A. V.   102, **132**
Kauzmann, W.   119, **131**
Kavanau, J. L.   102, **132**
Kawaguchi, K.   20, 30, 31, **54**
Kessler, Yu. M.   122, **132**
Kester, D. R.   103, 117, 125, **132**
Killworth, P. D.   59, 67, 73, 86, **90**
Kirkwood, J. G.   119, **132**
Koefoed, E.   14, 24, **54**
Koopmans, L. H.   143, **162**
Kortum, G.   116, **132**
Kraus, E. B.   62, **90**
Krause, J. T.   118, **131**
Krishnaswami, S.   167, 168, **238**
Kronick, P. L.   98, 100, **133**
Krueger, W. H.   10, 30, 31, 35, 36, **54**
Ku, T. L.   167, 168, **238**, **239**
Ku, T. W.   167-9, 171, **293**

Lacombe, H.   60, 61, 68, 72, 86, **90**
Larby, M. J.   62, 84, **89**
Laurs, R. M.   10, 34, 49, **54**
Lee, W. H.   101, 110, **131**
Lepple, F. K.   107, 108, 110, 111, **132**

Levine, S.   119, **133**
Lighthill, M. J.   72, **90**
Lin, C. C.   88, **90**
Long, F.   124, **132**
Longsworth, L. G.   109, **132**
Lu, C. C.   4, **53**
Luyten, J. R.   146, **162**

McDevitt, W.   124, **132**
MacDonald, R. W.   121, **132**
McIntyre, M. E.   77, 87, **90**
Margolis, S. V.   198, 199, **239**
Marshall, M. B.   12, 20, 22, 25, 31, 49, 50, 52, **54**
Marshall, W. L.   101, 106, 121, 125, **132**, **133**
Marumo, R.   20, 30, 31, **54**
Masson, D. O.   97, **132**
Masterton, W. L.   103, **132**
Mead, G. W.   5, **53**
MEDOC Group,   60, 61, 82, **90**
Merrett, N. R.   3, 8, 9, 34–6, 38, 40, 46, 52, **53**, **54**
Merrill, R. T.   183, **238**
Millero, F. J.   97, 98, 100, 103, 107, 108, 110, 111, 121, 124, **131**, **132**
Moelwyn-Hughes, E. A.   105, **133**
Moller, D. A.   138, **162**
Moore, T. C.   179, **239**
Moore, W. S.   167, 168, **238**
Morgan, J.   118, **133**
Morgenstein, M.   169, **239**
Moser, H. G.   41, **54**, 55, 56, 58
Muradova, G. A.   102, **132**
Murray, J.   4, 14, 20, 22, 25, **54**

Nafpaktitis, B. G.   45, **54**, 55
Nancollas, G. H.   122, **133**
Nayudu, Y. R.   166, **238**
Nemethy, G.   93, 106, 112, 120, **133**
North, N. A.   121, **132**, **133**

Ooyama, K.   64, 88, **90**
Orlanski, I.   73, **90**
Owen, B. B.   97, 98, 100, 106, 123, **132**, **133**

Padova, J.   100, 109, 110, 115, 116, **133**
Parin, N. V.   47, 52, **54**
Parr, A. E.   24, **54**
Passynski, A.   92, 110, **133**
Paulowich, S.   94, 96, 97, 100, 125, **131**
Paxton, J. R.   5, 10, 49, **54**
Pearcy, W. G.   10, 34, 49, **54**
Pedlosky, J.   73, 85, 86, **90**
Peixoto, J. P.   138, **162**
Pertseva-Ostroumova, T. A.   41, **54**, 55, 56
Pickford, G. E.   5, **54**
Povarova, Yu. M.   122, **132**
Pugh, P. R.   4, **55**
Pugh, W. L.   12, 16, 19, 45, 49, **54**
Pytkowicz, R. M.   103, **132**, **133**

Quist, A. S.   101, 106, **132**, **133**

Raab, W.   186, 187, 189, 197, 198, **239**
Reddy, A. K. N.   124, **131**
Regan, C. T.   40, **55**

# SUBJECT INDEX

# CONTENTS OF PREVIOUS VOLUMES

# VOLUME 4

# VOLUME 5

## Part 1

# VOLUME 6